Lecture Notes in Computer Science　　13461

More information about this series at https://link.springer.com/bookseries/558

Minming Li · Xiaoming Sun (Eds.)

Frontiers of Algorithmic Wisdom

International Joint Conference, IJTCS-FAW 2022
Hong Kong, China, August 15–19, 2022
Revised Selected Papers

Springer

Editors
Minming Li
Department of Computer Science
City University of Hong Kong
Hong Kong, China

Xiaoming Sun
Institute of Computing Technology
Chinese Academy of Sciences
Beijing, China

ISSN 0302-9743   ISSN 1611-3349   (electronic)
Lecture Notes in Computer Science
ISBN 978-3-031-20795-2   ISBN 978-3-031-20796-9   (eBook)
https://doi.org/10.1007/978-3-031-20796-9

This Springer imprint is published by the registered company Springer Nature Switzerland AG
The registered company address is: Gewerbestrasse 11, 6330 Cham, Switzerland

Preface

This volume contains the contributed, accepted papers presented at International Joint Conference on Theoretical Computer Science-Frontiers of Algorithmic Wisdom (IJTCS-FAW 2022), for the 16th International Conference on Frontiers of Algorithmic Wisdom (FAW) and the third International Joint Conference on Theoretical Computer Science (IJTCS), held in Hong Kong, China, during August 15–19, 2022. Due to COVID-19, the conference was run in a hybrid mode.

FAW started as the Frontiers of Algorithmic Workshop in 2007 at Lanzhou, China, and was held annually from 2007 to 2021 and published archival proceedings. IJTCS, the International joint theoretical Computer Science Conference, started in 2020, aimed to bring in presentations covering active topics in selected tracks in theoretical computer science.

To accommodate the diversified new research directions in theoretical computer science, FAW and IJTCS joined their forces together to organize an event for information exchange of new findings and work of enduring value in the field.

The conference had both contributed talks submitted to the four tracks in IJTCS-FAW 2022, namely, Track A: Algorithmic Game Theory, Track B: Game Theory in Blockchain, Track G: The 16th Conference on Frontiers of Algorithmic Wisdom (previously named Frontiers of Algorithmic Workshop), Track H: Computational and Network Economics, and invited talks in focused tracks on Algorithmic Game Theory; Game Theory in Blockchain; Multi-agent Learning, Multi-agent System, Multi-agent Games; Learning Theory; Quantum Computing; Machine Learning and Formal Method; and Conscious Turing Machine. Furthermore, a Female Forum, Young PhD Forum, Undergraduate Research Forum, and Young Faculty in TCS at CFCS session were also organized.

For the four tracks that accepted submissions, the Program Committee, consisting of 47 top researchers from the field, reviewed 25 submissions and decided to accept 18 regular papers and 1 short paper. These are presented in this proceedings volume. Each paper had at least three reviews, with additional reviews solicited as needed. The review process was conducted entirely electronically via Springer EquinOCS system. We are grateful to Springer for allowing us to handle the submissions and the review process and to the Program Committee for their insightful reviews and discussions, which made our job easier.

The best paper goes to "Optimally Integrating Ad Auction into E-Commerce Platforms" authored by Weian Li, Qi Qi, Changjun Wang and Changyuan Yu.

Besides the regular talks, IJTCS-FAW 2022 had keynote talks from Yang Cai (Yale University), Christian Catali (MIT), Hee-Kap Ahn (Pohang University of Science and Technology) and Zhiyi Huang (University of Hong Kong).

We are very grateful to all the people who made this meeting possible: the authors for submitting their papers, the Program Committee members and external reviewers for their excellent work, and all the keynote speakers and invited speakers. We also thank the Advisory Committee and Steering Committee for providing timely advice for running the conference. In particular, we would like to thank City University of Hong Kong and Peking University for providing organizational support. We would like to thank Algorand Foundation for their generous sponsorship for the event. Finally, we would like to thank Springer for their encouragement and cooperation throughout the preparation of this conference.

August 2022   Minming Li
                                                                   Xiaoming Sun

Organization

Program Committee Chairs

Li, Minming — City University of Hong Kong, Hong Kong, China
Sun, Xiaoming — Institute of Computing Technology, Chinese Academy of Sciences, China

Track Chairs

Algorithmic Game Theory

Yukun, Cheng — Suzhou University of Science and Technology, China
Zhengyang, Liu — Beijing Institute of Technology, China

Game Theory in Blockchain

Jing, Chen — Algorand, USA
Xiaotie, Deng — Peking University, China

Frontiers of Algorithmic Wisdom

Chung-Shou, Liao — National Tsinghua University, Taiwan
Haitao, Wang — Utah State University, USA

Computational and Network Economics

Biaoshuai, Tao — Shanghai Jiao Tong University, China
Jianwei, Huang — Chinese University of Hong Kong, Shenzhen, China

Program Committee

Chau, Vincent — Southeast University, China
Cheng, Yukun — Suzhou University of Science and Technology, China
Hung, Ling-Ju — National Taipei University of Business, Taiwan
Jin, Kai — Sun Yat-sen University, China
Kong, Yuqing — Peking University, China
Li, Shuai — Shanghai Jiao Tong University, China
Li, Bo — The Hong Kong Polytechnic University, Hong Kong, China
Li, Minming — City University of Hong Kong, Hong Kong, China
Liu, Jinyan — Beijing Institute of Technology, China
Liu, Shengxin — Harbin Institute of Technology, Shenzhen, China
Liu, Zhengyang — Beijing Institute of Technology, China
Luo, Kelin — University of Bonn, Germany

Sun, Xiaoming	Institute of Computing Technology, Chinese Academy of Sciences, China
Tang, Zhihao	Shanghai University of Finance and Economics, China
Tao, Biaoshuai	Shanghai Jiao Tong University, China
Wang, Zihe	Renmin University of China, China
Wu, Xiaowei	University of Macau, Macau, China
Xue, Jie	New York University Shanghai, China
Yang, Kuan	Shanghai Jiao Tong University, China
Zhang, Peng	Rutgers University, USA
Zhang, Yuhao	Shanghai Jiao Tong University, China
Zhang, Chihao	Shanghai Jiao Tong University, China
Zheng, Zhenzhe	Shanghai Jiao Tong University, China

Additional Reviewers

Chen, Binhui	Beijing Institute of Technology, China
Guo, Yongkang	Peking University, China
Lin, Chuang-Chieh (Joseph)	Tamkang University, Taiwan
Lu, Yuxuan	Peking University
Mei, Lili	Hangzhou Dianzi University, China
Nong, Qingqing	Ocean University of China, China
Qiu, Guoliang	Shanghai Jiao Tong University, China
Wang, Qian	Peking University, China
Xiao, Tao	Huawei, China
Yan, Xiang	Huawei, China
Yu, Wei	East China University of Science and Technology, China
Yu, Haoran	Beijing Institute of Technology, China

Invited Talks

Constrained Min-Max Optimization: Last-Iterate Convergence and Acceleration

Yang Cai

Yale University

Abstract. Min-max optimization plays a central role in numerous machine learning (ML) applications ranging from generative adversarial networks, adversarial examples, robust optimization, to multi-agent reinforcement learning. These ML applications pose new challenges in min-max optimization, requiring the algorithms to (i) exhibit last-iterate convergence as opposed to the more traditional average-iterate convergence, and (ii) to be robust in nonconvex-nonconcave environments. In this talk, we first focus on the last-iterate convergence rates of two most classical and popular algorithms for solving convex-concave min-max optimization – the extragradient (EG) method by Kopelevich (1976) and the optimistic gradient (OG) method by Popov (1980). Despite their long history and intensive attention from the optimization and machine learning community, the last-iterate convergence rates for both algorithms remained open. We resolve this open problem by showing both EG and OG exhibit a tight $O(1/\sqrt{T})$ last-iterate convergence rate under arbitrary convex constraints. In the second part of the talk, we discuss some recent progress on obtaining accelerated and optimal first-order methods for structured nonconvex-nonconcave min-max optimization. Our results are obtained via a new sum-of-squares programming based approach, which may be useful in analyzing other iterative methods.

How Crypto, Stablecoins, CBDCs and Web3 Will Reshape Competition

Christian Catalini

Founder, MIT Cryptoeconomics Lab and Research Scientist, MIT

Abstract. The talk will explore some of the recent developments in the crypto, payments and web3 space, as well as implications for competition and innovation.

Voronoi Diagrams in the Presence of Obstacles

Hee-Kap Ahn

Pohang University of Science and Technology

Abstract. A Voronoi diagram is a fundamental geometric structure that partitions space into regions based on the distance to points in a specific subset of the space. Due to the structural and combinatorial properties, Voronoi diagrams have applications in many fields, including geometry, informatics, biology, engineering, and architecture. In this talk, we will review the concept and properties of Voronoi diagrams and the recent progress in computing Voronoi diagrams, especially in the presence of obstacles.

Recent Progress in Online Matching

Zhiyi Huang

University of Hong Kong

Abstract. Originated from the seminal work by Karp, Vazirani, and Vazirani (1990), online matching has been established as one of the most fundamental topics in the literature of online algorithms. This talk presents the basics of online matching, and surveys the recent progress in two directions:

1. Open problems about online advertising: AdWords and Display Ads are generalizations of the online bipartite matching problem by Karp et al. These problems capture online advertising which generates tens of billions of US dollars annually. This year, we introduce a new technique called online correlated selection, and design the first online algorithms for the general cases of AdWords and Display Ads outperforming greedy, which has remained the state of the art for more than 10 years, despite many attempts to find better alternatives.

2. General arrival models: Traditional online matching models consider bipartite graphs and assume knowing one side of the bipartite graph upfront. The matching problems in many modern scenarios, however, do not fit into the traditional models. In the problem of matching rid of matching ride-sharing requests, for instance, the graph is not bipartite in general, and all vertices arrive online. There has been much progress in the past three years on online matching models beyond the traditional ones, including the fully online model, the general vertex arrival model, and the edge arrival model.

Contents

Computational and Network Economics

Papers and Talks Presented in IJTCS Tracks A-F and H-I, and the Forums

Invited Talks, Track A (Algorithmic Game Theory)

Practical Fixed-Parameter Algorithms for Defending Active Directory Style Attack Graphs [Mingyu Guo]

Invited Talks, Track B (Game Theory in Blockchain)

Token Design and Economic Incentives [Ye Li]
Markets for Crypto Tokens, and Security under Proof of Stake [Ravi Jagadeesan]
Eliciting Information without Verification [Yuqing Kong]
Insightful Mining Equilibria [Mengqian Zhang]
FileInsurer: A Scalable and Reliable Protocol for Decentralized File Storage in Blockchain [Hongyin Chen]
EIP-1559: Chaos and Efficiency in Ethereum's Transaction Fee Market [Stefanos Leonardos]
A Rational Protocol Treatment of 51% Attacks [Vassilis Zikas]

Invited Talks, Track C (Multi-agent Learning, Multi-agent System, Multi-agent Games)

A Continuum of Solutions to Cooperative Multi-Agent Reinforcement Learning [Yaodong Yang]
A Brief Introduction to Convergence Analysis of Reinforcement Learning Algorithms [Xingguo Chen]
Decision Structure in Decentralized Multi-Agent Learning [Yali Du]
Reverse Engineering Human Cooperation and Cultural Evolution [Joel Z Leibo]

Invited Talks, Track D (Learning Theory)

Unified Theory of Explaining Heuristic Findings in Attribution, Robustness, Generalization, Visual Features in a DNN [Quanshi Zhang, Huiqi Deng]
Demystifying (Deep) Reinforcement Learning with Optimism and Pessimism [Zhaoran Wang]
More Than a Toy: Random Matrix Models Predict How Real-World Neural Representations Generalize [Wei Hu]

Invited Talks, Track E (Quantum Computing)

Optimal Quantum Circuit Constructions for Quantum State Preparation and General Unitary Synthesis [Shengyu Zhang]
Quantum Adiabatic Theorem Revisited [Runyao Duan]
Winning Mastermind Overwhelmingly on Quantum Computers [Lvzhou Li]
Numerical Framework for Finite-size Security of Quantum Cryptography [Hongyi Zhou]

Invited Talks, Track F (Machine Learning and Formal Method)

QVIP: An ILP-based Formal Verification Approach for Quantized Neural Networks [Yedi Zhang]
Copy, Right? A Testing Framework for Copyright Protection of Deep Learning Models [Jingyi Wang]
Adaptive Fairness Improvement based on Causality Analysis [Mengdi Zhang]
Robustness Analysis for DNNs from the Perspective of Model Learning [Pengfei Yang]

Invited Talks, Track H (EconCS)

Escaping Saddle Points: from Agent-based Models to Stochastic Gradient Descent [Fang-Yi Yu]
Truthful Cake Sharing [Xinhang Lu]
An Efficient Algorithm for Approximating Nash Equilibrium in Zero-sum Imperfect-information Games [Ying Wen]

Invited Talks, Track I (Conscious AI)

Formalizing A Multimodal Language for Intelligence and Consciousness [Paul Liang]

Undergraduate Research Forum

Your Transformer May Not be as Powerful as You Expect [Shanda Li]
Element Distinctness, Birthday Paradox, and 1-out Pseudorandom Graphs [Hongxun Wu]
Nash Convergence of Mean-Based Learning Algorithms in First Price Auctions [Xinyan Hu]
Towards Data Efficiency in Offline Reinforcement Learning [Baihe Huang]
On the Feasibility of Unclonable Encryption, and More [Xingjian Li]
Characterizing Parametric and Convergence Stability in Nonconvex Nonsmooth Optimization [Hanyu Li]

Female Forum

Quantum Copy Protection and Unclonable Cryptography [Jiahui Liu]
Hardness Results for Weaver's Discrepancy Problem [Peng Zhang]
Graph limits and graph homomorphism inequalities [Fan Wei]
A Faster Interior-Point Method for Sum-of-Squares Optimization [Shunhua Jiang]

Young PhD Forum

Optimal Rates of (Locally) Differentially Private Heavy-tailed Multi-Armed Bandit
    [Yulian Wu]
Optimal Private Payoff Manipulation against Commitment in Extensive-form Games
    [Yurong Chen]
Beyond Characteristic Function Method to Get Exponential Concentration Bound
    [Zhide Wei]
Multiwinner Voting Using Favorite-Candidate Rule [Zeyu Ren]
The Power of Multiple Choices in Online Stochastic Matching [Xinkai Shu]

Young Faculty in TCS

An Optimal Separation between Two Property Testing Models for Bounded Degree
    Directed Graphs [Pan Peng]
Ordinal Approximation Algorithms for MMS Allocation of Chores [Xiaowei Wu]
The Power of Uniform Sampling for Coresets [Xuan Wu]
A Polynomial Time Algorithm for Submodular 4-Partition [Chao Xu]
Graphical Resource Allocation with Matching-Induced Utilities [Bo Li]
Survivable Network Design: Group-Connectivity [Yuhao Zhang]

CSIAM Forum

DAMYSUS: Streamlined BFT Consensus Leveraging Trusted Components [Jiangshan
    Yu]
Blockchain-Based Private Provable Data Possession [Huaqun Wang]

Algorithmic Game Theory

EFX Under Budget Constraint

Sijia Dai[1], Guichen Gao[1], Shengxin Liu[2], Boon Han Lim[3], Li Ning[1],
Yicheng Xu[1], and Yong Zhang[1(✉)]

[1] Shenzhen Institute of Advanced Technology, Chinese Academy of Sciences,
Shenzhen, People's Republic of China
{sj.dai,gc.gao,li.ning,yc.xu,zhangyong}@siat.ac.cn

[2] Harbin Institute of Technology (Shenzhen), Shenzhen, People's Republic of China
sxliu@hit.edu.cn

[3] Tunku Abdul Rahman University, Kajang, Malaysia
limbhan@utar.edu.my

Abstract. Fair division captures many real-world scenarios and plays
an important role in many research fields including computer science,
economy, operations research, etc. For the problem of indivisible goods
allocation, it is well-known that an allocation satisfying the *maximum
Nash Social Welfare* (Max-NSW) is *envy-free up to one good* (EF1), which
is an important fairness concept. However, EF1 is often considered to be
somewhat weak since one may remove the most valuable good. In this
paper, we investigate the relationship between the Max-NSW allocation
and a stronger fairness concept, *envy-free up to any good* (EFX), which
means the envyness disappears after the removal of the least valuable
good. We focus on the budget-feasible variant in which each agent has a
budget to cover the total cost of goods she gets. We show that a Max-
NSW allocation guarantees $\frac{1}{4}$-EFX when agents' value are in the *Binary*
$\{0, 1\}$. Moreover, we provide an algorithm to find a budget-feasible EFX
allocation for the *Binary* variant.

Keywords: Fair division · Maximum nash social welfare · Budget
constraint · Envy-free up to any good

1 Introduction

The theory of fair division addresses the fundamental problem of allocating limited goods or resources among agents in a fair and/or efficient manner. It has
widespread applications in areas like vaccine distribution, kidney matching, property inheritance, government auction, and so on.

In fair division model, the owner of a resource wants to allocate limited
resources efficiently and the recipients of resources want to receive it fairly. In
economics, Pareto Optimality (PO) is usually used to measure the efficiency of

Supported by the National Key Research and Development Project of China (Grant No.
2019YFB2102500), NSFC (Grant No. 12071460, 62102117), the Fundamental Research
Project of Shenzhen City (No. JCYJ20210324102012033) and the Shenzhen Science and
Technology Program (Grant No. RCBS20210609103900003).

M. Li and X. Sun (Eds.): IJTCS-FAW 2022, LNCS 13461, pp. 3–14, 2022.
https://doi.org/10.1007/978-3-031-20796-9_1

allocation, which means no other allocation can make some agents strictly better off without making any agent strictly worse off. A lot of researches focuses on finding effective algorithms to balance the efficiency and the fairness in real-world applications [11,15,16,19,23]. Mamoru and Kenjiro [17] propose the concept of Nash Social Welfare, which represents the multiplicative form of all agents' valuations w.r.t. their assigned bundles. The property of the maximizing the Nash Social Welfare leads it to be regarded as a balance on both efficiency and fairness [17].

The Nash Social Welfare is well-studied for divisible goods [21]. Eisenberg and Gale [12] show that the maximum Nash Social Welfare allocation can be computed in polynomial time by using the convex program. However, computing the maximum Nash Social Welfare allocation for indivisible goods is proved to be APX-hard [18] even when agents have additive valuations of goods, i.e., no polynomial-time approximation schemes (PTAS) unless $P = NP$. Thus, it is natural to consider how to approximate the maximum Nash Social Welfare in polynomial time. Cole et al. [10] first propose an algorithm with the approximation ratio of 2.89 and they then improve the bound to 2 by a modified algorithm [9]. Barman et al. [2] introduce the concept of Fisher market equilibrium and they design an algorithm with the approximate ratio 1.45 and show that the output of algorithm could achieve Pareto Optimality. There are also some works on the Nash Social Welfare maximization problem for some particular cases, e.g., the greedy maximum Nash Social Welfare algorithm for identical valuations and binary valuations [3], an approximation maximum Nash Social Welfare algorithm under submodular valuations [14].

Note that the maximum Nash Social Welfare allocation always exists. It is interesting to investigate which kind of fairness can be achieved when an allocation reaches the maximum Nash Social Welfare. Previous studies prove that the allocation satisfying the maximum Nash Social Welfare can reach π_n-Maximin Share (MMS) (where π_n is a value that desearses as the number of agents increases) and 0.618-pairwise Maximin Share (PMMS) [6]. MMS and PMMS mentioned above are important measurements of fairness. MMS means that everyone should at least get the goods to maximize the worst distribution of the share while PMMS consider fairness between any pair of agents.

Envy-free (EF) is one of the criteria of fair division, which states that when resources are distributed among agents, each agent should receive a share which at least as good as the share received by other agents in his opinion. But when the resources to be allocated are indivisible, even if there is one resource and two agents, it is impossible to achieve an EF allocation. Envy-free up to one good (EF1) is proposed by Lipton et al. [20]. It is known that the maximum Nash Social Welfare allocation is also an EF1 allocation [6]. Another well-known fairness notion, envy-free up to any good (EFX), is introduced by Caragiannis et al. [6], which is weaker than EF but stronger than EF1. They prove that a PMMS allocation implies an EFX allocation if all agents have positive values for all goods. Thus, the maximum Nash Social Welfare allocation can also reach 0.618-EFX.

There are some interesting works to show the relationship between the allocation satisfying the maximum Nash Social Welfare and the fairness under some conditions [1,4,5,7,8]. For example, the maximum Nash Social Welfare allocation is always EFX as long as there are at most two possible values for the goods [1]. The maximum Nash Social Welfare allocation is always EF1 if the agents have identical valuations with matroid constraint [4].

In some realities, agents must pay for the goods allocated to her. Thus, a natural generalization of the fair division problem is to consider the budget constraint. Each agent is associated with some fixed budget and the cost of the assigned bundle must be not exceed her budget. Wu et al. [22] introduce the budget constraint for EF1 allocation in the fair division problem. As far as we know, it is the first result considering the budget constraint in fairness allocation. They show that any maximum Nash Social Welfare allocation is at least $\frac{1}{4}$-EF1 and the ratio is tight. Besides, they prove that any maximum Nash Social Welfare allocation guarantees $\frac{1}{2}$-EF1 if the budget of each agent is sufficiently large. Gan et al. [13] study EF1 allocation in some special cases and show that a $\frac{1}{2}$-EF1 allocations can be computed in polynomial time if the valuations of all agents are identical. Meanwhile, the maximum Nash Social Welfare allocation can be computed in polynomial-time and the approximation ratio w.r.t. EF1 is asymptotically close to 1 when agents have identical valuations and the budget of each agent is sufficiently large.

In this paper, we focus on the fair division under the budget constraint, that means each agent can only afford the bundle with the total cost no more than her budget. We investigate the properties of the maximum Nash Social Welfare allocation with the budget constraint and find a budget-feasible EFX allocation in this setting. The remainder of this paper is structured as follows. Section 2 presents the model of fair division and some classical fairness concepts. Section 3 proves that the maximum Nash Social Welfare allocation guarantees the approximation ratio of EFX under the budget constraint. Section 4 provides an algorithm to compute a budget-feasible EFX allocation for the *Binary* variant. Section 5 gives the concluding remark and some possible future research for the fair division problem.

2 Preliminaries

Let $M = \{g_1, \ldots, g_m\}$ denotes the set of goods and $N = \{1, 2, \ldots, n\}$ denotes the set of agents. Throughout this paper, we assume the goods are indivisible, i.e., each good must be entirely allocated to one agent. Each agent $i \in N$ has a valuation function $v_i(\cdot) : 2^M \to \mathbb{R}_{\geq 0}$ that measures the utility of each agent on different subsets of goods. In this paper, we assume that the valuation function $v_i(\cdot)$ of each agent is additive, i.e., $v_i(X_i) = \sum_{g \in X_i} v_i(\{g\})$ for any subset (or bundle) X_i of M. For simplicity, we write $v_i(g)$ instead of $v_i(\{g\})$ for good $g \in M$.

Every good $g \in M$ is associated with a cost $c(g) \geq 0$ and we assume that the cost of goods is the same and additive for each agent, i.e., $c(X_i) = \sum_{g \in X_i} c(g), \forall i \in N$, $X_i \subseteq M$. In addition, each agent has a different budget $B_i \geq 0$ for any $i \in N$.

Let X_0 denotes the set of unallocated goods. X_0 can be regarded as a charity, who has unbounded budget and $v_0(g) = 0$ for $g \in M$. We consider such an allocation $\mathbf{X} = \{X_0, X_1, \ldots, X_n\}$ which is a partition of goods set M, where $X_i \subseteq M$ is the bundle of goods assigned to agent i and $X_i \cap X_j = \emptyset$ for any $0 \leq i < j \leq n$.

When considering the budget constraint, we say an allocation $\mathbf{X}$ is *budget-feasible* if each agent can afford the allocated bundle, i.e., $c(X_i) \leq B_i$ for any $i \in N$.

We investigate the following three kinds of valuation functions, which are reasonable and well-adopted in many applications.

- *Binary*: $v_i(g) \in \{0, 1\}$ for every $i \in N$ and $g \in M$;
- *Identical*: $v(\cdot)$: each agent has the same valuation on any good $g \in M$;
- *Interval*: $v_i(g) \in [b, a]$, where $0 < b < a$ for every $i \in N$ and $g \in M$.

Now we generalize the fairness concepts by considering the budget constraint. Note that when each agent has sufficiently large budget and they can afford any subsets of the good set M, the problem will be degenerated to the traditional one without the budget constraint. Again, the objective is to find an allocation which is fair to all agents. Under the budget constraint, we assume that the agent will only envy the bundle she can afford [22].

Definition 1. *(Budget-Feasible-Envy-freeness and Its Relaxations)* For any $\alpha \in [0, 1]$, a budget-feasible allocation $\mathbf{X} = (X_i)_{i \in N}$ is:

- *α-approximate budget-feasible envy-free (α-BFEF): if for any $i, j \in N$ and any $S \subseteq X_j$ with $c(S) \leq B_i$ such that*

$$v_i(X_i) \geq \alpha \cdot v_i(S).$$

  Note that a 1-BFEF allocation is an BFEF allocation.
- *α-approximate budget-feasible envy-free up to one good (α-BFEF1): if for any $i, j \in N$ and any $S \subseteq X_j$ with $c(S) \leq B_i$ such that*

$$\exists\, g \in S,\ v_i(X_i) \geq \alpha \cdot v_i(S \setminus g).$$

  Note that a 1-BFEF1 allocation is an BFEF1 allocation.
- *α-approximate budget-feasible envy-free up to any good (α-BFEFX): if for any $i, j \in N$, any $S \subseteq X_j$ with $c(S) \leq B_i$ such that*

$$\forall\, g \in S,\ v_i(X_i) \geq \alpha \cdot v_i(S \setminus g).$$

  Note that a 1-BFEFX allocation is an BFEFX allocation.

Our goal is to analyze the fairness property when the Nash Social Welfare (NSW) is maximized while the budget constraint is satisfied. This gives rise to the following definitions.

Definition 2. *(PO) An allocation* $\mathbf{X}$ *is Pareto Optimal (PO) if there exist no allocation* X' *such that* $v_i(X_i') \geq v_i(X_i)$ *for all* $i \in N$ *and* $v_j(X_j') > v_j(X_j)$ *for some* $j \in N$.

Definition 3. *(Budget-Feasible-Max-NSW) An allocation* $\mathbf{X}^*$ *is a budget-feasible Max-NSW allocation if it is budget-feasible and maximizes the Nash Social Welfare, i.e.,*

$$\prod_{i \in N} v_i(X_i^*) \geq \prod_{i \in N} v_i(X_i)$$

for any other budget-feasible allocation $\mathbf{X}$, *where* $X_i^* \cap X_j^* = \emptyset$ *and* $X_i \cap X_j = \emptyset$ *for any* $i, j \in [n]$.

In this paper, we mainly consider BFEFX allocations in some cases. For simplicity, we say agent j x-*envies* agent i with respect to her budget B_j if there exists $S \subseteq X_i$ such that $c(S) \leq B_j$ and $v_j(X_j) < v_j(S \setminus g)$ for some $g \in S$. In other words, no x-envy with respect to the budget among all agents is the necessary and sufficient condition of satisfying BFEFX.

3   Max-NSW Allocation and EFX Under Budget Constraint

In this section, we investigate the approximated fairness guarantee for the Max-NSW allocation under the budget constraint.

We claim that if $v_i(g) = 0$ for some $i \in N$ and $g \in M$, to achieve the Max-NSW allocation, good g will not be allocated to agent i. Otherwise, the budget of agent i might be decreased while her valuation will not increase.

Remark 1. In a Max-NSW allocation, if $v_i(g) = 0$ for some $i \in N$ and $g \in M$, good g will not be allocated to agent i.

Proof. If there is a good whose valuation is 0 for all agents, i.e., $v_k(g) = 0$, for all $k \in N$, a reasonable idea is allocating it to the charity instead of an agent. $\square$

Lemma 1. *For the Binary variant, any Max-NSW allocation* $\mathbf{X}^*$ *is* $\frac{1}{4}$-*BFEFX and PO.*

Proof. If $\mathbf{X}^*$ is not PO, there exists another allocation $\mathbf{X}'$ that the value of some agent is strictly higher than the corresponding value in $\mathbf{X}^*$ while the values of other agents are non-decreasing. It means the NSW of $\mathbf{X}'$ is larger, which contradicts $\mathbf{X}^*$ is a Max-NSW allocation.

We show the correctness of the bound $\frac{1}{4}$ by contradiction. Assume a Max-NSW allocation $\mathbf{X}^*$ is not $\frac{1}{4}$-BFEFX, there exists the case that an agent i x-*envies* another agent j, which can be denoted as

$$\exists\, S \subseteq X_j^*, \ c(S) \leq B_i$$

and

$$\exists \, g' \in S, \; v_i(S \setminus g') > 4 \cdot v_i(X_i^*)$$

For the *Binary* variant, the value of a Max-NSW allocation depends on the valuation of all agents. If at least one agent gets 0, i.e., the valuation of all goods allocated to this agent is 0 and the value of Max-NSW will be 0, too. Otherwise, the value of Max-NSW will be a positive number.

Let $NW(\mathbf{X})$ be the value of Nash Social Welfare w.r.t. the allocation $\mathbf{X}$. We analyze two cases, $NW(\mathbf{X}^*) > 0$ and $NW(\mathbf{X}^*) = 0$, separately.

Case 1. $NW(\mathbf{X}^*) > 0$.

– If $\min_{g \in X_j^*} v_i(g) = 1$.

Due to Theorem 3.1 in [22] and the *Binary*-value property, the Max-NSW allocation is $\frac{1}{4}$-BFEFX, i.e.,

$$\forall \, S \subseteq X_j^* \text{ satisfying } c(S) \le B_i, \forall \, g \in S, \; v_i(X_i^*) \ge 1/4 \cdot v_i(S \setminus g).$$

– If $\min_{g \in X_j^*} v_i(g) = 0$.

We may assume at least one good in X_j^* whose valuation of agent i is 1. Otherwise agent i would never *x-envy* agent j. Hence, $|S| \ge 2$.

If S contains $\ell(> 1)$ goods with valuation $v_i(\cdot) = 0$, we may remove $\ell - 1$ of them from S and remain g'. In this way, the valuation $v_i(S)$ and $v_i(S \setminus g')$ do not change. Thus, we mayconsider that S contains only one good with zero valuation.

Combining the assumption, an Max-NSW allocation $\mathbf{X}^*$ is not $\frac{1}{4}$-BFEFX and Remark 1, we have $|S| > 4 \cdot |X_i^*| + 1$.

Evenly partition $S \setminus g'$ into two sets, S_1 and S_2, i.e., $||S_1| - |S_2|| \le 1$. Construct a new allocation $\mathbf{X}' = \{X_1', X_2', ..., X_n'\}$, in which $X_k' = X_k^*$ for all $k \ne i, j$, $X_i' = \arg\max_{S_\ell \in \{S_1, S_2\}} v_i(S_\ell)$, $X_j' = X_j^* \setminus X_i'$ and $X_0' = X_0 \cup X_i^*$.

Obviously, we have

$$v_i(X_i') \ge \frac{1}{2} \cdot v_i(S \setminus g') > 2 \cdot v_i(X_i^*).$$

Agent j does not *x-envy* agent i no matter she chooses S_1 or S_2, since agent j gets another one and g' with the valuation at least $\frac{1}{2} \cdot v_j(S)$, i.e.,

$$v_j(X_j') \ge \frac{1}{2} \cdot v_j(X_j^*).$$

Since $v_k(X_k') = v_k(X_k^*)$ for all $k \ne i, j$, it follows that

$$\prod_{\ell \in N} v_\ell(X_\ell') > \prod_{\ell \in N} v_\ell(X_\ell^*),$$

which contradicts the fact that $\mathbf{X}^*$ is Max-NSW.

Case 2. $NW(\mathbf{X}^*) = 0$.

If $X_i^* = \emptyset$ and $X_j^* = \emptyset$, they are envy-free each other.

If $X_i^* \neq \emptyset$ and $X_j^* \neq \emptyset$, similar to Case 1,

$$v_i(X_i^*) \geq \frac{1}{4} \cdot v_i(S \setminus g), \ \forall \, g \in S.$$

Now we consider the remaining case, $X_i^* = \emptyset$ and $X_j^* \neq \emptyset$.

– If $|X_j^*| = 1$

Since there is only one good in X_j^*, it is obviously true that

$$v_i(X_i^*) \geq \frac{1}{4} \cdot v_i(S \setminus g), \ \forall \, g \in S.$$

– $|X_j^*| \geq 2$

We claim that there exists at least one good $g' \in X_j^*$ such that $v_i(g') = 1$. Otherwise, agent i will not x-envy agent j. If agent i is the only agent with zero valuation in the allocation, switch g' from X_j^* to X_i^* leading to a new allocation with a positive Nash Social Welfare, which also contradicts the fact that $\mathbf{X}^*$ is the Max-NSW allocation.

If there are some other agent $k \in N \setminus \{i, j\}$ and $v_k(X_k^*) = 0$, the envyness can be eliminated by moving all goods which leads to envyness to the charity X_0, while keep the valuation of Nash Social Welfare unchanged.   □

Lemma 2. *There is a particular instance where every Max-NSW allocation fails to be better than $(\frac{1}{2} + \epsilon)$-BFEFX for any $\epsilon > 0$ if the valuation is Binary.*

Proof. This negative result can be shown via the following counterexample.

Example 1. There are 4 goods and two agents $\{1, 2\}$ with $B_1 = B_2 = 4$. The valuations of agents and costs of goods are shown in Table 1.

Table 1. Illustration to show a $(\frac{1}{4} + \epsilon)$-BFEFX allocation is unreachable.

	g_1	g_2	g_3	g_4
v_1	0	1	1	1
v_2	1	1	1	1
c	4	1	1	1

The only maximum Nash Social Welfare allocation $\mathbf{X}^* = \{X_1^* = \{g_2, g_3, g_4\}, X_2^* = \{g_1\}\}$. In this case,

$$v_2(X_2^*) < (\frac{1}{2} + \epsilon) \cdot v_2(X_1^* \setminus g_4),$$

where $\epsilon > 0$. We find that the Max-NSW allocation may not satisfy $(\frac{1}{2} + \epsilon)$-BFEFX for any $\epsilon > 0$ even if agents have the same budget.   □

Theorem 1. *A Max-NSW allocation* $\mathbf{X}^*$ *is* $\frac{1}{4}$-*BFEFX and PO if the valuation is Binary under the budget constraint. Moreover, no PO allocation can be* $(\frac{1}{2}+\epsilon)$-*BFEFX for any* $\epsilon > 0$ *under the above setting.*

Proof. The correctness of this theorem can be directly obtained from Lemma 1 and Lemma 2. □

4   Computing a BFEFX Allocation

In this section, we give an algorithm to compute a BFEFX allocation for the *Binary* variant in polynomial time. Intuitively, the cheaper cost of goods, the more value an agent may get. To allocate more valuable good sets to agents, a simple idea is assigning good with smaller cost to agent with smaller budget. However, it may violate BFEFX since an agent with larger budget may *x-envies* some other agent w.r.t. the budget.

In the algorithm, goods will be assigned to agents in turn with respect to the non-decreasing order of their original budgets. Let B'_i be the remaining budget of agent i, initially, $B'_i = B_i$. Let M' be the remaining unassigned good set, initially, $M' = M$. We say an agent is *active* if she still can afford some goods with positive value in the remaining unassigned goods. Let $\mathcal{A}$ denotes the set of active agents. Formally speaking, agent $i \in \mathcal{A}$ if there exists $g \in M'$ such that $c(g) \le B'_i$ and $v_i(g) = 1$.

The algorithm is implemented round by round (the outer **while** loop in Algorithm 1). In each round ℓ, an agent i may get at most one good g^ℓ_i. Note that BFEFX may violate at some step. We will show later (in Lemma 3) that such case only happens when assigning some good to an agent i and another agent j with larger budget *x-envies* agent i w.r.t. the budget B_j. When this happens, we reallocate some assigned good from agent i to agent j so as to eliminate the violation. According to Lemma 3, the assignment and reallocation will not affect the *x-envy* property of agent ℓ with $\ell < i$.

Lemma 3. *In processing Algorithm 1, agent k_j x-envies agent k_i w.r.t. B_{k_j} only when $j > i$.*

Proof. According to the algorithm, any assigned good is with value 1 for the agent who gets it. If agent i gets a good from M' in round ℓ, she must get one good in all previous round $\ell' < \ell$. That is because agent i is still *active* in round ℓ.

When agent k_i gets good $g^\ell_{k_i}$ in round ℓ, $v_{k_i}(X_{k_i}) = \ell$. Assuming that agent k_j *x-envies* agent k_i w.r.t. B_{k_j} at this time step and this does not happen in previous steps. That means assigning $g^\ell_{k_i}$ to agent k_i leads to such *x-envy*. So, there exists $S \subseteq X_{k_i}$ with $c(S) \le B_{k_j}$ and $g^\ell_{k_i} \in S$, satisfying

$$|S| - 2 = v_{k_j}(S \setminus g^\ell_{k_i}) - 1 \le v_{k_j}(X_{k_j}) < v_{k_j}(S \setminus g^\ell_{k_i}) = |S| - 1.$$

Algorithm 1: Finding a BFEFX allocation when the valuation function is *Binary*

---

1 **Input:** agents set N, goods set M, budgets set $\mathcal{B} = \{B_1, B_2, \ldots, B_n\}$ and costs
   set $\mathcal{C} = \{c(g_1), c(g_2), \ldots, c(g_m)\}$;

2 Initialize: $X_i \leftarrow \emptyset$, $B_i' \leftarrow B_i$ for all $i \in N$;

3 Initialize: $\mathcal{A} \leftarrow N$; $\triangleright$ Assume any agent can afford at least one good in M.

4 Initialize: $M' \leftarrow M$;

5 Sorting agents according to their budgets, i.e., $\{k_1, k_2, ..., k_n\}$ such that
   $B_{k_i} \leq B_{k_j}$ for $i < j$;

6 **while** $\mathcal{A} \neq \emptyset$ and $M' \neq \emptyset$ **do**

7 | **for** $i = 1$ *to* n **do**

8 | | **if** agent $k_i \in \mathcal{A}$ **then**

9 | | Pick the good $g \leftarrow \arg\min_{\{g \in M'\}} c(g)$ to agent k_i such that
                       $v_{k_i}(g) = 1$ and $c(X_{k_i}) + c(g) \leq B_{k_i}$;

10 | | $X_{k_i} \leftarrow X_{k_i} \cup g$, $M' \leftarrow M' \setminus g$;

11 | | **Procedure:** Eliminate *x-envy* of agent k_i; $\triangleright$ Eliminate other agent
                  x-envy k_i.

12 $X_0 \leftarrow M'$;
   Return X = $\{X_0, X_1, X_2, ..., X_n\}$;

We have

$$v_{k_j}(X_{k_j}) = |S| - 2.$$

If $j < i$, k_j should be considered earlier than k_i in each round. According to the algorithm, $g_{k_i}^\ell$ must be assigned to agent k_j, not k_i.

Therefore, this lemma holds. $\square$

Suppose the case mentioned in Lemma 3 happens, i.e., agent k_j *x-envies* agent k_i at some time step, the **Procedure** can eliminate the *x-envy* by switching a good from k_i to k_j. To justify the correctness of the procedure, two facts must be verified. (1) The switched good has positive value for agent k_j, and (2) agent k_j can afford the switched good.

Remark 2. If agent k_j *x-envies* agent k_i at round ℓ and S is the subset chosen by the procedure, we have $v_{k_j}(g_{k_i}^\ell) = 0$ and for any $g \in S \setminus g_{k_i}^\ell$, $v_{k_j}(g) = 1$.

Remark 3. If agent k_j *x-envies* agent k_i at round ℓ and S is the subset chosen by the procedure, we have $c(X_{k_j}) \leq c(S \setminus g^*)$ and $c(X_{k_j}) + c(g^*) \leq c(S) \leq B_{k_j}$.

The full proof version of Remark 2 and Remark 3 can be found in Appendices.

To show the correctness of the algorithm, we need to prove the following two lemmas.

Lemma 4. *In each round, the elimination of x-envy is one-way, i.e., after handling the pair (k_i, k_j) $(k_i < k_j)$, agent k_j does not x-envy agent k_i w.r.t. B_{k_j} and moreover, the x-envy property of other agent k_ℓ $(\ell < j)$ does not affected.*

Procedure: Eliminate x-$envy$ of agent k_i

---

while some agents x-$envy$ agent k_i w.r.t. her budget **do**

    Let k_j be the minimal index of j violating the BFEFX condition; $\triangleright\ i < j$
    according to Lemma 3

        $S \leftarrow \arg\max_{S \subseteq X_{k_i}} v_{k_j}(S);$ $\triangleright\ |S| - 1 \leq v_{k_j}(S) \leq |S|$

        $g^* \leftarrow \arg\max_{g' \in S \setminus g_{k_i}^{\ell}} c(g');$

        $X_{k_i} \leftarrow X_{k_i} \setminus g^*,\ X_{k_j} \leftarrow X_{k_j} \cup g^*;$

        if $v_{k_i}(g) = 0$ for all $g \in M'$ or $c(X_{k_i}) + c(g) > B_{k_i}$ for all $g \in M'$ such
    that $v_{k_i}(g) = 1$ **then**

            $\mathcal{A} \leftarrow \mathcal{A} \setminus k_i;$

        else

            Pick the good $g \leftarrow \arg\min_{g \in M'}\{c(g) | v_{k_i}(g) = 1\}$

            $X_{k_i} \leftarrow X_{k_i} \cup g,\ M' \leftarrow M' \setminus g;$

    Procedure: Eliminate x-$envy$ of agent k_j.

Proof. After the first round, no matter whether or not an agent gets a good, no agent x-$envies$ others.

In some following step, x-$envy$ may happen, w.l.o.g., agent k_j x-$envies$ agent k_i w.r.t. B_{k_j}. As analyzed in Lemma 3, $|X_{k_i}| \geq |S| > |X_{k_j}| + 1$ and $|X_{k_i}| - |X_{k_j}| \geq |S| - |X_{k_j}| = 2$, where S is the goods set selected by the procedure.

According to the algorithm, good g^* will be reallocated from agent k_i to agent k_j and satisfying $c(X_{k_j}) + c(g^*) \leq B_{k_j}$. Let X'_{k_j} and X'_{k_i} denote the updated bundle of agents k_j and k_i, respectively. Since

$$v_{k_j}(X'_{k_j}) = |X_{k_j}| + 1 = |S| - 1 > v_{k_j}(S \setminus g^*) = |S| - 2$$

and

$$v_{k_i}(X'_{k_i}) = |X_{k_i}| - 1 \geq |X_{k_j}| + 1 \geq v_{k_i}(X'_{k_j}),$$

the x-$envy$ between agents k_i and k_j is eliminated.

When agent k_j gets the good g^*, a new x-$envy$ may happen between agent k_j and some other agent k_ℓ. We claim that $\ell > j$. Otherwise, assume that $\ell < j$. When allocating g^* to agent k_i, agent k_ℓ x-$envies$ k_i. Since g^* is assigned to agent k_i in some previous round, such x-$envy$ should be already eliminated, contradicting to the assumption. Thus, the newly introduced x-$envy$ only happens on some agent k_ℓ with $\ell > j$. Therefore, the procedure will be implemented in a one-way fashion. $\qquad\square$

Lemma 5. *When the algorithm stops, no agent x-$envies$ the charity X_0.*

The full proof version can be found in Appendices.

Lemma 6. *The algorithm will be stopped in $O(mn^2)$ time.*

Proof. According to Lemma 4, when assigning good to an agent, the **procedure** to eliminate x-$envy$ will be executed at most $O(n)$ times. Combining with the outer **while** loop (repeat $O(m)$ times) and the **for** loop (repeat $O(n)$ times), the running time of the algorithm is $O(mn^2)$. $\qquad\square$

Theorem 2. *The proposed algorithm correctly computes a BFEFX allocation for Binary variant in $O(mn^2)$ time.*

Proof. Combining Lemma 4, Lemma 5 and Lemma 6, this theorem can be proved. □

5   Concluding Remarks

Fair allocation is important in many fields and has been well studied during recent years. The concept of the maximum Nash Social Welfare is fundamental while budget feasible is also a reasonable constraint. The combination of these three notions, *fairness, Nash Social Welfare and budget-feasible*, are interesting and valuable in both theory and practice. In this work, we investigate the relationship between the maximum Nash Social Welfare allocation and BFEFX allocations in the budget feasible setting.

However, some important issues in this area are still not clear. Is it possible to find a budget-feasible allocation which exactly or approximately satisfies the maximum Nash Social Welfare? For the budget-feasible setting, the best guarantee so far is a $\frac{1}{2}$-BFEF1 allocation for *Identical* variant [13]. Does there always exist a BFEF1 allocation in general case? If the answer is no, can we find an efficient algorithm to approximate the property of BFEF1? In addition, it is not clear whether EFX allocation exists when the number of agents is greater than 3 [7]. For the budget-feasible setting, whether the BFEFX allocation in general case can be approximated with a good ratio. Any positive or negative answer of these questions may lead to a better understanding for the fair allocation.

References

1. Amanatidis, G., Birmpas, G., Filos-Ratsikas, A., Hollender, A., Voudouris, A.A.: Maximum Nash welfare and other stories about EFX. In: Proceedings of the 29th International Joint Conference on Artificial Intelligence, pp. 24–30. IJCAI 2020, AAAI Press, Yokohama, Japan (2020)
2. Barman, S., Krishnamurthy, S.K., Vaish, R.: Finding fair and efficient allocations. In: Proceedings of the 19th ACM Conference on Economics and Computation, pp. 557–574. EC 2018, ACM, New York, NY, USA (2018)
3. Barman, S., Krishnamurthy, S.K., Vaish, R.: Greedy algorithms for maximizing Nash social welfare. In: Proceedings of the 17th International Conference on Autonomous Agents and MultiAgent Systems, pp. 7–13. AAMAS 2018, ACM, Stockholm, Sweden (2018)
4. Biswas, A., Barman, S.: Fair division under cardinality constraints. In: Proceedings of the 27th International Joint Conference on Artificial Intelligence, pp. 91–97. IJCAI 2018, AAAI Press, Stockholm, Sweden (2018)
5. Caragiannis, I., Gravin, N., Huang, X.: Envy-freeness up to any item with high Nash welfare: The virtue of donating items. In: Proceedings of the 20th ACM Conference on Economics and Computation, pp. 527–545. EC 2019, ACM, New York, NY, USA (2019)

6. Caragiannis, I., Kurokawa, D., Moulin, H., Procaccia, A.D., Shah, N., Wang, J.: The unreasonable fairness of maximum Nash welfare. ACM Trans. Econ. Comput. **7**(3), 1–32 (2019)

7. Chaudhury, B.R., Garg, J., Mehlhorn, K.: EFX exists for three agents. In: Proceedings of the 21st ACM Conference on Economics and Computation, pp. 1–19. EC 2020, ACM, New York, NY, USA (2020)

8. Chaudhury, B.R., Kavitha, T., Mehlhorn, K., Sgouritsa, A.: A little charity guarantees almost envy-freeness. SIAM J. Comput. **50**(4), 1336–1358 (2021)

9. Cole, R., Devanur, N., Gkatzelis, V., Jain, K., Mai, T., Vazirani, V.V., Yazdanbod, S.: Convex program duality, fisher markets, and Nash social welfare. In: Proceedings of the 18th ACM Conference on Economics and Computation, pp. 459–460. EC 2017, ACM, New York, NY, USA (2017)

10. Cole, R., Gkatzelis, V.: Approximating the Nash social welfare with indivisible items. In: Proceedings of the Forty-Seventh Annual ACM Symposium on Theory of Computing, pp. 371–380. STOC 2015, ACM, New York, NY, USA (2015)

11. Dolev, D., Feitelson, D.G., Halpern, J.Y., Kupferman, R., Linial, N.: No justified complaints: on fair sharing of multiple resources. In: Innovations in Theoretical Computer Science, pp. 68–75. ITCS 2012, ACM, New York, NY, USA (2012)

12. Eisenberg, E., Gale, D.: Consensus of subjective probabilities: the pari-mutuel method. Ann. Math. Statist. **30**(1), 165–168 (1959)

13. Gan, J., Li, B., Wu, X.: Approximately envy-free budget-feasible allocation (2021). https://arxiv.org/abs/2106.14446

14. Garg, J., Kulkarni, P., Kulkarni, R.: Approximating Nash social welfare under submodular valuations through (un)matchings. In: Proceedings of the 2020 ACM-SIAM Symposium on Discrete Algorithms, pp. 2673–2687. SOSA 2020, SIAM, Salt Lake City, UT, USA (2020)

15. Ghodsi, A., Zaharia, M., Hindman, B., Konwinski, A., Shenker, S., Stoica, I.: Dominant resource fairness: Fair allocation of multiple resource types. In: Proceedings of the 8th USENIX Conference on Networked Systems Design and Implementation, pp. 323–336. NSDI 2011, USENIX Association, USA (2011)

16. Grandl, R., Ananthanarayanan, G., Kandula, S., Rao, S., Akella, A.: Multi-resource packing for cluster schedulers. In: ACM SIGCOMM 2014 Conference, pp. 455–466. SIGCOMM 2014, ACM, Chicago, IL, USA (2014)

17. Kaneko, M., Nakamura, K.: The Nash social welfare function. Econometrica **47**(2), 423–435 (1979)

18. Lee, E.: Apx-hardness of maximizing Nash social welfare with indivisible items. Inf. Process. Lett. **122**, 17–20 (2017)

19. Li, P., Hua, Q., Hu, Z., Ting, H.-F., Zhang, Y.: Approximation algorithms for the selling with preference. J. Combinat. Optim. **40**(2), 366–378 (2020). https://doi.org/10.1007/s10878-020-00602-3

20. Lipton, R., Markakis, E., Mossel, E., Saberi, A.: On approximately fair allocations of indivisible goods. In: Proceedings of the 5th ACM Conference on Electronic Commerce, pp. 125–131. EC 2004, ACM, New York, NY, USA (2004)

21. Moulin, H.: Fair Division and Collective Welfare, vol. 1, 1st edn. MIT Press (2003)

22. Wu, X., Li, B., Gan, J.: Budget-feasible maximum Nash social welfare allocation is almost envy-free (2020). https://arxiv.org/abs/2012.03766

23. Zhang, Y., Chin, F.Y., Poon, S.H., Ting, H.F., Xu, D., Yu, D.: Offline and online algorithms for single-minded selling problem. Theoret. Comput. Sci. **821**, 15–22 (2020)

Two-Facility Location Games
with Distance Requirement

Ling Gai[1][✉] , Dandan Qian[1] , and Chenchen Wu[2]

[1] Glorious Sun School of Business and Management, Donghua University,
Shanghai 200051, China
lgai@dhu.edu.cn, qiandandan@mail.dhu.edu.cn
[2] College of Science, Tianjin University of Technology, Tianjin 300384, China

Abstract. We consider the game of locating two homogeneous facilities
in the interval $[0, 1]$ with maximum distance requirement. In this game,
n agents report their preferred locations, then the designed mechanism
outputs the locations of two homogeneous facilities such that the total
utility of agents is maximized or the total cost is minimized. The location
information of agents is private and could be misreported to influence
the output. A strategy-proof mechanism with good performance ratio
need to be designed. We first show that for the desirable facilities case,
there is no deterministic strategy-proof mechanism can reach a constant
approximation ratio comparing with the optimal solution without private
information. Then we focus on the obnoxious facilities case. We propose
four group strategy-proof mechanisms and prove their approximation
ratios, separately. The performance of mechanisms are compared under
different maximum distance requirement.

Keywords: Facility location game · Strategy-proof mechanism ·
Distance requirement · Approximation ratio

1 Introduction

Facility location games have been in the center of research at the intersection
of artificial intelligence and social choice due to their practical importance. In
its original motivation, it is applied to determine the geographical locations of
some public facilities. The government first announces the mechanism employed,
then collects the private information of users (agents) and outputs the final locations. Generally, the government tries to adopt a good mechanism to minimize
the social cost or maximize the social welfare, while this objective relies heavily on the authenticity of data collected. The reason that agents may report
untruthfully is for better rewards, being closer to a preferred facility or farther
away from an obnoxious one. In order to incentivize the agents to be honest, a
truthful mechanism has to be designed. This guarantee of real data comes with
a cost on the objective value. So, a strategy-proof mechanism with good performance becomes the desire solution in the computational social choice area.

M. Li and X. Sun (Eds.): IJTCS-FAW 2022, LNCS 13461, pp. 15–24, 2022.
https://doi.org/10.1007/978-3-031-20796-9_2

Chan et al. [1] gave a thorough and comprehensive survey on this topic. Facility location game also has excellent application in other non-geographical fields, like the temperature choosing for a classroom, or committee voting for people with different political reviews.

In this paper, we focus on the games of locating two homogeneous facilities in the interval $[0, 1]$ with the maximum distance requirement included. It is meaningful in reality especially when the facilities belong to one same government or company, who does not want to pay too much operating and transportation cost on connecting facilities. Both desirable facility and obnoxious facility are considered. Without specified statement, the facilities are assumed to be homogenous, which means they are both desirable or both obnoxious to all the agents.

Here are some motivations for the problem with distance requirement. Take locating the 5G mobile communication base stations as an example, the distance between stations should not be too large due to the limited service interval. For the band of 2.6 GHz and the edge rate of 2 Mbps, the maximum distance between two stations is about 450 m in dense urban areas, about 700 m in general urban areas, and about 1300 m in suburban areas. For the band of 3.5 GHz, due to the high frequency and poor coverage, the maximum distance between stations is about 100 m smaller than that at 2.6 GHz. This spacing standard is currently occupied by the three major operators of China (China Mobile, China Unicom and China Telecom). Another example is the locating of vertical garbage stations and waste disposal site. Generally the garbages are collected by sanitation workers and dumped into the vertical garbage station, then they are compressed to blocks and delivered to large waste disposal site. The distance between garbage station and waste disposal site should be in some reasonable range to guarantee the operating efficiency. Other cases include the distance between metro stations, the political views in election campaign, etc.

Related Works. There are huge number of literature concerning about facility location games. The most related results are in [6,17], several mechanisms with minimum distance requirement are designed. Note that some of the output facilities satisfies the boundary condition, which means these mechanisms are also applicable in our case with maximum distance requirement. We will show that they are not always the good option and the choice of mechanisms is correlated to the distance requirement parameter heavily. Procaccia and Tennenholtz [15] first studied strategy-proof mechanism design for one facility on a line. Golomb and Tzamos [10] studied the output of a facility location in the range of $[0, 1]$, focusing on minimizing the worst-case additive errors by using additive approximation ratio. Zhang and Li [18] discussed the strategy-proof mechanism when locating one facility on a line and all agents have their own weights. [13] improved the lower bound from $n-2$ to $(n-1)/2$ for the deterministic strategy-proof mechanism. Fotakis and Tzamos [9] showed that any strategy-proof mechanism either allows a unique dictator or always places the facilities at the leftmost and rightmost positions is feasible when the number of agents $n \geq 5$ in the game. Fotakis et al. [8] studied the multistage K-facility reallocation problem on a line, where the facility location is based on the stage-dependent locations of n agents.

Mei et al. [14] created a happiness function to measure the satisfaction of agents and studied the upper and lower bounds of the mechanism based on the preference and location. Krogmann et al. [11] considered facility location games where the strategic facilities and clients influence each other. Li et al. [12] studied the facility location games with payments and assume that facilities are strategic players on a line.

The study for obnoxious facility location with deterministic and randomized mechanisms was firstly proposed by Cheng et al. [3]. Zou and Li [19] proposed strategy-proof mechanisms for the facility location game with dual preference when both location and preference were private information. Dokow et al. [5] studied strategy-proof mechanisms on a discrete graph. Feldman et al. [7] restricted the location of facilities to a set of fixed candidate points in a certain interval to study the mechanism design. Besides these discrete constraints, distance requirement on the facility location has also been studied. Cheng et al. [4] put forward the model in which all agents want to stay away from the facilities within the service range $1/2 \leq r \leq 1$. Chen et al. [2] presented the problem of dispersing n agents in a k–dimensional polytope meanwhile keeping a suitable distance away from each other. Wu et al. [16] focused on a circle network to design strategy-proof mechanisms with the minimum distance requirement.

Our Results. For the desirable facility location game with maximum distance requirement, we prove the negative result that no deterministic strategy-proof mechanism exists with constant approximation ratio. Then we consider the obnoxious facility location game. Four mechanisms are designed separately and proved to be strategy-proof. Their approximation ratios are showed in Table 1.

Table 1. Summary of the approximation ratio results under distance requirement b, $0 < b \leq 1$

$0 < |y_1 - y_2| \leq b$		
Obnoxious	UB	LB
M1	$\frac{4}{1-2b}$	3
M2	$max\{\frac{8}{b}, \frac{3}{1-b}\}$	5
M3	$max\{\frac{16}{3-2b}, \frac{6}{2b-1}\}$	$\frac{3}{2}$
M4	$max\{\frac{4}{1-b}, \frac{3}{2b-1}\}$	$\frac{3}{2}$
Desirable	$+\infty$	/

Organization of the Paper. The rest of paper is organized as follows. In Sect. 2, the definitions and terminologies are presented. In Sect. 3 the negative result of desirable two-facility location game with maximum distance requirement is analyzed. In Sect. 4 the obnoxious two-facility location game is studied. Four strategy-proof mechanisms are presented with approximation ratio proved. Section 5 presents the conclusion and sketch of our future work.

2   Preliminaries

Let $N = \{1, 2, \ldots, n\}$ be the set of agents. Each agent has a private information of his location x_i on a line interval $[0, 1]$. Let $\mathbf{x} = (x_1, ..., x_n)$ be the location profile. Two facilities need to be output by a mechanism $f(\mathbf{x})$ are denoted as $(y_1, y_2) \in I^2$. The distance requirement is denoted as $0 < |y_1 - y_2| \leq b$, where $b \in (0, 1]$.

For the desirable facilities case, the cost of an agent i is the minimum distance to one of the two facilities. That is

$$c_i(f(\mathbf{x}), x_i) = min\{|x_i - y_1|, |x_i - y_2|\}. \tag{1}$$

Following utilitarian objective, the social cost of a mechanism $f(\mathbf{x})$ with respect to x is the sum of costs of n agents, i.e.

$$SC(f(\mathbf{x}), x_i) = \sum_{i=1}^{n} c_i(f(\mathbf{x}), x_i). \tag{2}$$

In the obnoxious facility location game, the utility of an agent i is defined as the minimum distance to one of the two facilities which should be maximized, i.e.,

$$u_i(f(\mathbf{x}), x_i) = min\{|x_i - y_1|, |x_i - y_2|\}. \tag{3}$$

Following utilitarian objective, the social utility of a mechanism $f(\mathbf{x})$ with respect to $\mathbf{x}$ is

$$SU(f(\mathbf{x}), x_i) = \sum_{i=1}^{n} u_i(f(\mathbf{x}), x_i). \tag{4}$$

Following the definitions in literatures, a mechanism f is strategy-proof if any agent cannot increase the benefit by misreporting his location information. Specifically, given agent i, profile $\mathbf{x} = \{x_i, x_{-i}\} \in I^n$, and any location $x_i' \in I^n$, it holds that

$$u_i(f(x_i, \mathbf{x}_{-i}), x_i) \geq u_i(f(x_i', \mathbf{x}_{-i}), x_i).$$

A mechanism f is group strategy-proof if for any coalition of agents, at least one of them cannot benefit if they misreport simultaneously. Specifically, given a non-empty set $S \subseteq N$, profile $\mathbf{x} = \{x_S, x_{-S}\} \in I^n$, and any location $x_S' \in I^n$, it holds that

$$u_i(f(x_S, \mathbf{x}_{-S}), x_i) \geq u_i(f(x_S', \mathbf{x}_{-S}), x_i),$$

where $\mathbf{x}_{-i} = (x_1, x_2, ..., x_{i-1}, x_{i+1}, ..., x_n)$ is the location profile of all agents without agent i.

A mechanism f with maximizing objective has an approximation ratio $r \geq 1$ if for any location profile $\mathbf{x}$, the ratio between the optimal solution $OPT(\mathbf{x})$ and the mechanism solution $SU(f, \mathbf{x})$ is at most r, i.e., $OPT(\mathbf{x}) \leq rSU(f, \mathbf{x})$. The version with minimizing objective is defined similarly.

3   Desirable Two-Facility Location Game with Maximum Distance Requirement

In the desirable two-facility location game, all agents want to get close to one of the two homogeneous facilities. The cost of an agent i is $c_i((y_1, y_2), x_i) = min\{|x_i - y_1|, |x_i - y_2|\}$ and the social cost is $SC((y_1, y_2), \mathbf{x}) = \sum_{i=1}^{n} c_i((y_1, y_2), x_i)$. So, the problem with maximum distance requirement can be shown as

$$min_{(y_1, y_2) \in D} \sum_{i=1}^{n} min\{|x_i - y_1|, |x_i - y_2|\},$$

where $D = \{(y_1, y_2)|0 < |y_1 - y_2| \leq b, 0 \leq y_1 \leq y_2 \leq 1, 0 < b \leq 1\}$.

Fotakis and Tzamos [9] has proved that for instances with more than 5 agents, any deterministic strategy-proof mechanism either admits a unique dictator, or always places the facilities at the leftmost and the rightmost location of the instance. In the following we will show that these two mechanisms do not fit the case with maximum distance requirement. Furthermore, there is no other strategy-proof mechanism could reach a constant approximation ratio.

Theorem 1. *Given $n \geq 5$, no strategy-proof mechanism has bounded approximation ratio for the desirable two-facility location game with maximum distance requirement.*

The proof is omitted in this version.

4   Obnoxious Two-Facility Location Game with Maximum Distance Requirement

In this section, we study the obnoxious homogeneous two-facility location game, where all agents want to be far away from both homogeneous facilities. The utility of agent i is defined as $u_i((y_1, y_2), x_i) = min\{|x_i - y_1|, |x_i - y_2|\}$ and the social utility is $SU((y_1, y_2), \mathbf{x}) = \sum_{i=1}^{n} u_i((y_1, y_2), x_i)$. The problem can be shown as below

$$max_{(y_1, y_2) \in D} \sum_{i=1}^{n} min\{|x_i - y_1|, |x_i - y_2|\},$$

where $D = \{(y_1, y_2)|0 < |y_1 - y_2| \leq b, 0 \leq y_1 \leq y_2 \leq 1, 0 < b \leq 1\}$.

Mechanism M1. Define $T_1 = \{i|x_i \in [0, \frac{1}{2}], x_i \in \mathbf{x}\}, T_2 = \{i|x_i \in (\frac{1}{2}, 1], x_i \in \mathbf{x}\}, N = T_1 \bigcup T_2$.

Given $0 < b < \frac{1}{2}$, if $|T_1| \geq |T_2|$, return $(y_1, y_2) = (1 - b, 1)$; if $|T_1| < |T_2|$, return $(y_1, y_2) = (0, b)$.

Theorem 2. *The mechanism M1 is group strategy-proof with an approximation ratio $r = \frac{4}{1-2b}$ with $0 < b < \frac{1}{2}$.*

Proof. Given $0 < b < \frac{1}{2}$, for any agent i in T_1,

$$u_i((1-b,1),x_i) = (1-b) - x_i \geq u_i((0,b),x_i) = min\{x_i, |b - x_i|\}\}; \qquad (5)$$

for any agent i in T_2,

$$u_i((1-b,1),x_i) = min\{x_i, (1-b) - x_i\} \leq u_i((0,b),x_i) = x_i - b; \qquad (6)$$

Let $S \subseteq N$ be the collation of the agents. When all agents cannot gain together by misreporting, the mechanism is group strategy-proof. We denote n_1, n_2 as the number of agents in the interval T_1, T_2 with truthful report, respectively; n_1', n_2' as the number of agents in the interval T_1, T_2 with misreporting. The new false location profile is denoted as $\mathbf{x'}$ and the facility location is (y_1', y_2'). There are two cases.

Case 1. $n_1 \geq n_2$, thus $(y_1, y_2) = (1-b, 1)$.

Case 1.1. If $n_1' \geq n_2'$, then $(y_1', y_2') = (1-b, 1)$ and $u_i(f(\mathbf{x}), x_i) = u_i(f(\mathbf{x'}), x_i)$ for any agent $i \in N$.

Case 1.2. If $n_1' < n_2'$, then $(y_1', y_2') = (0, b)$. We can observe that at least one agent i in $[0, \frac{1}{2}]$ misreports his location to $x_i' \in (\frac{1}{2}, 1]$. Because of (5) and (6), $u_i(f(\mathbf{x}), x_i) \geq u_i(f(\mathbf{x'}), x_i)$.

Case 2. $n_1 < n_2$, thus $(y_1, y_2) = (0, b)$. The analysis of strategy-proofness is similar to that of Case 1.

The Mechanism is group strategy-proof.

We next consider the approximation ratio. Without loss of generality, assume that $|T_1| \geq |T_2|$, and the mechanism outputs $(y_1, y_2) = (1-b, 1)$. The approximation ratio

$$
\begin{aligned}
r &= \frac{OPT(\mathbf{x})}{SU((1-b,1),\mathbf{x})} \\
&= \frac{max_{y_1,y_2 \in D} \sum_{i \in N} min\{|y_1 - x_i|, |y_2 - x_i|\}}{\sum_{i \in T_1} min\{|x_i - 1 + b|, |1 - x_i|\} + \sum_{i \in T_2} min\{|x_i - 1 + b|, |1 - x_i|\}} \\
&\leq \frac{\sum_{i \in N} 1}{\sum_{i \in T_1}(1 - b - \frac{1}{2}) + \sum_{i \in T_2} 0} \\
&= \frac{n}{|T_1|(\frac{1}{2} - b)} \\
&\leq \frac{4}{1 - 2b}.
\end{aligned}
$$

$\square$

We can get the lower bound of mechanism M1 when $n/2$ agents are located at point $1/2$ and $n/2$ agents are at point 1. Under this case, $OPT(x) = 3n/4$, and by mechanism M1, the output is $(1 - b, 1)$, the social utility is $\frac{n}{2}(\frac{1}{2} - b)$. Since $\frac{OPT(x)}{SU((1-b,1),x_i)} = 3 + \frac{6b}{n - 2b}$, the lower bound is 3.

Mechanism 2. Define $T_3 = \{i | x_i \in [0, b/4], x_i \in \mathbf{x}\}$, $T_4 = \{i | x_i \in (b/4, 1/2], x_i \in \mathbf{x}\}$, $T_5 = \{i | x_i \in (1/2, \frac{4-b}{4}], x_i \in \mathbf{x}\}$ and $T_6 = \{i | x_i \in (\frac{4-b}{4}, 1], x_i \in \mathbf{x}\}$, $N =$

$T_3 \cup T_4 \cup T_5 \cup T_6$. Given $2/3 < b < 1$, if $|T_3| + |T_5| \geq |T_4| + |T_6|$, return $(y_1, y_2) = (b/2, 1)$; if $|T_3| + |T_5| < |T_4| + |T_6|$, return $(y_1, y_2) = (0, \frac{2-b}{2})$.

Before proving the performance of mechanism M2, we give the following conclusion.

Lemma 1. *For any agent* i,

$$|x_i - y_1| \leq |x_i - b| + 1, |x_i - y_2| \leq |1 - x_i| + 1, |x_i - y_1| = |x_i| + |y_1| \leq x_i + 1 \quad (7)$$

$$|x_i - y_1| \leq |x_i - \frac{3 - 2b}{4}| + 1 \quad (8)$$

$$|x_i - y_1| \leq |x_i - \frac{1 - b}{2}| + 1, |x_i - y_2| \leq |x_i - \frac{1 + b}{2}| + 1 \quad (9)$$

The proof of is omitted due to page constraint.

Theorem 3. *The mechanism is group strategy-proof with an approximation ratio* $r = max\{\frac{8}{b}, \frac{3}{1-b}\}$ *with* $2/3 < b < 1$.

Proof. Given $\frac{2}{3} < b < 1$, for any agent $i \in T_3$, $u_i((\frac{b}{2}, 1), x_i) = \frac{b}{2} - x_i \geq u_i((0, \frac{2-b}{2}), x_i) = x_i$;

for any agent $i \in T_4$, $u_i((\frac{b}{2}, 1), x_i) = |\frac{b}{2} - x_i| \leq u_i((0, \frac{2-b}{2}), x_i) = min\{x_i, \frac{2-b}{2} - x_i\}$;

for any agent $i \in T_5$, $u_i((\frac{b}{2}, 1), x_i) = min\{x_i - \frac{b}{2}, 1 - x_i\} \geq u_i((0, \frac{2-b}{2}), x_i) = |x_i - \frac{2-b}{2}|$;

for any agent $i \in T_6$, $u_i((\frac{b}{2}, 1), x_i) = 1 - x_i \leq u_i((0, \frac{2-b}{2}), x_i) = x_i - \frac{2-b}{2}$.

The proof of strategy-proofness is similar to that of Theorem 1.

Without loss of generality, assume that $|T_3| + |T_5| \geq |T_4| + |T_6|$, and the mechanism outputs $(y_1, y_2) = (b/2, 1)$.

$$SU((\frac{b}{2}, 1), x) = \sum_{i \in N} min\{|x_i - \frac{b}{2}|, |1 - x_i|\}$$

$$\geq |T_3|\frac{b}{4} + \sum_{i \in T_5} min\{|x_i - \frac{b}{2}|, |1 - x_i|\} + \sum_{i \in T_4 \cup T_6} min\{|x_i - \frac{b}{2}|, |1 - x_i|\}$$

From Lemma 1, we know that

$$OPT(x) = max_{(y_1, y_2) \in D} \sum_{i \in N} min\{|y_1 - x_i|, |y_2 - x_i|\}$$

$$\leq max_{(y_1, y_2) \in D} \sum_{i \in T_3} u_i(f(x), x_i) + max_{(y_1, y_2) \in D} \sum_{i \in T_5} u_i(f(x), x_i)$$

$$+ max_{(y_1, y_2) \in D} \sum_{i \in T_4 \cup T_6} u_i(f(x), x_i)$$

$$\leq |T_3|(1 - x_i) + |T_5|(1 - x_i) + \sum_{i \in T_4 \cup T_6} min\{|x_i - \frac{b}{2}|, |1 - x_i|\} + 1)$$

$$\leq |T_3|(2 - x_i) + |T_5|(2 - x_i) + \sum_{i \in T_4 \cup T_6} min\{|x_i - \frac{b}{2}|, |1 - x_i|\}$$

The approximation ratio is

$$r = \frac{OPT(x)}{SU(\frac{b}{2},1),x)}$$

$$\leq \frac{|T_3|(2-x_i)+|T_5|(2-x_i)}{|T_3|\frac{b}{4}+\sum_{i \in T_5} min\{|x_i-\frac{b}{2}|,|1-x_i|\}}$$

$$\leq max\{\frac{|T_3|(2-x_i)}{|T_3|\frac{b}{4}}, max\{\frac{|T_5|(2-x_i)}{|T_5||x_i-\frac{b}{2}|}, \frac{|T_5|(2-x_i)}{|T_5||1-x_i|}\}\}$$

$$= max\{\frac{8}{b}, \frac{3/2}{\frac{1}{2}-\frac{b}{2}}, \frac{4+b}{b}\} = max\{\frac{8}{b}, \frac{3}{1-b}\}$$

Lower Bound. Suppose there are $n/4$ agents located at point $b/4$, $n/4$ agents located at point $1/2$, $n/4$ agents located at point $\frac{4-b}{4}$ and $n/4$ agents located at point 1. Under this case, $OPT(x) = \frac{5n}{8}$, and by mechanism M2, the output is $(b/2,1)$, the social utility is $\frac{n}{8}$. So $OPT(x)/SU((b/2,1),x_i) = 5$, the lower bound of mechanism M2 is 5.

Mechanism M3. Define $T_7 = \{i|x_i \in [0,\frac{3-2b}{8}], x_i \in \mathbf{x}\}, T_8 = \{i|x_i \in (\frac{3-2b}{8},\frac{1}{2}], x_i \in \mathbf{x}\}, T_9 = \{i|x_i \in (\frac{1}{2},\frac{5+2b}{8}], x_i \in \mathbf{x}\}$ and $T_{10} = \{i|x_i \in (\frac{5+2b}{8},1], x_i \in \mathbf{x}\}, N = T_7 \cup T_8 \cup T_9 \cup T_{10}$.

Given $\frac{1}{2} \leq b \leq 1$, if $|T_7|+|T_9| \geq |T_8|+|T_{10}|$, return $(y_1,y_2) = (\frac{3-2b}{4},1)$; if $|T_7|+|T_9| < |T_8|+|T_{10}|$, return $(y_1,y_2) = (0,\frac{1+2b}{4})$.

Theorem 4. *The mechanism M3 is group strategy-proof with an approximation ratio* $r = max\{\frac{16}{3-2b}, \frac{6}{2b-1}\}$.

The proof is omitted here and stated in the extended version.

Lower Bound. Suppose there are $n/4$ agents located at point $\frac{3-2b}{8}$, $n/4$ agents located at point $1/2$, $n/4$ agents located at point $(5+2b)/8$ and $n/4$ agents located at point 1. In this case, $OPT(x) = 3n/8$, and the output of mechanism M3 is $(\frac{3-2b}{4},1)$ with the social utility of $\frac{n}{8(1+b)}$. Since $OPT(x)/SU((\frac{3-2b}{4},1),x_i) = \frac{3}{1+b} \in [3/2,2)$, the lower bound of mechanism M3 is 3/2.

Mechanism M4. Define $T_{11} = \{i|x_i \in [0,\frac{1-b}{2}], x_i \in \mathbf{x}\}, T_{12} = \{i|x_i \in (\frac{1-b}{2},1/2], x_i \in \mathbf{x}\}, T_{13} = \{i|x_i \in (1/2,\frac{1+b}{2}], x_i \in \mathbf{x}\}$ and $T_{14} = \{i|x_i \in (\frac{1+b}{2},1], x_i \in \mathbf{x}\}, N = T_{11} \cup T_{12} \cup T_{13} \cup T_{14}$.

Given $1/2 \leq b \leq 1$, if $|T_{11}|+|T_{13}| \geq |T_{12}|+|T_{14}|$, return $(y_1,y_2) = (1-b,1)$; if $|T_{11}|+|T_{13}| < |T_{12}|+|T_{14}|$, return $(y_1,y_2) = (0,b)$.

Theorem 5. *The mechanism M4 is group strategy-proof with an approximation ratio* $r = max\{\frac{4}{1-b}, \frac{3}{2b-1}\}$.

The proof is omitted due to page constraint.

Lower Bound. Suppose there are $n/4$ agents located at point 0, $n/4$ agents located at point $\frac{1-b}{2}$, $n/4$ agents located at point $\frac{1+b}{2}$ and $n/4$ agents located at point 1. In this case, $OPT(x) = \frac{n}{4(1+b)}$. The output of mechanism M4 is $(1-b, 1)$, the social utility is $\frac{n}{2(1-b)}$. Since $\frac{OPT(x)}{SU((1-b,1),x_i)} = \frac{1+b}{2-2b} \in [3/2, +\infty)$, the lower bound of mechanism M4 is 3/2.

Given above four mechanisms with approximation ratios related to distance requirement b, we present the Fig. 1 to show their comparison. The curves reflect the change of approximation ratios.

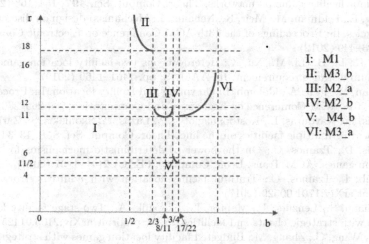

I: M1
II: M3_b
III: M2_a
IV: M2_b
V: M4_b
VI: M3_a

Fig. 1. The mechanism with better performance under different distance requirement

5 Conclusions

In this paper, we consider the two-facility location game with maximum distance requirement. For the game of locating two desirable facilities with maximum distance requirement on a line interval, we show that no deterministic strategy-proof mechanism have a bounded approximation ratio comparing with the optimal solution. For the game of locating two obnoxious facilities with maximum distance requirement, we propose group strategy-proof mechanisms and prove their performance ratios. Since the performance ratio is highly correlated to distance requirement parameter b, we recommend Fig. 1 to show the proper choice of mechanisms under different circumstances.

Future work includes the randomized mechanism design for the obnoxious two-facility location game with distance requirement and the case with more facilities. We will also study the facility location game with both minimum and maximum distance requirement.

References

1. Chan, H., Filos-Ratsikas, A., Li, B., Li, M., Wang, C.: Mechanism design for facility location problems: a survey. arXiv preprint arXiv:2106.03457 (2021)
2. Chen, J., Li, B., Li, Y.: Efficient approximations for the online dispersion problem. SIAM J. Comput. **48**(2), 373–416 (2019)
3. Cheng, Y., Yu, W., Zhang, G.: Mechanisms for obnoxious facility game on a path. In: Wang, W., Zhu, X., Du, D.-Z. (eds.) COCOA 2011. LNCS, vol. 6831, pp. 262–271. Springer, Heidelberg (2011). https://doi.org/10.1007/978-3-642-22616-8_21
4. Cheng, Y., Yu, W., Zhang, G.: Strategy-proof approximation mechanisms for an obnoxious facility game on networks. Theor. Comput. Sci. **497**, 154–163 (2013)
5. Dokow, E., Feldman, M., Meir, R., Nehama, I.: Mechanism design on discrete lines and cycles. In: Proceedings of the 13th ACM Conference on Electronic Commerce, pp. 423–440 (2012)
6. Duan, L., Li, B., Li, M., Xu, X.: Heterogeneous two-facility location games with minimum distance requirement. In: AAMAS, pp. 1461–1469 (2019)
7. Feldman, M., Fiat, A., Golomb, I.: On voting and facility location. In: Proceedings of the 2016 ACM Conference on Economics and Computation, pp. 269–286 (2016)
8. Fotakis, D., Kavouras, L., Kostopanagiotis, P., Lazos, P., Skoulakis, S., Zarifis, N.: Reallocating multiple facilities on the line. Theor. Comput. Sci. **858**, 13–34 (2021)
9. Fotakis, D., Tzamos, C.: On the power of deterministic mechanisms for facility location games. ACM Trans. Econ. Comput. (TEAC) **2**(4), 1–37 (2014)
10. Golomb, I., Tzamos, C.: Truthful facility location with additive errors. arXiv preprint arXiv:1701.00529 (2017)
11. Krogmann, S., Lenzner, P., Molitor, L., Skopalik, A.: Two-stage facility location games with strategic clients and facilities. arXiv preprint arXiv:2105.01425 (2021)
12. Li, M., Wang, C., Zhang, M.: Budgeted facility location games with strategic facilities. In: Proceedings of the Twenty-Ninth International Conference on International Joint Conferences on Artificial Intelligence, pp. 400–406 (2021)
13. Lu, P., Sun, X., Wang, Y., Zhu, Z.A.: Asymptotically optimal strategy-proof mechanisms for two-facility games. In: Proceedings of the 11th ACM Conference on Electronic Commerce, pp. 315–324 (2010)
14. Mei, L., Li, M., Ye, D., Zhang, G.: Facility location games with distinct desires. Discrete Appl. Math. **264**, 148–160 (2019)
15. Procaccia, A.D., Tennenholtz, M.: Approximate mechanism design without money. ACM Trans. Econ. Comput. (TEAC) **1**(4), 1–26 (2013)
16. Wu, X., Mei, L., Zhang, G.: Two-facility location games with a minimum distance requirement on a circle. In: Du, D.-Z., Du, D., Wu, C., Xu, D. (eds.) COCOA 2021. LNCS, vol. 13135, pp. 497–511. Springer, Cham (2021). https://doi.org/10.1007/978-3-030-92681-6_39
17. Xu, X., Li, B., Li, M., Duan, L.: Two-facility location games with minimum distance requirement. J. Artif. Intell. Res. **70**, 719–756 (2021)
18. Zhang, Q., Li, M.: Strategyproof mechanism design for facility location games with weighted agents on a line. J. Comb. Optim. **28**(4), 756–773 (2014). https://doi.org/10.1007/s10878-013-9598-8
19. Zou, S., Li, M.: Facility location games with dual preference. In: Proceedings of the 2015 International Conference on Autonomous Agents and Multiagent Systems, pp. 615–623 (2015)

Constrained Heterogeneous Two-Facility Location Games with Max-Variant Cost

Qi Zhao, Wenjing Liu(✉)(iD), Qizhi Fang, and Qingqin Nong(iD)

Ocean University of China, Qingdao 266100, China
zq1012@stu.ouc.edu.cn, {liuwj,qfang,qqnong}@ouc.edu.cn

Abstract. In this paper, we propose a constrained heterogeneous facility location model where a set of alternative locations are feasible for building facilities and the number of facilities built at each location is limited. Supposing that a set of agents on the real line can strategically report their locations and each agent's cost is her distance to the further facility that she is interested in, we study deterministic mechanism design without money for constrained heterogeneous two-facility location games.

Depending on whether agents have optional preference, the problem is considered in two settings: the compulsory setting and the optional setting. In the compulsory setting where each agent is served by the two heterogeneous facilities, we provide a 3-approximate deterministic group strategyproof mechanism for the sum/maximum cost objective respectively, which is also the best deterministic strategyproof mechanism under the corresponding social objective. In the optional setting where each agent can be interested in one of the two facilities or both, we propose a deterministic group strategyproof mechanism with approximation ratio of at most $2n+1$ for the sum cost objective and a deterministic group strategyproof mechanism with approximation ratio of at most 9 for the maximum cost objective.

Keywords: Mechanism design · Facility location · Strategyproof · Constrained

1 Introduction

In the origin mechanism design problem for heterogeneous facility location games, there are a set of strategic agents who are required to report their private information and a social planner intends to locate several heterogeneous facilities by a mechanism based on the reported information, with the purpose of optimizing some social objective. In this paper, we study the problem of locating two heterogeneous facilities under a constrained setting, which means a set of alternative locations are feasible for building facilities and the number of facilities built at each location is limited.

Compared with the origin setting where facilities can be built anywhere in a specific metric space and there is no limit on the number of facilities at each

© The Author(s), under exclusive license to Springer Nature Switzerland AG 2022
M. Li and X. Sun (Eds.): IJTCS-FAW 2022, LNCS 13461, pp. 25–43, 2022.
https://doi.org/10.1007/978-3-031-20796-9_3

location, our constrained setting models well many practical applications. For example, in the realistic urban planning, facilities can only be built at designated sites and the number of facilities at each site is limited. To accommodate these constraints, we propose a multiset of feasible locations and at most one facility is permitted to build at each location. Further, we focus on the Max-variant where the cost of each agent depends on her distance to the farthest one if she is served by two or more heterogeneous facilities. The Max-variant can be found applications in natural scenarios [26]. For example, a local authority plans to locate different raw material warehouses for several processing plants. Assuming each plant has multiple transport trucks having the same speed, the time that the plant has to wait depends on its distance to the farthest one if it requires raw materials from different sites.

We discuss the mechanism design problem for constrained heterogeneous two-facility location games with Max-variant cost in two settings: the first is the compulsory setting, where each agent is served by the two heterogeneous facilities; the second is the optional setting, where each agent is served by either one of the two facilities or both. Considering that agents may manipulate the facility locations by misreporting their private information, we concentrate on mechanisms that can perform well under some social objective (e.g., minimizing the sum/maximum cost) while guaranteeing truthful report from agents (i.e., strategyproof or group strategyproof).

1.1   Our Contribution

This paper studies deterministic mechanism design without money for constrained heterogeneous two-facility location games with Max-variant cost under the objective of minimizing the sum/maximum cost.

Our key innovations and results are summarized as follows.

In Sect. 2, we formulate the constrained heterogeneous facility location game with Max-variant cost. We propose a finite multiset of alternative locations which are feasible for building facilities and require that at most one facility can be built at each location. Thus, by adjusting the number of same elements in the multiset, the model can accommodate different scenarios where the number of facilities at the same location is limited.

In Sect. 3, we focus on deterministic mechanism design in the compulsory setting. We propose a set of adjacent alternative location pairs, which all agents have single peaked preferences over and the optimal solution under the sum/maximum cost objective can always be found in. We prove that any deterministic strategyproof mechanism has an approximation ratio of at least 3 under the sum/maximum cost objective. In addition, we present 3-approximate deterministic group strategyproof mechanisms for both social objectives, which implies that the best deterministic strategyproof mechanisms have been obtained.

In Sect. 4, we discuss the optional setting. For the sum cost objective, we propose a deterministic group strategyproof mechanism with approximation ratio of at most $2n + 1$. For the maximum cost objective, we design a deterministic group strategyproof mechanism with approximation ratio of at most 9.

1.2 Related Work

Mechanism design without money for facility location games has been extensively studied in recent years. Early studies focused on the characterization of strategyproof mechanisms. Moulin [19] identified all the possible strategyproof mechanisms for one-facility location on the line with single peaked preferences, whose results were extended by Schummer and Vohra [21] and Dokow et al. [9] to tree and cycle networks.

Approximate mechanism design without money was initiated by Procaccia and Tennenholtz [20], who studied deterministic and randomized strategyproof mechanisms with constant approximation ratio for facility location games under the sum cost and the maximum cost in three settings: one-facility, two-facility and multiple facilities per agent. Following this research agenda, numerous studies have emerged, including improvements on the lower/upper bound of approximation [14,17] and further variants.

Cheng et al. [7] introduced approximate mechanism design for obnoxious facility location games where the facility is not desirable to each agent. Zou and Li [29] studied the dual preference setting where the facility can be desirable or undesirable for different agents. Zhang and Li [27] introduced weights to agents and Filos-Ratsikas et al. [12] studied one-facility location problem with double-peaked preferences. Serafino and Ventre [22] introduced heterogeneous two-facility location games where each agent cares about either one facility or both and her cost depends on the sum of distances to her interested facilities (referred to as the Sum-variant). Later, Yuan et al. [26] considered the Min-variant and Max-variant instead and Anastasiadis and Deligkas [1] studied heterogeneous k-facility setting with Min-variant. Besides, various individual and social objectives were also studied. Mei et al. [18] introduced a happiness factor to measure each agent's individual utility. Feigenbaum and Sethuraman [10] considered the L_p-form of the vector of agent-costs instead of the classic sum cost. Cai et al. [4] and Chen et al. [5] studied facility location problems under the objective of minimizing the maximum envy. Ding et al. [8] and Liu et al. [16] considered the envy ratio objective. Zhou et al. [28] studied group-fair facility location problems.

Further, motivated by real-world applications, researchers have begun to study the mechanism design problem with constraints on the facilities. Aziz et al. [2,3] studied facility location problems with capacity constraints. Chen et al. [6] studied the two-opposite-facility location problem with maximum distance constraint by imposing a penalty. Xu et al. [25] studied minimum distance requirement for the heterogeneous two-facility location problem. In addition, considering that in reality the feasible locations that facilities could be built at are usually limited, mechanism design for facility location games with limited locations were also studied. Sui and Boutilier [23] studied approximately strategyproof mechanisms for facility location games with constraints on the feasible placement of facilities. Feldman et al. [11] studied the one-facility location setting under the sum cost objective in the context of voting embedded in some underlying metric space. Tang et al. [24] further considered the maximum cost objective and the two-facility setting. Li et al. [15] studied the heterogeneous two-facility setting with optional preference, which is also the most related to our work among all studies on the constrained heterogeneous facility location problem. However, there are at least three differences between us: (1) our model requires a limit on the number of facilities at each feasible location and [15] does not; (2) each agent's location is private and her preference on facilities is public in our model while it is the opposite in [15]; (3) we consider the Max-variant cost while [15] considers the Min-variant where the cost of each agent depends on her distance to the closest facility within her acceptable set.

2 Model

Let $N = \{1, 2, \ldots, n\}$ be a set of agents located on the real line $\mathcal{R}$ and $\mathcal{F} = \{F_1, F_2\}$ be the set of two heterogeneous facilities to be built. Each agent $i \in N$ has a location $x_i \in \mathcal{R}$ and a facility preference $p_i \subseteq \mathcal{F}$, where x_i is i's private information and p_i is public. Denote $\mathbf{x} = (x_1, x_2, \ldots, x_n)$ and $\mathbf{p} = (p_1, p_2, \ldots, p_n)$ as the n agents' location profile and facility preference profile, respectively. For $i \in N$, let $\mathbf{x}_{-i} = (x_1, \ldots, x_{i-1}, x_{i+1}, \ldots, x_n)$ be the location profile without agent i, then $\mathbf{x} = (x_i, \mathbf{x}_{-i})$. For $S \subseteq N$, denote $\mathbf{x}_S = (x_i)_{i \in S}$, $\mathbf{p}_S = (p_i)_{i \in S}$, and $\mathbf{x}_{-S} = (x_i)_{i \notin S}$, then $\mathbf{x} = (\mathbf{x}_S, \mathbf{x}_{-S})$.

Let $A = \{a_1, a_2, \ldots, a_m\} \in \mathcal{R}^m$ be a multiset of alternative locations which are feasible for building facilities and at most one facility can be built at each location. Assume without loss of generality that $a_1 \leq a_2 \leq \ldots \leq a_m$. Denote an instance of the n agents by $I(\mathbf{x}, \mathbf{p}, A)$ or simply by I without confusion.

Individual and Social Objectives. When locating F_1, F_2 at $y_1 \in A, y_2 \in A \backslash \{y_1\}$ respectively, denote the facility location profile by $\mathbf{y} = (y_1, y_2)$. Under Max-variant, the cost of agent i is denoted by $c_i(\mathbf{y}, (x_i, p_i)) = \max_{F_j \in p_i} |y_j - x_i|$. While each agent seeks to minimize her individual cost, the social planner aims to minimize the sum cost or maximum cost of the n agents. For a location and facility preference profile $(\mathbf{x}, \mathbf{p}) \in \mathcal{R}^n \times (2^{\mathcal{F}})^n$, the sum cost and the maximum cost under $\mathbf{y}$ are denoted by $sc(\mathbf{y}, (\mathbf{x}, \mathbf{p})) = \sum_{i \in N} c_i(\mathbf{y}, (x_i, p_i))$ and $mc(\mathbf{y}, (\mathbf{x}, \mathbf{p})) = \max_{i \in N} c_i(\mathbf{y}, (x_i, p_i))$, respectively. Let $OPT_{sc}(\mathbf{x}, \mathbf{p})$ and $OPT_{mc}(\mathbf{x}, \mathbf{p})$ be the optimal solution under the sum cost and the maximum cost, respectively.

Considering the limit on facility locations, the mechanism in our constrained setting is defined as follows.

Definition 1. *A **deterministic mechanism** f is a function that maps the n agents' location profile $\mathbf{x}$ and facility preference profile $\mathbf{p}$ to a location profile of the two facilities, i.e., $f(\mathbf{x}, \mathbf{p}) = \mathbf{y} = (y_1, y_2), \forall (\mathbf{x}, \mathbf{p}) \in \mathcal{R}^n \times (2^{\mathcal{F}})^n$, where $\mathbf{y} = (y_1, y_2)$ should satisfy $y_1 \in A$ and $y_2 \in A \backslash \{y_1\}$.*

Given a mechanism f and a reported location profile $\mathbf{x}' \in \mathcal{R}^n$, the cost of agent $i \in N$ under f is $c_i(f(\mathbf{x}',\mathbf{p}),(x_i,p_i))$. The sum cost and maximum cost of f are $sc(f(\mathbf{x}',\mathbf{p}),(\mathbf{x},\mathbf{p})) = \sum_{i \in N} c_i(f(\mathbf{x}',\mathbf{p}),(x_i,p_i))$ and $mc(f(\mathbf{x}',\mathbf{p}),(\mathbf{x},\mathbf{p})) = \max_{i \in N} c_i(f(\mathbf{x}',\mathbf{p}),(x_i,p_i))$, respectively. Since agents may misreport their locations to benefit themselves, strategyproofness of mechanisms becomes necessary.

Definition 2. *A mechanism f is **strategyproof** if each agent can never benefit from misreporting her location, regardless of the others' strategies, i.e., for every location and facility preference profile $(\mathbf{x},\mathbf{p}) \in \mathcal{R}^n \times \left(2^{\mathcal{F}}\right)^n$, every agent $i \in N$, and every $x_i' \in \mathcal{R}$, $c_i(f(\mathbf{x},\mathbf{p}),(x_i,p_i)) \leq c_i(f((x_i',\mathbf{x}_{-i}),\mathbf{p}),(x_i,p_i))$.*

Definition 3. *A mechanism f is **group strategyproof** if for any group of agents misreporting their locations, at least one of them cannot benefit regardless of the others' strategies, i.e., for every location and facility preference profile $(\mathbf{x},\mathbf{p}) \in \mathcal{R}^n \times \left(2^{\mathcal{F}}\right)^n$, every group of agents $S \subseteq N$ and every $\mathbf{x}_S' \in \mathcal{R}^{|S|}$, there exists $i \in S$ such that $c_i(f(\mathbf{x},\mathbf{p}),(x_i,p_i)) \leq c_i(f((\mathbf{x}_S',\mathbf{x}_{-S}),\mathbf{p}),(x_i,p_i))$.*

We aim at deterministic strategyproof or group strategyproof mechanisms that can perform well under the sum/maximum cost objective. The worst-case approximation ratio is used to evaluate a mechanism's performance. Without confusion, denote $sc(f(\mathbf{x}',\mathbf{p}),(\mathbf{x},\mathbf{p}))$, $sc(OPT_{sc}(\mathbf{x},\mathbf{p}),(\mathbf{x},\mathbf{p}))$, $mc(f(\mathbf{x}',\mathbf{p}),(\mathbf{x},\mathbf{p}))$ and $mc(OPT_{mc}(\mathbf{x},\mathbf{p}),(\mathbf{x},\mathbf{p}))$ by $sc(f,(\mathbf{x},\mathbf{p}))$, $sc(OPT,(\mathbf{x},\mathbf{p}))$, $mc(f,(\mathbf{x},\mathbf{p}))$ and $mc(OPT,(\mathbf{x},\mathbf{p}))$ respectively for simplicity. The approximation ratio under the sum cost objective is defined as follows and it is similar under the maximum cost objective.

Definition 4. *A mechanism f is said to have an **approximation ratio** of $\rho(\rho \geq 1)$ under the sum cost objective, if*

$$\rho = \sup_{I(\mathbf{x},\mathbf{p},A)} \frac{sc(f,(\mathbf{x},\mathbf{p}))}{sc(OPT,(\mathbf{x},\mathbf{p}))}. \tag{1}$$

In this paper, we are interested in deterministic strategyproof or group strategyproof mechanisms with small approximation ratio under the sum/maximum cost objective.

Notations. For a location profile $\mathbf{x} \in \mathcal{R}^n$, denote the median location in $\mathbf{x}$ by $med(\mathbf{x})$, the leftmost location in $\mathbf{x}$ by $lt(\mathbf{x}) = \min_{i \in N}\{x_i\}$, the rightmost location by $rt(\mathbf{x}) = \max_{i \in N}\{x_i\}$, and the center location by $cen(\mathbf{x}) = \frac{lt(\mathbf{x})+rt(\mathbf{x})}{2}$. For a facility preference profile $\mathbf{p} \in \left(2^{\mathcal{F}}\right)^n$, denote $N_k = \{i \in N \mid p_i = \{F_k\}\}$ for $k \in \{1,2\}$, and $N_{1,2} - \{i \subset N \mid p_i - \{F_1,F_2\}\}$.

3   Compulsory Setting

In this section, we study the compulsory setting where each agent is served by the two heterogeneous facilities, i.e., $p_i = \{F_1,F_2\}, \forall i \in N$. For simplicity, we omit p_i or $\mathbf{p}$ in this section. For example, replace $(\mathbf{x},\mathbf{p})$ by $\mathbf{x}$ and the cost of agent $i \in N$ under the facility location profile $\mathbf{y} = (y_1, y_2)$ is denoted by $c_i(\mathbf{y}, x_i) = \max_{j \in \{1,2\}} |y_j - x_i|$.

For the multiset of alternative locations $A = \{a_1, \ldots, a_m\}$ with $a_1 \leq \ldots \leq a_m$, denote $AP = \{(a_1, a_2), (a_2, a_3), \ldots, (a_{m-1}, a_m)\}$. Then the real line can be partitioned into $m-1$ zones where the kth zone (denoted by $Z_k, k = 1, \ldots, m-1$) represents the set of points whose favorite location pair in AP is (a_k, a_{k+1}). We refer to Z_k as the zone of location pair (a_k, a_{k+1}). Obviously, it holds that

$$
Z_k = \begin{cases}
\left(-\infty, \frac{a_k + a_{k+2}}{2} \right], & k = 1 \\
\left(\frac{a_{k-1} + a_{k+1}}{2}, \frac{a_k + a_{k+2}}{2} \right], & 2 \leq k \leq m-2 \\
\left(\frac{a_{k-1} + a_{k+1}}{2}, +\infty \right), & k = m-1
\end{cases}
\tag{2}
$$

The preferences of all agents over AP are (not strictly) *single peaked*: for each agent $i \in N$ with location $x_i \in Z_l$, her *peak* (or favorite) in AP is (a_l, a_{l+1}) and her cost under (a_k, a_{k+1}) monotonically increases as $|k - l|$ increases. Based on the single peaked preference, locating at the peak of $\mathbf{x}$'s any ith statistic order (denoted by $\mathbf{x}_{(i)}$) is group strategyproof.

Lemma 1. *Given a location profile $\mathbf{x}$, locating at the peak of $\mathbf{x}_{(i)}$ in AP for any $i \in \{1, 2, \ldots, n\}$ is group strategyproof.*

Proof. Given any i, the set of agents N can be divided into $L(i) = \{j \in N \mid x_j < \mathbf{x}_{(i)}\}$, $R(i) = \{j \in N \mid x_j > \mathbf{x}_{(i)}\}$, and $M(i) = \{j \in N \mid x_j = \mathbf{x}_{(i)}\}$.

To show group strategyproofness, we need to prove that for every nonempty $S \subseteq N$ with deviation $\mathbf{x}'_S \in \mathcal{R}^{|S|}$, there exists $j \in S$ who cannot benefit from the coalitional deviation. Denote $\mathbf{x}' = (\mathbf{x}'_S, \mathbf{x}_{-S})$ and the mechanism by f.

Case 1: $M(i) \cap S \neq \emptyset$. Then the cost of any agent $j \in M(i) \cap S$ cannot decrease by the deviation since $f(\mathbf{x})$ is her favorite.

Case 2: $M(i) \cap S = \emptyset$. If $\mathbf{x}'_{(i)} < \mathbf{x}_{(i)}$, there must exist some agent $j \in R(i) \cap S$ who prefers the peak of $\mathbf{x}_{(i)}$ to that of $\mathbf{x}'_{(i)}$ since $x_j > \mathbf{x}_{(i)}$, which implies that agent j cannot benefit from the deviation. Similarly, if $\mathbf{x}'_{(i)} > \mathbf{x}_{(i)}$, there must exist some agent $j \in L(i) \cap S$ who prefers the peak of $\mathbf{x}_{(i)}$ to that of $\mathbf{x}'_{(i)}$ since $x_j < \mathbf{x}_{(i)}$ and cannot benefit from the deviation. □

Lemma 1 provides a class of group strategyproof mechanisms for the compulsory setting where all agents are served by two facilities. Next we will select proper mechanisms from this class for the sum/maximum cost objective respectively.

3.1   Sum Cost

For the sum cost objective, we first show that there exists an optimal solution where the two facilities are located at adjacent alternatives.

Lemma 2. *Given a location profile $\mathbf{x} \in \mathcal{R}^n$, there exists an optimal solution in AP under the sum cost objective.*

Proof. Let $OPT_{sc}(\mathbf{x}) = (y_1^\star, y_2^\star)$ be an optimal solution. Without loss of generality, assume that $y_1^\star \leq y_2^\star$. Supposing there exists some $a \in A$ such that $y_1^\star \leq a \leq y_2^\star$, we only need to show that $sc((y_1^\star, a), \mathbf{x}) \leq sc((y_1^\star, y_2^\star), \mathbf{x})$.

For each agent $i \in N$, $c_i((y_1^\star, a), x_i) = \max\{|y_1^\star - x_i|, |a - x_i|\}$, $c_i((y_1^\star, y_2^\star), x_i) = \max\{|y_1^\star - x_i|, |y_2^\star - x_i|\}$. If $x_i \leq (a + y_2^\star)/2$, obviously $c_i((y_1^\star, a), x_i) \leq c_i((y_1^\star, y_2^\star), x_i)$; otherwise, $c_i((y_1^\star, a), x_i) = |y_1^\star - x_i| = c_i((y_1^\star, y_2^\star), x_i)$.

Thus, we have

$$sc((y_1^\star, a), \mathbf{x}) = \sum_{i \in N} c_i((y_1^\star, a), x_i) \tag{3}$$

$$= \sum_{i: x_i \leq (a+y_2^\star)/2} c_i((y_1^\star, a), x_i) + \sum_{i: x_i > (a+y_2^\star)/2} c_i((y_1^\star, a), x_i) \tag{4}$$

$$\leq \sum_{i: x_i \leq (a+y_2^\star)/2} c_i((y_1^\star, y_2^\star), x_i) + \sum_{i: x_i > (a+y_2^\star)/2} c_i((y_1^\star, y_2^\star), x_i) \tag{5}$$

$$= sc((y_1^\star, y_2^\star), \mathbf{x}) \tag{6}$$

$\square$

Intuitively, each agent always prefers the two facilities located as close as possible, since her cost depends on her distance to the farther one. By Lemma 2, an optimal solution (or mechanism) can always be found in $m - 1$ steps. However, it may be not strategyproof. Consider an instance $I(\mathbf{x}, A)$ with $\mathbf{x} = (0, 2), A = \{-1 - 2\varepsilon, -1, 1 + 3\varepsilon\}$ where $\varepsilon > 0$ is sufficiently small. It holds that $OPT_{sc}(\mathbf{x}) = (-1, 1 + 3\varepsilon), c_1(OPT_{sc}(\mathbf{x}), x_1) = 1 + 3\varepsilon$. Replacing $x_1 = 0$ by $x_1' = -1$, we have $OPT_{sc}(\mathbf{x}') = (-1 - 2\varepsilon, -1), c_1(OPT_{sc}(\mathbf{x}'), x_1) = 1 + 2\varepsilon$. Thus, agent 1 with $x_1 = 0$ can strictly decrease her cost by reporting $x_1' = -1$.

Theorem 1. *Under the sum cost objective, any deterministic strategyproof mechanism has an approximation ratio of at least 3.*

Proof. Suppose f is a deterministic strategyproof mechanism with approximation ratio of $3 - \delta$ for some $\delta > 0$.

Consider an instance $I(\mathbf{x}, A)$ with $\mathbf{x} = (-\varepsilon, \varepsilon)$ and $A = \{-1, -1, 1, 1\}$, where $\varepsilon > 0$ is sufficiently small. $f(\mathbf{x})$ can be $(-1, -1)$, $(1, 1)$, $(-1, 1)$, or $(1, -1)$ and assume w.l.o.g. that $f(\mathbf{x}) = (1, 1)$ or $(-1, 1)$. Then the cost of agent 1 is $c_1(f(\mathbf{x}), x_1) = 1 + \varepsilon$.

For another instance $I(\mathbf{x}', A)$ with $\mathbf{x}' = (-1, \varepsilon)$, it holds that $OPT_{sc}(\mathbf{x}') = (-1, -1)$ and $sc(OPT, \mathbf{x}') = 1 + \varepsilon$. If $f(\mathbf{x}') = (1, 1), (-1, 1)$, or $(1, -1)$, then $sc(f, \mathbf{x}') \geq 3 - \varepsilon$. This implies that

$$\frac{sc(f, \mathbf{x}')}{sc(OPT, \mathbf{x}')} \geq \frac{3 - \varepsilon}{1 + \varepsilon} > 3 - \delta \tag{7}$$

for sufficiently small $\varepsilon > 0$, which is a contradiction. Thus, $f(\mathbf{x}') = (-1, -1)$.

Note that $c_1(f(\mathbf{x}'), x_1) = 1 - \varepsilon$. This indicates that agent 1 can decrease her cost by misreporting her location as $x_1' = -1$, which contradicts f's strategyproofness. $\square$

Mechanism 1. Given a location profile $\mathbf{x} \in \mathcal{R}^n$, output the peak of med($\mathbf{x}$) in AP, i.e., the location pair $(y_1, y_2) \in \underset{(s_1,s_2)\in AP}{\arg\min} \max_{j\in\{1,2\}} |s_j - \text{med}(\mathbf{x})|$, breaking ties in any deterministic way.

Theorem 2. *Mechanism 1 is group strategyproof and has an approximation ratio of 3 under the sum cost objective.*

Proof. By Lemma 1, *Mechanism 1* is group strategyproof. We now turn to its approximation ratio.

Given a location profile $\mathbf{x} \in \mathcal{R}^n$, let $OPT_{sc}(\mathbf{x}) = (y_1^\star, y_2^\star) \in AP$ be an optimal solution. Denote *Mechanism 1* by f and $f(\mathbf{x}) = (y_1, y_2)$.

Considering that both $f(\mathbf{x})$ and $OPT_{sc}(\mathbf{x})$ are adjacent location pairs in AP, assume w.l.o.g. that $(y_1^\star, y_2^\star)$ is on the right of (y_1, y_2).

Let $y_2' \in A$ be the location adjacent to the right of y_2 and $y' = (y_1 + y_2')/2$ be the right border of the zone of (y_1, y_2). We first give two claims, then compare $sc(f, \mathbf{x})$ with $sc(OPT, \mathbf{x})$.

Claim 1. $|\{i \in N \mid x_i \leq y'\}| \geq |\{i \in N \mid x_i > y'\}|$, since med($\mathbf{x}$) $\leq y'$.

Claim 2. For any agent i with $x_i \leq y'$, it holds that $c_i(f(\mathbf{x}), x_i) \leq c_i(OPT_{sc}(\mathbf{x}), x_i)$, since the peak of agent i in AP is (y_1, y_2) or to the left.

The sum cost of *Mechanism 1* is

$$sc(f, \mathbf{x}) = \sum_{i\in N} c_i\left((y_1, y_2), x_i\right) = \sum_{i\in N} \max_{j\in\{1,2\}} |x_i - y_j| \tag{8}$$

$$= \sum_{x_i \leq y'} \max_{j\in\{1,2\}} |x_i - y_j| + \sum_{x_i > y'} \max_{j\in\{1,2\}} |x_i - y_j|, \tag{9}$$

where the first term is denoted by α and the second by β.

The optimal sum cost is

$$sc(OPT, \mathbf{x}) = \sum_{i\in N} c_i\left((y_1^\star, y_2^\star), x_i\right) = \sum_{i\in N} \max_{j\in\{1,2\}} |x_i - y_j^\star| \tag{10}$$

$$= \sum_{x_i \leq y'} \max_{j\in\{1,2\}} |x_i - y_j^\star| + \sum_{x_i > y'} \max_{j\in\{1,2\}} |x_i - y_j^\star|, \tag{11}$$

where the first term is denoted by γ and the second by δ.

Note that

$$\beta = \sum_{x_i > y'} \max_{j \in \{1,2\}} |x_i - y_j| \le \sum_{x_i > y'} \max_{j \in \{1,2\}} \{|x_i - y^\star| + |y^\star - y_j|\} \tag{12}$$

$$\le \sum_{x_i > y'} |x_i - y^\star| + \sum_{x_i > y'} \max_{j \in \{1,2\}} |y^\star - y_j| \tag{13}$$

$$\le \sum_{x_i > y'} |x_i - y^\star| + \sum_{x_i \le y'} \max_{j \in \{1,2\}} |y^\star - y_j| \tag{14}$$

$$\le \sum_{x_i > y'} |x_i - y^\star| + \sum_{x_i \le y'} \max_{j \in \{1,2\}} \{|y^\star - x_i| + |x_i - y_j|\} \tag{15}$$

$$\le \sum_{i \in N} |x_i - y^\star| + \sum_{x_i \le y'} \max_{j \in \{1,2\}} |x_i - y_j| \tag{16}$$

$$\le \sum_{i \in N} \max_{j \in \{1,2\}} |x_i - y_j^\star| + \sum_{x_i \le y'} \max_{j \in \{1,2\}} |x_i - y_j| = \gamma + \delta + \alpha. \tag{17}$$

Here, the third inequality holds by Claim 1. Besides, we have $\alpha \le \gamma$ by Claim 2. Thus,

$$\frac{sc(f, \mathbf{x})}{sc(OPT, \mathbf{x})} = \frac{\alpha + \beta}{\gamma + \delta} \le \frac{\alpha + \gamma + \delta + \alpha}{\gamma + \delta} \le \frac{3\gamma + \delta}{\gamma + \delta} \le 3 \tag{18}$$

Combining with Theorem 1, the approximation ratio of *Mechanism 1* is 3.   □

3.2   Maximum Cost

Compared with the sum cost objective, there is a more precise statement on the optimal solution under the maximum cost objective.

Lemma 3. *Given a location profile* $\mathbf{x} \in \mathcal{R}^n$, *the peak of* $\mathrm{cen}(\mathbf{x})$ *in AP is exactly an optimal solution under the maximum cost objective.*

Proof. Let $\mathbf{a} = (a_k, a_{k+1})$ be the peak of $\mathrm{cen}(\mathbf{x})$ in *AP*. If there exists $s(\in A) < a_k$, then

$$(s + a_{k+1})/2 \le (a_{k-1} + a_{k+1})/2 \le \mathrm{cen}(\mathbf{x}). \tag{19}$$

If there exists $s(\in A) > a_{k+1}$, then

$$(a_k + s)/2 \ge (a_k + a_{k+2})/2 \ge \mathrm{cen}(\mathbf{x}). \tag{20}$$

Let $\mathbf{y} = (y_1, y_2)$ be any feasible solution that is different from (a_k, a_{k+1}). Assume w.l.o.g. that $y_1 \le y_2$, then either $y_1 < a_k$ or $a_{k+1} < y_2$. By symmetry, we only need to compare $mc(\mathbf{y}, \mathbf{x})$ with $mc(\mathbf{a}, \mathbf{x})$ through the following two cases.

Case 1: $a_k \le a_{k+1} \le \mathrm{cen}(\mathbf{x})$. In this case, $mc(\mathbf{a}, \mathbf{x}) = \mathrm{rt}(\mathbf{x}) - a_k$. If $y_1 < a_k$, then

$$mc(\mathbf{y}, \mathbf{x}) \ge \mathrm{rt}(\mathbf{x}) - y_1 > \mathrm{rt}(\mathbf{x}) - a_k = mc(\mathbf{a}, \mathbf{x}). \tag{21}$$

If $a_{k+1} < y_2$, then $y_2 - \mathrm{cen}(\mathbf{x}) \ge \mathrm{cen}(\mathbf{x}) - a_k$ by Eq. (20). Thus, we have

$$mc(\mathbf{y}, \mathbf{x}) \ge y_2 - \mathrm{lt}(\mathbf{x}) = y_2 - \mathrm{cen}(\mathbf{x}) + \mathrm{cen}(\mathbf{x}) - \mathrm{lt}(\mathbf{x}) \tag{22}$$

$$\ge \mathrm{cen}(\mathbf{x}) - a_k + \mathrm{rt}(\mathbf{x}) - \mathrm{cen}(\mathbf{x}) = mc(\mathbf{a}, \mathbf{x}). \tag{23}$$

Case 2: $a_k \leq \text{cen}(\mathbf{x}) < a_{k+1}$. In this case, $mc(\mathbf{a}, \mathbf{x}) = \max\{\text{rt}(\mathbf{x}) - a_k, a_{k+1} - \text{lt}(\mathbf{x})\}$. If $y_1 < a_k$, then $\text{rt}(\mathbf{x}) - y_1 > \text{rt}(\mathbf{x}) - a_k$ and by Eq. (19), it holds that

$$\text{rt}(\mathbf{x}) - y_1 = \text{rt}(\mathbf{x}) - \text{cen}(\mathbf{x}) + \text{cen}(\mathbf{x}) - y_1 \tag{24}$$

$$\geq \text{cen}(\mathbf{x}) - \text{lt}(\mathbf{x}) + a_{k+1} - \text{cen}(\mathbf{x}) = a_{k+1} - \text{lt}(\mathbf{x}). \tag{25}$$

Thus, we have $mc(\mathbf{y}, \mathbf{x}) \geq \text{rt}(\mathbf{x}) - y_1 \geq mc(\mathbf{a}, \mathbf{x})$. Similarly if $a_{k+1} < y_2$, then $y_2 - \text{lt}(\mathbf{x}) > a_{k+1} - \text{lt}(\mathbf{x})$ and $y_2 - \text{lt}(\mathbf{x}) = y_2 - \text{cen}(\mathbf{x}) + \text{cen}(\mathbf{x}) - \text{lt}(\mathbf{x}) \geq \text{cen}(\mathbf{x}) - a_k + \text{rt}(\mathbf{x}) - \text{cen}(\mathbf{x}) = \text{rt}(\mathbf{x}) - a_k$. Thus, we have $mc(\mathbf{y}, \mathbf{x}) \geq y_2 - \text{lt}(\mathbf{x}) \geq mc(\mathbf{a}, \mathbf{x})$. □

However, the optimal mechanism is not strategyproof. Consider an instance $I(\mathbf{x}, A)$ with $\mathbf{x} = (-\varepsilon, \varepsilon)$ and $A = \{-1, 1, 1 + \varepsilon\}$. It holds that $OPT_{mc} = (-1, 1)$ and $c_2(OPT(\mathbf{x}), x_2) = 1 + \varepsilon$ for sufficiently small $\varepsilon > 0$. Replacing $x_2 = \varepsilon$ by $x_2' = 2$, we have $OPT_{sc}(\mathbf{x}') = (1, 1 + \varepsilon), c_2(OPT_{sc}(\mathbf{x}'), x_2) = 1$. Thus, agent 2 with $x_2 = \varepsilon$ can strictly decrease her cost by misreporting $x_2' = 2$.

Theorem 3. *Under the maximum cost objective, any deterministic strategyproof mechanism has an approximation ratio of at least 3.*

Proof. Suppose f is a deterministic strategyproof mechanism with approximation ratio of $3 - \delta$ for some $\delta > 0$.

Consider an instance $I(\mathbf{x}, A)$ with $\mathbf{x} = (-\varepsilon, \varepsilon)$ and $A = \{-1, -1, 1, 1\}$, where $\varepsilon > 0$ is sufficiently small. $f(\mathbf{x})$ can be $(-1, -1)$, $(1, 1)$, $(-1, 1)$, or $(1, -1)$ and assume w.l.o.g. that $f(\mathbf{x}) = (1, 1)$ or $(-1, 1)$. Then the cost of agent 1 is $c_1(f(\mathbf{x}), x_1) = 1 + \varepsilon$.

For another instance $I(\mathbf{x}', A)$ with $\mathbf{x}' = (-2 - \varepsilon, \varepsilon)$, it holds that $OPT_{mc}(\mathbf{x}') = (-1, -1)$ and $mc(OPT, \mathbf{x}') = 1 + \varepsilon$. If $f(\mathbf{x}') = (1, 1)$, $(-1, 1)$, or $(1, -1)$, then $mc(f, \mathbf{x}') = 3 + \varepsilon$. This implies that

$$\frac{mc(f, \mathbf{x}')}{mc(OPT, \mathbf{x}')} = \frac{3 + \varepsilon}{1 + \varepsilon} > 3 - \delta \tag{26}$$

for sufficiently small $\varepsilon > 0$, which is a contradiction. Thus, $f(\mathbf{x}') = (-1, -1)$.

Considering that $c_1(f(\mathbf{x}'), x_1) = 1 - \varepsilon$, agent 1 can decrease her cost by misreporting her location as $x_1' = -2 - \varepsilon$, which contradicts f's strategyproofness. □

Mechanism 2. Given a location profile $\mathbf{x} \in \mathcal{R}^n$, output the peak of $\text{lt}(\mathbf{x})$ in AP, i.e., the location pair $(y_1, y_2) \in \underset{(s_1, s_2) \in AP}{\arg\min} \max_{j \in \{1,2\}} |s_j - \text{lt}(\mathbf{x})|$, breaking ties in any deterministic way.

Theorem 4. *Mechanism 2 is group strategy-proof and has an approximation ratio of 3 under the maximum cost objective.*

Proof. By Lemma 1, *Mechanism 2* is group strategyproof. We now turn to its approximation ratio.

Given a location profile $\mathbf{x} \in \mathcal{R}^n$, let $OPT_{mc}(\mathbf{x}) = (y_1^\star, y_2^\star)$ be the peak of $cen(\mathbf{x})$ in AP which is also an optimal solution. Denote *Mechanism 2* by f and $f(\mathbf{x}) = (y_1, y_2)$. Assume without loss of generality that $rt(\mathbf{x}) - lt(\mathbf{x}) = 1$. It is easy to see that $mc(OPT, \mathbf{x}) \geq \frac{1}{2}$, and

$$mc(OPT, \mathbf{x}) \geq \max_{j \in \{1,2\}} |lt(\mathbf{x}) - y_j^\star| \geq \max_{j \in \{1,2\}} |lt(\mathbf{x}) - y_j|. \tag{27}$$

We compare $mc(f, \mathbf{x})$ with $mc(OPT, \mathbf{x})$ through the following analysis.
Case 1: $y_1 \leq y_2 \leq lt(\mathbf{x}) \leq rt(\mathbf{x})$, or $y_1 \leq lt(\mathbf{x}) \leq y_2 \leq rt(\mathbf{x})$.

$$mc(f, \mathbf{x}) = |rt(\mathbf{x}) - y_1| = 1 + |lt(\mathbf{x}) - y_1| \leq 3mc(OPT, \mathbf{x}). \tag{28}$$

Case 2: $lt(\mathbf{x}) \leq y_1 \leq y_2 \leq rt(\mathbf{x})$.

$$mc(f, \mathbf{x}) \leq 1 \leq 2mc(OPT, \mathbf{x}). \tag{29}$$

Case 3: $lt(\mathbf{x}) \leq y_1 \leq rt(\mathbf{x}) \leq y_2$, or $lt(\mathbf{x}) \leq rt(\mathbf{x}) \leq y_1 \leq y_2$.
In this case, the right border of the zone of (y_1, y_2) is no less than $(y_1 + y_2)/2 \geq cen(\mathbf{x}) \geq lt(\mathbf{x})$. Combining with the fact that $lt(\mathbf{x})$ lies in the zone of (y_1, y_2), it holds that $cen(\mathbf{x})$ also lies in the zone of (y_1, y_2). This implies that $f(\mathbf{x}) = OPT_{mc}(\mathbf{x})$. Thus, we have

$$mc(f, \mathbf{x}) = mc(OPT, \mathbf{x}). \tag{30}$$

Case 4: $y_1 \leq lt(\mathbf{x}) \leq rt(\mathbf{x}) \leq y_2$. Note that

$$|lt(\mathbf{x}) - y_2| \leq \max_{j \in \{1,2\}} |y_j - lt(\mathbf{x})| \leq \max_{j \in \{1,2\}} |y_j^\star - lt(\mathbf{x})| \leq mc(OPT, \mathbf{x}), \tag{31}$$

and

$$|rt(\mathbf{x}) - y_1| = 1 + |lt(\mathbf{x}) - y_1| \leq 1 + mc(OPT, \mathbf{x}) \leq 3mc(OPT, \mathbf{x}). \tag{32}$$

Thus, it holds that

$$mc(f, \mathbf{x}) = \max \{|lt(\mathbf{x}) - y_2|, |rt(\mathbf{x}) - y_1|\} | \leq 3mc(OPT, \mathbf{x}). \tag{33}$$

Above all, $mc(f, \mathbf{x}) \leq 3mc(OPT, \mathbf{x})$. Combining with Theorem 3, *Mechanism 2* has an approximation ratio of 3.   □

4   Optional Setting

In this section, we discuss the optional setting where each agent can be interested in either one of the two heterogeneous facilities or both. The cost of agent $i \in N$ is $c_i(\mathbf{y}, (x_i, p_i)) = \max_{F_k \in p_i} |y_k - x_i|$.

Note that even in the optional setting, each agent $i \in N$ has some kind of single peaked preference: if $p_i = \{F_1\}$ or $\{F_2\}$, she has single peaked preference over A; if $p_i = \{F_1, F_2\}$, she has single peaked preference over AP. Our mechanisms will be proposed based on the single peaked preference.

In the following subsections, two mechanisms for one-facility location games will be used as subroutines in our mechanisms. Supposing that a set of n agents have single peaked preference over the set of alternative locations A, the related results are listed as follows.

SC-Mechanism [11]. Given $\mathbf{x} \in \mathcal{R}^n$ and A, output $y \in \arg\min_{a \in A}|a - \text{med}(\mathbf{x})|$, breaking ties in any deterministic way.

Proposition 1 ([11]). *SC-Mechanism is group strategyproof and has an approximation ratio of 3 under the sum cost objective.*

MC-Mechanism [24]. Given $\mathbf{x} \in \mathcal{R}^n$ and A, output $y \in \arg\min_{a \in A}|a - \text{lt}(\mathbf{x})|$, breaking ties in any deterministic way.

Proposition 2 ([24]). *MC-Mechanism is group strategyproof and has an approximation ratio of 3 under the maximum cost objective.*

4.1   Sum Cost

Mechanism 3. Given a location and facility preference profile $(\mathbf{x}, \mathbf{p}) \in \mathcal{R}^n \times \left(2^{\mathcal{F}}\right)^n$, output the facility location profile $\mathbf{y} = (y_1, y_2)$ as follows:

- if $|N_{1,2}| > 0$, select $(y_1, y_2) \in \arg\min_{(s_1,s_2) \in AP} \max_{j \in \{1,2\}} \left|s_j - \text{med}\left(\mathbf{x}_{N_{1,2}}\right)\right|$, breaking ties in any deterministic way;
- if $|N_{1,2}| = 0$ and $|N_1| \geq |N_2|$, select $y_1 \in \arg\min_{y \in A} |y - \text{med}\left(\mathbf{x}_{N_1}\right)|$, and $y_2 \in \arg\min_{y \in A \backslash \{y_1\}} |y - \text{med}\left(\mathbf{x}_{N_2}\right)|$ (if $N_2 \neq \emptyset$), breaking ties in any deterministic way;
- if $|N_{1,2}| = 0$ and $|N_1| < |N_2|$, select $y_2 \in \arg\min_{y \in A} |y - \text{med}\left(\mathbf{x}_{N_2}\right)|$, and $y_1 \in \arg\min_{y \in A \backslash \{y_2\}} |y - \text{med}\left(\mathbf{x}_{N_1}\right)|$ (if $N_1 \neq \emptyset$), breaking ties in any deterministic way.

Theorem 5. *Mechanism 3 is group strategyproof and has an approximation ratio of at most $2n + 1$ under the sum cost objective.*

Proof. **Group Strategyproofness.** Given $(\mathbf{x}, \mathbf{p}) \in \mathcal{R}^n \times \left(2^{\mathcal{F}}\right)^n$, *Mechanism 3* outputs the facility location profile according to the public information $\mathbf{p}$. To show group strategyproofness, we need to prove that for every nonempty $S \subseteq N$ with deviation $\mathbf{x}'_S \in \mathcal{R}^{|S|}$, there exists $j \in S$ who cannot benefit from the coalitional deviation. Denote $\mathbf{x}' = (\mathbf{x}'_S, \mathbf{x}_{-S})$, *Mechanism 3* by f, *Mechanism 1* by f_1, and *SC-Mechanism* by f_2.

 Case 1: $|N_{1,2}| > 0$, then $f(\mathbf{x}, \mathbf{p}) = f_1(\mathbf{x}_{N_{1,2}})$ and $f(\mathbf{x}', \mathbf{p}) = f_1(\mathbf{x}'_{N_{1,2} \cap S}, \mathbf{x}_{N_{1,2} \backslash S})$. If $N_{1,2} \cap S \neq \emptyset$, any agent in $N_{1,2} \cap S$ cannot benefit from the deviation $\mathbf{x}'_{N_{1,2} \cap S}$ by f_1's group strategyproofness. If $N_{1,2} \cap S = \emptyset$, $f(\mathbf{x}', \mathbf{p}) = f_3(\mathbf{x}_{N_{1,2}})$, which implies that any agent in $S \subseteq N_1 \cup N_2$ cannot benefit from the deviation.

Case 2: $|N_{1,2}| = 0$ and $|N_1| \geq |N_2|$. It holds that $f(\mathbf{x}, \mathbf{p}) = (f_2(\mathbf{x}_{N_1}), f_2(\mathbf{x}_{N_2}))$ and $f(\mathbf{x}', \mathbf{p}) = (f_2(\mathbf{x}'_{N_1 \cap S}, \mathbf{x}_{N_1 \setminus S}), f_2(\mathbf{x}'_{N_2 \cap S}, \mathbf{x}_{N_2 \setminus S}))$, with $f_2(\mathbf{x}_{N_2}) \in A \backslash f_2(\mathbf{x}_{N_1})$ and $f_2(\mathbf{x}'_{N_2 \cap S}, \mathbf{x}_{N_2 \setminus S}) \in A \backslash f_2(\mathbf{x}'_{N_1 \cap S}, \mathbf{x}_{N_1 \setminus S})$. If $N_1 \cap S \neq \emptyset$, any agent in $N_1 \cap S$ cannot benefit from the deviation $\mathbf{x}'_{N_1 \cap S}$ by f_2's group strategyproofness. If $N_1 \cap S = \emptyset$, $f(\mathbf{x}', \mathbf{p}) = (f_2(\mathbf{x}_{N_1}), f_2(\mathbf{x}'_{N_2 \cap S}, \mathbf{x}_{N_2 \setminus S}))$ with $f_2(\mathbf{x}'_{N_2 \cap S}, \mathbf{x}_{N_2 \setminus S}) \in A \backslash f_2(\mathbf{x}_{N_1})$. Still by f_2's group strategyproofness, any agent in $N_2 \cap S$ cannot benefit from the deviation $\mathbf{x}'_{N_2 \cap S}$.

Case 3: $|N_{1,2}| = 0$ and $|N_1| < |N_2|$. This case is similar to *Case 2*.

Approximation Ratio. Given $(\mathbf{x}, \mathbf{p}) \in \mathcal{R}^n \times (2^{\mathcal{F}})^n$, let $OPT_{sc}(\mathbf{x}, \mathbf{p}) = \mathbf{y}^\star = (y_1^\star, y_2^\star)$ be an optimal solution and $f(\mathbf{x}, \mathbf{p}) = \mathbf{y} = (y_1, y_2)$. We now compare $sc(f, (\mathbf{x}, \mathbf{p}))$ with $sc(OPT, (\mathbf{x}, \mathbf{p}))$.

Case 1: If $|N_{1,2}| > 0$, the output of *Mechanism 3* on $I(\mathbf{x}, \mathbf{p}, A)$ equals to that of *Mechanism 1* on $I(\mathbf{x}_{N_{1,2}}, \mathbf{p}_{N_{1,2}}, A)$. Denote the optimal solution on $I(\mathbf{x}_{N_{1,2}}, \mathbf{p}_{N_{1,2}}, A)$ as $\mathbf{y}^{opt}$.

By Theorem 2, it holds that

$$\sum_{i \in N_{1,2}} c_i(\mathbf{y}, (x_i, p_i)) \leq 3 \sum_{i \in N_{1,2}} c_i(\mathbf{y}^{opt}, (x_i, p_i)) \leq 3 \sum_{i \in N_{1,2}} c_i(\mathbf{y}^\star, (x_i, p_i)). \quad (34)$$

Thus, we have

$$sc(f, (\mathbf{x}, \mathbf{p})) = \sum_{i \in N_1 \cup N_2 \cup N_{1,2}} c_i(\mathbf{y}, (x_i, p_i)) \quad (35)$$

$$\leq \sum_{i \in N_1} |x_i - y_1| + \sum_{i \in N_2} |x_i - y_2| + 3 \sum_{i \in N_{1,2}} c_i(\mathbf{y}^\star, (x_i, p_i)) \quad (36)$$

$$\leq \sum_{i \in N_1} |x_i - y_1^\star| + \sum_{i \in N_2} |x_i - y_2^\star| + 3 \sum_{i \in N_{1,2}} c_i(\mathbf{y}^\star, (x_i, p_i)) \quad (37)$$

$$+ |N_1| \cdot |y_1 - y_1^\star| + |N_2| \cdot |y_2 - y_2^\star| \quad (38)$$

$$\leq 3sc(OPT, (\mathbf{x}, \mathbf{p})) + |N_1 \cup N_2| \cdot 2sc(OPT, (\mathbf{x}, \mathbf{p})) \quad (39)$$

$$\leq (2n + 1)sc(OPT, (\mathbf{x}, \mathbf{p})). \quad (40)$$

Here, the above third inequality holds because for $j = 1, 2$,

$$|y_j - y_j^\star| \leq |y_j - \text{med}(\mathbf{x}_{N_{1,2}})| + |\text{med}(\mathbf{x}_{N_{1,2}}) - y_j^\star| \quad (41)$$

$$\leq \max_{k \in \{1,2\}} |y_k - \text{med}(\mathbf{x}_{N_{1,2}})| + |\text{med}(\mathbf{x}_{N_{1,2}}) - y_j^\star| \quad (42)$$

$$\leq \max_{k \in \{1,2\}} |y_k^\star - \text{med}(\mathbf{x}_{N_{1,2}})| + |\text{med}(\mathbf{x}_{N_{1,2}}) - y_j^\star| \quad (43)$$

$$\leq 2sc(OPT, (\mathbf{x}, \mathbf{p})). \quad (44)$$

Case 2: If $|N_{1,2}| = 0$ and $|N_1| \geq |N_2|$. Without loss of generality, assume that $N_2 \neq \emptyset$. y_1 equals to the output of *SC-Mechanism* on instance $I_1 = I(\mathbf{x}_{N_1}, \mathbf{p}_{N_1}, A)$, and y_2 equals to the output of *SC-Mechanism* on instance $I_2 = I(\mathbf{x}_{N_2}, \mathbf{p}_{N_2}, A \backslash \{y_1\})$. Denote by y_1^{opt} the optimal solution on instance I_1 and y_2^{opt} the optimal solution on instance I_2.

For $k = 1, 2$, let $sc(y, I_k) = \sum_{i \in N_k} |x_i - y|$, then

$$sc(OPT, (\mathbf{x}, \mathbf{p})) = \sum_{i \in N_1} |x_i - y_1^\star| + \sum_{i \in N_2} |x_i - y_2^\star| = sc\,(y_1^\star, I_1) + sc\,(y_2^\star, I_2) \quad (45)$$

$$sc(f, (\mathbf{x}, \mathbf{p})) = \sum_{i \in N_1} |x_i - y_1| + \sum_{i \in N_2} |x_i - y_2| = sc\,(y_1, I_1) + sc\,(y_2, I_2) \quad (46)$$

For I_1, by Proposition 1, it holds that

$$sc\,(y_1, I_1) \le 3sc\,(y_1^{opt}, I_1) \le 3sc\,(y_1^\star, I_1) \quad (47)$$

For I_2, we consider the following two cases.

Case 2.1: If $y_2^\star \in A \setminus \{y_1\}$, by Proposition 1, it holds that

$$sc\,(y_2, I_2) \le 3sc\,(y_2^{opt}, I_2) \le 3sc\,(y_2^\star, I_2) \quad (48)$$

Case 2.2: $y_2^\star \notin A \setminus \{y_1\}$, then $y_1 = y_2^\star$ and $y_1^\star \in A \setminus \{y_1\}$. On the one hand, by Proposition 1, we have

$$sc\,(y_2, I_2) \le 3sc\,(y_2^{opt}, I_2) \le 3sc\,(y_1^\star, I_2). \quad (49)$$

On the other hand,

$$sc\,(y_1^\star, I_2) = \sum_{i \in N_2} |x_i - y_1^\star| \le \sum_{i \in N_2} |x_i - y_2^\star| + \sum_{i \in N_1} |y_2^\star - y_1^\star| \quad (50)$$

$$\le \sum_{i \in N_2} |x_i - y_2^\star| + \sum_{i \in N_1} |y_1 - x_i| + \sum_{i \in N_1} |x_i - y_1^\star| \quad (51)$$

$$= sc\,(y_2^\star, I_2) + sc\,(y_1, I_1) + sc\,(y_1^\star, I_1) \quad (52)$$

$$\le sc\,(y_2^\star, I_2) + 4sc\,(y_1^\star, I_1), \quad (53)$$

where the first inequality holds because $|N_1| \ge |N_2|$ and the third holds by Eq. (47).

Combining Eq. (49) and Eq. (53), we have

$$sc\,(y_2, I_2) \le 3sc\,(y_2^\star, I_2) + 12sc\,(y_1^\star, I_1) \quad (54)$$

Thus, by Eq. (47) and Eq. (54), it holds that

$$sc(f, (\mathbf{x}, \mathbf{p})) = sc\,(y_1, I_1) + sc\,(y_2, I_2) \quad (55)$$

$$\le 3sc\,(y_1^\star, I_1) + 3sc\,(y_2^\star, I_2) + 12sc\,(y_1^\star, I_1) \quad (56)$$

$$\le 15sc(OPT, (\mathbf{x}, \mathbf{p})) \quad (57)$$

Case 3: $|N_{1,2}| = 0$ and $|N_1| < |N_2|$. This case is similar to *Case 2*.

Above all, *Mechanism 3* has an approximation ratio of at most $2n + 1$. $\quad\square$

4.2   Maximum Cost

Mechanism 4. Given a location and facility preference profile $(\mathbf{x}, \mathbf{p}) \in \mathcal{R}^n \times (2^{\mathcal{F}})^n$, output the facility location profile $\mathbf{y} = (y_1, y_2)$ as follows:

- if $|N_{1,2}| > 0$, select $(y_1, y_2) \in \underset{(s_1, s_2) \in AP}{\arg\min} \ \max_{j \in \{1,2\}} \left| s_j - \mathrm{lt}\left(\mathbf{x}_{N_{1,2}}\right)\right|$, breaking ties in any deterministic way;
- if $|N_{1,2}| = 0$, select $y_1 \in \underset{y \in A}{\arg\min} |y - \mathrm{lt}\left(\mathbf{x}_{N_1}\right)|$ (if $N_1 \neq \emptyset$), and $y_2 \in \underset{y \in A \setminus \{y_1\}}{\arg\min} |y - \mathrm{lt}\left(\mathbf{x}_{N_2}\right)|$ (if $N_2 \neq \emptyset$), breaking ties in any deterministic way.

Theorem 6. *Mechanism 4 is group strategyproof and has an approximation ratio of at most 9 under the maximum cost objective.*

Proof. The proof of *Mechanism 4*'s group strategyproofness is similar to that of *Mechanism 3*'s, which is omitted here. Now we focus on the approximation ratio of *Mechanism 4*.

Denote *Mechanism 4* by f. Given $(\mathbf{x}, \mathbf{p}) \in \mathcal{R}^n \times (2^{\mathcal{F}})^n$, let $OPT_{mc}(\mathbf{x}, \mathbf{p}) = \mathbf{y}^\star = (y_1^\star, y_2^\star)$ be an optimal solution and $f(\mathbf{x}, \mathbf{p}) = \mathbf{y} = (y_1, y_2)$. We now compare $mc(f, (\mathbf{x}, \mathbf{p}))$ with $mc(OPT, (\mathbf{x}, \mathbf{p}))$.

Case 1: If $|N_{1,2}| > 0$, the output of *Mechanism 4* on $I(\mathbf{x}, \mathbf{p}, A)$ equals to that of *Mechanism 2* on $I\left(\mathbf{x}_{N_{1,2}}, \mathbf{p}_{N_{1,2}}, A\right)$. Denote by $\mathbf{y}^{opt} = (y_1^{opt}, y_2^{opt})$ the optimal solution on $I\left(\mathbf{x}_{N_{1,2}}, \mathbf{p}_{N_{1,2}}, A\right)$.

By Theorem 4, it holds that

$$\max_{i \in N_{1,2}} c_i\left(\mathbf{y}, (x_i, p_i)\right) \leq 3 \max_{i \in N_{1,2}} c_i\left(\mathbf{y}^{opt}, (x_i, p_i)\right) \leq 3 \max_{i \in N_{1,2}} c_i\left(\mathbf{y}^\star, (x_i, p_i)\right). \quad (58)$$

Thus, we have

$$mc(f, (\mathbf{x}, \mathbf{p})) \qquad\qquad\qquad\qquad\qquad\qquad\qquad\qquad (59)$$

$$= \max_{i \in N_1 \cup N_2 \cup N_{1,2}} \left\{c_i\left(\mathbf{y}, (x_i, p_i)\right)\right\} \qquad\qquad\qquad\qquad (60)$$

$$= \max\left\{\max_{i \in N_1}\left\{|y_1 - x_i|\right\}, \max_{i \in N_2}\left\{|y_2 - x_i|\right\}, \max_{i \in N_{1,2}}\left\{c_i(\mathbf{y}, (x_i, p_i))\right\}\right\} \quad (61)$$

$$\leq \max\left\{\max_{i \in N_1}\left\{|y_1^\star - x_i| + |y_1 - y_1^\star|\right\}, \max_{i \in N_2}\left\{|y_2^\star - x_i|\right.\right. \qquad (62)$$

$$\left. + |y_2 - y_2^\star|\right\}, 3 \max_{i \in N_{1,2}} c_i\left(\mathbf{y}^\star, (x_i, p_i)\right)\bigg\} \qquad\qquad (63)$$

$$\leq \max\left\{\max_{i \in N_1}\left\{|y_1^\star - x_i| + 2mc(OPT, (\mathbf{x}, \mathbf{p}))\right\}, \max_{i \in N_2}\left\{|y_2^\star - x_i|\right.\right. \quad (64)$$

$$\left. + 2mc(OPT, (\mathbf{x}, \mathbf{p}))\right\}, 3 \max_{i \in N_{1,2}} c_i\left(\mathbf{y}^\star, (x_i, p_i)\right)\bigg\} \qquad (65)$$

$$\leq 3mc(OPT, (\mathbf{x}, \mathbf{p})). \qquad\qquad\qquad\qquad\qquad\qquad (66)$$

Here, the above second inequality holds because for $j = 1, 2$,

$$|y_j - y_j^\star| \leq |y_j - \mathrm{lt}\,(\mathbf{x}_{N_{1,2}})| + |\mathrm{lt}\,(\mathbf{x}_{N_{1,2}}) - y_j^\star| \tag{67}$$

$$\leq \max_{k \in \{1,2\}} |y_k - \mathrm{lt}\,(\mathbf{x}_{N_{1,2}})| + |\mathrm{lt}\,(\mathbf{x}_{N_{1,2}}) - y_j^\star| \tag{68}$$

$$\leq \max_{k \in \{1,2\}} |y_k^\star - \mathrm{lt}\,(\mathbf{x}_{N_{1,2}})| + |\mathrm{lt}\,(\mathbf{x}_{N_{1,2}}) - y_j^\star| \tag{69}$$

$$\leq 2mc(OPT, (\mathbf{x}, \mathbf{p})). \tag{70}$$

Case 2: $|N_{1,2}| = 0$. Assume w.l.o.g. that $N_1 \neq \emptyset, N_2 \neq \emptyset$. y_1 equals to the output of *MC-Mechanism* on instance $I_1 = I(\mathbf{x}_{N_1}, \mathbf{p}_{N_1}, A)$, and y_2 equals to the output of *MC-Mechanism* on instance $I_2 = I(\mathbf{x}_{N_2}, \mathbf{p}_{N_2}, A \backslash \{y_1\})$. Denote by y_1^{opt} the optimal solution on instance I_1 and y_2^{opt} the optimal solution on instance I_2. For $k = 1, 2$, let $mc(y, I_k) = \max_{i \in N_k} |x_i - y|$, then

$$mc(OPT, (\mathbf{x}, \mathbf{p})) = \max \left\{ \max_{i \in N_1} |x_i - y_1^\star|, \max_{i \in N_2} |x_i - y_2^\star| \right\} \tag{71}$$

$$= \max \{ mc\,(y_1^\star, I_1), mc\,(y_2^\star, I_2) \} \tag{72}$$

$$mc(f, (\mathbf{x}, \mathbf{p})) = \max \{ mc\,(y_1, I_1), mc\,(y_2, I_2) \} \tag{73}$$

For I_1, by Proposition 2, it holds that

$$mc\,(y_1, I_1) \leq 3mc\,(y_1^{opt}, I_1) \leq 3mc\,(y_1^\star, I_1) \tag{74}$$

For I_2, we consider the following two cases.
Case 2.1: If $y_2^\star \in A \backslash \{y_1\}$, by Proposition 2, it holds that

$$mc\,(y_2, I_2) \leq 3mc\,(y_2^{opt}, I_2) \leq 3mc\,(y_2^\star, I_2) \tag{75}$$

Case 2.2: $y_2^\star \notin A \backslash \{y_1\}$, then $y_1 = y_2^\star$ and $y_1^\star \in A \backslash \{y_1\}$. On the one hand, by Proposition 2, we have

$$mc\,(y_2, I_2) \leq 3mc\,(y_2^{opt}, I_2) \leq 3mc\,(y_1^\star, I_2). \tag{76}$$

On the other hand,

$$mc\,(y_1^\star, I_2) = \max_{i \in N_2} |y_1^\star - x_i| \leq \max_{i \in N_2} |y_2^\star - x_i| + |y_1^\star - y_1| \tag{77}$$

$$\leq \max_{i \in N_2} |y_2^\star - x_i| + |y_1^\star - \mathrm{lt}\,(\mathbf{x}_{N_1})| + |\mathrm{lt}\,(\mathbf{x}_{N_1}) - y_1| \tag{78}$$

$$\leq mc\,(y_2^\star, I_2) + 2mc\,(y_1^\star, I_1) \tag{79}$$

Combining Eq. (76) and Eq. (79), we have

$$mc\,(y_2, I_2) \leq 3mc\,(y_2^\star, I_2) + 6mc\,(y_1^\star, I_1) \tag{80}$$

Thus, by Eq. (74) and Eq. (80), it holds that

$$mc(f, (\mathbf{x}, \mathbf{p})) = \max \{ mc\,(y_1, I_1), mc\,(y_2, I_2) \} \tag{81}$$

$$\leq \max \{ 3\,mc\,(y_1^\star, I_1), 3mc\,(y_2^\star, I_2) + 6mc\,(y_1^\star, I_1) \} \tag{82}$$

$$\leq 9\,mc(OPT, (\mathbf{x}, \mathbf{p})) \tag{83}$$

Above all, *Mechanism 4* has an approximation ratio of at most 9. □

5   Conclusion

In this paper, we considered the mechanism design problem for constrained heterogeneous two-facility location games where a set of alternatives are feasible for building facilities and the number of facilities built at each alternative is limited. We studied deterministic mechanisms design without money under the Max-variant cost where the cost of each agent depends on the distance to the further facility. In the compulsory setting where each agent is served by two facilities, we showed that the optimal solution under the sum/maximum cost objective is not strategyproof and proposed a 3-approximate deterministic group strategyproof mechanism which is also the best deterministic strategyproof mechanism for the corresponding social objective. In the optional setting where each agent can be interested in either one of the two facilities or both, we designed a deterministic group strategyproof mechanism with approximation ratio with at most $2n + 1$ for the sum cost objective and a deterministic group strategyproof mechanism with approximation ratio with at most 9 for the maximum cost objective.

There are several directions for future research. First, the bounds for approximation ratio of deterministic strategyproof mechanisms in the optional setting do not match yet. Are there more desirable bounds in this setting? Second, randomized mechanism design for constrained heterogeneous facility location games remains an open question. Third, the cost of each agent served by two facilities here is simply the sum of her distances from facilities. How about mechanism design for constrained facility location games in more general settings, such as agents having weighted preference for facilities [13]? Further, our model can be extended to include more than two facilities or in more general metric spaces.

Acknowledgements. This research was supported in part by the National Natural Science Foundation of China (12171444, 11971447, 11871442), the Natural Science Foundation of Shandong Province of China (ZR2019MA052).

References

1. Anastasiadis, E., Deligkas, A.: Heterogeneous facility location games. In: Proceedings of the 17th International Conference on Autonomous Agents and Multiagent Systems, pp. 623–631 (2018)
2. Aziz, H., Chan, H., Lee, B., Li, B., Walsh, T.: Facility location problem with capacity constraints: algorithmic and mechanism design perspectives. In: Proceedings of the 34th AAAI Conference on Artificial Intelligence, vol. 34, no. 2, pp. 1806–1813 (2020). https://doi.org/10.1609/aaai.v34i02.5547
3. Aziz, H., Chan, H., Lee, B., Parkes, D.C.: The capacity constrained facility location problem. Games Econ. Behav. **124**, 478–490 (2020). https://doi.org/10.1016/j.geb.2020.09.001
4. Cai, Q., Filos-Ratsikas, A., Tang, P.: Facility location with minimax envy. In: Proceedings of the 25th International Joint Conference on Artificial Intelligence, pp. 137–143 (2016)

5. Chen, X., Fang, Q., Liu, W., Ding, Y.: Strategyproof mechanisms for 2-facility location games with minimax envy. In: Zhang, Z., Li, W., Du, D.-Z. (eds.) AAIM 2020. LNCS, vol. 12290, pp. 260–272. Springer, Cham (2020). https://doi.org/10.1007/978-3-030-57602-8_24

6. Chen, X., Hu, X., Tang, Z., Wang, C.: Tight efficiency lower bounds for strategyproof mechanisms in two-opposite-facility location game. Inf. Process. Lett. **168**, 106098 (2021). https://doi.org/10.1016/j.ipl.2021.106098

7. Cheng, Y., Yu, W., Zhang, G.: Strategy-proof approximation mechanisms for an obnoxious facility game on networks. Theor. Comput. Sci. **497**, 154–163 (2013). https://doi.org/10.1016/j.tcs.2011.11.041

8. Ding, Y., Liu, W., Chen, X., Fang, Q., Nong, Q.: Facility location game with envy ratio. Comput. Ind. Eng. **148**(3), 106710 (2020). https://doi.org/10.1016/j.cie.2020.106710

9. Dokow, E., Feldman, M., Meir, R., Nehama, I.: Mechanism design on discrete lines and cycles. In: Proceedings of the 13rd ACM Conference on Electronic Commerce, pp. 423–440 (2012). https://doi.org/10.1145/2229012.2229045

10. Feigenbaum, I., Sethuraman, J., Ye, C.: Approximately optimal mechanisms for strategyproof facility location: minimizing L_p norm of costs. Comput. Sci. Game Theory **42**(2), 277–575 (2017). https://doi.org/10.1287/moor.2016.0810

11. Feldman, M., Fiat, A., Golomb, I.: On voting and facility location. In: Proceedings of the 2016 ACM Conference on Economics and Computation, pp. 269–286 (2016). https://doi.org/10.1145/2940716.2940725

12. Filos-Ratsikas, A., Li, M., Zhang, J., Zhang, Q.: Facility location with double-peaked preferences. Auton. Agents Multi-Agent Syst. **31**(6), 1209–1235 (2017). https://doi.org/10.1007/s10458-017-9361-0

13. Fong, K., Li, M., Lu, P., Todo, T., Yokoo, T.: Facility location games with fractional preferences. In: Proceedings of the 32nd AAAI Conference on Artificial Intelligence, pp. 1039–1046 (2018). https://doi.org/10.1609/aaai.v32i1.11458

14. Fotakis, D., Tzamos, C.: On the power of deterministic mechanisms for facility location games. ACM Trans. Econ. Comput. **2**(4), 1–37 (2014). https://doi.org/10.1145/2665005

15. Li, M., Lu, P., Yao, Y., Zhang, J.: Strategyproof mechanism for two heterogeneous facilities with constant approximation ratio. In: Proceedings of the 29th International Joint Conference on Artificial Intelligence, pp. 238–245 (2020). https://doi.org/10.24963/ijcai.2020/34

16. Liu, W., Ding, Y., Chen, X., Fang, Q., Nong, Q.: Multiple facility location games with envy ratio. Theor. Comput. Sci. **864**, 1–9 (2021). https://doi.org/10.1016/j.tcs.2021.01.016

17. Lu, P., Sun, X., Wang, Y., Zhu, Z.A.: Asymptotically optimal strategy-proof mechanisms for two-facility games. In: Proceedings of the 11th ACM Conference on Electronic Commerce, pp. 315–324 (2010). https://doi.org/10.1145/1807342.1807393

18. Mei, L., Li, M., Ye, D., Zhang, G.: Facility location games with distinct desires. Discrete Appl. Math. **264**, 148–160 (2019). https://doi.org/10.1016/j.dam.2019.02.017

19. Moulin, H.: On strategy-proofness and single peakedness. Public Choice **35**(4), 437–455 (1980). https://doi.org/10.1007/BF00128122

20. Procaccia, A.D., Tennenholtz, M.: Approximate mechanism design without money. In: Proceedings of the 10th ACM Conference on Electronic Commerce, pp. 177–186 (2009). https://doi.org/10.1145/1566374.1566401

21. Schummer, J., Vohra, R.: Strategy-proof location on a network. J. Econ. Theory **104**(2), 405–428 (2002). https://doi.org/10.1006/jeth.2001.2807

22. Serafino, P., Ventre, C.: Heterogeneous facility location without money. Theor. Comput. Sci. **636**, 27–46 (2016). https://doi.org/10.1016/j.tcs.2016.04.033
23. Sui, X., Boutilier, C.: Approximately strategy-proof mechanisms for (constrained) facility location. In: Proceedings of the 14th International Conference on Autonomous Agents and Multiagent Systems, pp. 605–613 (2015)
24. Tang, Z., Wang, C., Zhang, M., Zhao, Y.: Mechanism design for facility location games with candidate locations. In: Wu, W., Zhang, Z. (eds.) COCOA 2020. LNCS, vol. 12577, pp. 440–452. Springer, Cham (2020). https://doi.org/10.1007/978-3-030-64843-5_30
25. Xu, X., Li, B., Li, M., Duan, L.: Two-facility location games with minimum distance requirement. J. Artif. Intell. Res. **70**, 719–756 (2021). https://doi.org/10.1613/jair.1.12319
26. Yuan, H., Wang, K., Fong, K.C., Zhang, Y., Li, M.: Facility location games with optional preference. In: Proceedings of the 22nd European Conference on Artificial Intelligence, pp. 1520–1527 (2016). https://doi.org/10.3233/978-1-61499-672-9-1520
27. Zhang, Q., Li, M.: Strategyproof mechanism design for facility location games with weighted agents on a line. J. Comb. Optim. **28**(4), 756–773 (2013). https://doi.org/10.1007/s10878-013-9598-8
28. Zhou, H., Li, M., Chan, H.: Strategyproof mechanisms for group-fair facility location problems. arXiv preprint arXiv:2107.05175 (2021)
29. Zou, S., Li, M.: Facility location games with dual preference. In: Proceedings of the 14th International Conference on Autonomous Agents and Multiagent Systems, pp. 615–623 (2015)

Optimally Integrating Ad Auction into E-Commerce Platforms

Weian Li[1], Qi Qi[2(✉)], Changjun Wang[3], and Changyuan Yu[4]

[1] Center on Frontiers of Computing Studies, Peking University, Beijing, China
weian_li@pku.edu.cn
[2] Gaoling School of Artificial Intelligence, Renmin University of China,
Beijing, China
qi.qi@ruc.edu.cn
[3] Academy of Mathematics and Systems Science, Chinese Academy of Sciences,
Beijing, China
wcj@amss.ac.cn
[4] Baidu Inc., Beijing, China
yuchangyuan@baidu.com

Abstract. Advertising becomes one of the most popular ways of monetizing an online transaction platform. Usually, sponsored advertisements are posted on the most attractive positions to enhance the number of clicks. However, multiple e-commerce platforms are aware that this action may hurt the search experience of users, even though it can bring more incomes. To balance the advertising revenue and the user experience loss caused by advertisements, most e-commerce platforms choose fixing some areas for advertisements and adopting some simple restrictions on the number of ads, such as a fixed number K of ads on the top positions or one advertisement for every N organic searched results. Different from these common rules of treating the allocation of ads separately (from the arrangements of the organic searched items), in this work we build up an integrated system with mixed arrangements of advertisements and organic items. We focus on the design of truthful mechanisms to properly list the advertisements and organic items and optimally trade off the instant revenue and the user experience. Furthermore, for different settings and practical requirements, we extend our optimal truthful allocation mechanisms to cater for these realistic conditions. Finally, we exert several experiments to verify the improvement of our mechanism compared to the common-used advertising mechanism.

Keywords: E-Commerce platforms · Ad auctions · Optimal mixed arrangements

This work is supported by Beijing Outstanding Young Scientist Program No. BJJWZYJH012019100020098, and Intelligent Social Governance Platform, Major Innovation & Planning Interdisciplinary Platform for the "Double-First Class" Initiative, Renmin University of China, and National Natural Science Foundation of China (NSFC) (No. 11971046).

1   Introduction

With the development of Internet and mobile devices, e-commerce platforms have become the most fashionable online marketplace to bridge merchants and consumers, whose popularity is originated from its convenient shopping mode. That is, whenever and wherever consumers want to pick up their desirable products, they only require to enter a relevant keyword into the search bar of Apps of e-commerce platforms by mobile devices. Then, platforms will return a list of the matching products in the search result pages provided to be selected. Hence, how to recommend products accurately and effectively is a considerable problem of platforms, which also affects both the short-run revenue and the long-run prosperity.

Nowadays, in one standard webpage of search results, based on a given keyword, it usually presents two types of items, *organic search results* and *sponsored advertisements*[1]. The platform calculates the displayed sequence of two types of items followed by certain rules. Traditionally, organic search results and sponsored advertisements are totally separate components in the process of arranging items. First, e-commerce platforms will decide how many top positions are used to display advertisements, and then run a preset auction mechanism, like generalized second price auction, to output the ads. For the rest positions, they are used to show the organic items ranked by the relevance of items or other criteria which depend on different platforms (e.g., see [4,5,29]).

The first step of the above process is so-called *Sponsored Search Auction*, which is first launched by Google at the end of 20th century. Several common pricing models can be adopted in sponsored search auctions, like pay-per-click, pay-per-impression and pay-per-action. In this paper, we focus on the pay-per-click model which means that, for the winners of auction, they only pay when their advertisements are clicked. Due to the dissatisfaction with the ranking of organic items, some merchants hope to join the auction (becoming the advertisers) and charge some extra money for improving the position and drawing more attention from customers. Because of the millions of search everyday, sponsored search auction has become the vital instant incoming source of e-commerce platform today. Different from the first step, ranking organic items aims to enhance the efficiency of search and user experience, so its criterion is usually the relevance of items. In this investigation, we consider exploiting a relatively comprehensive but very simple indicator to reflect relevance, that is, the expected merchandise volume, or the expected sale amount to rank the organic items with two reasons. The first reason is that the expected merchandise volume includes the user's interest in browsing the product (click rate), the detailed purchase tendency (conversion rate), and the decision after the balance between price and product quality. In another word, the merchandise volume can directly or indirectly reflect the user experience. The second reason is that, nowadays, most e-commerce platforms regard the gross merchandise volume (GMV) as a crite-

---

[1] Sponsored advertisements are usually labeled by "Sponsored" or "Ad" to be distinguished with organic search results.

rion to measure how much they take over the whole market, and compete for GMV during some shopping ceremonies.

However, due to some specialty of e-commerce platforms, like that the sponsored advertisements are also the organic items essentially, still insisting on the conventional pattern and regarding sponsored advertisements and organic results as independent sections may expose some downsides: **(a)** obviously, it is not appropriate for different keywords to use the same number of positions providing for ads. The conventional pattern cannot automatically solve the above problem; **(b)** to optimize the user experience, the platform will list the organic results according to their expected transaction volume or their expected sale quantities (these volumes or quantities are well known by the platforms), i.e., the item with the highest expected "volume" will be shown in the first slot (exclude the slots for advertisements), the item with the second highest expected "volume" will be shown in the second slot, and so on. Nevertheless, during the sponsored search auction, if some items with low volume would rather pay high price to get more clicks, by the goal of sponsored search auction, maximizing the instant revenue, these items will be picked up, which will harm the platforms' expected GMV and the long-run profits. In view of the above points, is it still a great idea to fix the ad positions in advance? Can the platforms do better if they weighing the allocation of the sponsored advertisements and the organic results at the same time, not separately? If yes[2], how to balance the two conflicting objectives of user experience (i.e., GMV) and instant advertising revenue in designing mechanisms? Furthermore, how to design the optimal mechanisms with the integrated allocation?

In reality, on some platforms such as Tmall, Taobao etc., the advertisements and organic items have already been combined together to display. However, the currently used mixed allocation rules are still very heuristic. For example, always take out one slot for advertisement of every N displaying slots. Motivated by this, in this work, we will, from the theoretical aspect, try to study how to design truthful mechanisms to optimally trade off the expected revenue and volume for this new integrated setting (answering the third question and fourth question proposed above). To the best of our knowledge, our work is the first attempt on studying auction mechanism design with multiple objectives in the new layout of mixed arrangements.

1.1   Our Contribution

In this paper, our contributions can be summarized as follows.

- **Novelty**—We initially build up an integrated model to describe the layout of mixed arrangements of organic results and sponsored advertisements, called an integrated ad system (IAS), where two main objectives, revenue and GMV, are considered. We propose two general kinds of problems with two different

---

[2] We give an example to illustrate that the GMV and revenue will be better if we consider organic results and sponsored advertisements at the same time in full version.

ways to trade off the revenue and GMV: unconstrained problem of linearly combining the volume and revenue as a single objective; constrained problem of bounding the volume and maximizing the revenue (see Sect. 2.3).

- **Techniques**—For the unconstrained problem, we prove that the optimal truthful mechanism (Theorem 1) can be obtained by a transformation on objectives and then allocating the items by their "revised virtual values". For the constrained problem, using variables relaxation and Lagrangian dual methods, we show that the constrained problem can be transformed into an unconstrained problem by properly choosing a parameter (Theorem 2). Then we give a numerical algorithm to compute the desired parameter, thus deriving the optimal truthful mechanism for the constrained problem.

- **Extensions**—We extend our study to three general settings with practical restrictions: requiring an upper bound of the total number of advertisements, and requiring certain sparsity of the allocation of advertisements where for the later, we divide it into two models with different restriction on the sparsity. However, under these restrictions, some good properties will not exist anymore, but by some more delicate treatments, we can still design the optimal mechanisms for both constrained and unconstrained problems.

- **Practicability**—We first verify that our mechanisms can be implemented as planned. Then comparing with the currently used mechanism, we show the superiority of our mechanisms. At last, we take the correlation between value and weight of advertisers into account, our mechanisms still perform better.

1.2 Related Work

Sponsored Search Auction. Auction lies in the core of mechanism design research, while sponsored search auction has been an area of great focus in computer science in the last twenty years. For the traditional revenue optimization, [20] solved the problem for single item in Bayesian-Nash equilibrium setting. [19] studied multi-unit optimal auction. From the first launch of sponsored search auction in 1997, a series of auction mechanisms are put forward, like generalized first price (GFP) auction by Overture and generalized second price (GSP) by Google, which makes sponsored search auction become a vital incoming source of variable online platforms in practise. [2] showed that the lower bound of revenue produced by GSP, the most popular mechanism is equal to the revenue of classic VCG mechanism [8,12,28]. Due to the untruthfulness of GSP, [11,27] independently examined the behavior of bidders and proposed locally envy-free equilibrium (LEFE) and symmetric Nash equilibrium (SNE), respectively. Another idea of a squashing parameter of GSP to improve the revenue was proposed by [13]. [21] applied these works and studied its effects on Yahoo! auction using the optimal reserve price. [26] also studied several different techniques for increasing the revenue, including via several different reserve prices, squashing, combinations of a reserve price and squashing, etc. Rather than using the reserve price simply as a minimum bid, [23] presented the idea of incorporating the reserve price into ranking score and showed that this mechanism may increase the revenue compared to the squashing mechanism.

For the topic of user experience, [1] introduced a concept of hidden cost, which was advertiser-specific and represented the quality of the ads' landing page. [3] studied a model that introduced the users' search cost, and showed that reserve price can improve users' welfare. [14] defined the shadow cost, which is revenue reduction in long run and depends on both advertiser and slot.

As for the trade-offs among different objectives, there are also some related studies. [16] first studied the problem of designing optimal mechanisms to balance revenue and welfare. [25] considered the convex combination of revenue and welfare to improve the prediction. [6] applied the linear combination of different objectives as their objective function. [24] introduced a class of parameterized mechanisms to balance different objectives.

Rank of Organic Results. Besides the methods [4,5,29] used in practise, there are still multiple theoretical investigations about ranking organic results, recently. [9] first incorporated bias of search results into consideration, when ranking the items. In the next few years, this topic is still studied by many researchers, e.g., [10,18,30]. [17] showed how to optimally rank search results to maximize an objective that combines search-result relevance and sales revenue. [7] studied the optimal ranking rule of multiple objectives includes consumer and seller surplus, as well as the sales revenue, taking consumers' choice into consideration.

However, all these work treats the sponsored search auction separately and does not consider the mixed arrangement of organic results and advertisements. We design an integrated system that takes both organic items and paid advertisements into consideration together.

Our Organizations. In Sect. 2, we give all the necessary notations and definitions, and build up the modelling framework of the integrated ad system. Two core optimization problems are also proposed in this section. In Sect. 3, we consider designing optimal mechanisms for these two problems. In Sect. 4, we generalize our study to more practical settings and design the corresponding optimal mechanisms. Finally, we summarize our works and put forward several open problems in Sect. 5. The most of proofs, tables, algorithm and numerical experiments can refer to our full version [15].

2   Notations and Preliminaries

2.1   Integrated Ad System

In the classic ad auction, advertisers bid for the keywords, and the pre-set advertising slots, which are separated from organic search result, are allocated to the ads with the highest bids. In this paper, we propose a new model called *integrated ad system* which caters for modern e-commerce online platforms, where both the positions and total number of advertising slots are not determined in advance. After collecting bids from advertisers, the platform decides how to rank all items (including sponsored advertisements and organic results) and how much each advertiser should pay.

We start with a formal description of the IAS. In an IAS, there are n_1 advertisers and n_2 organic results competing for K available slots of one search-result page simultaneously. For simplicity, we call advertisements and organic results as ad items and organic items, respectively. Denote A as the set of ad items and O as the set of organic items, and exploit $i \in A$ or $i \in O$ to represent an ad item or an organic item. Let $k \in \{1, 2, \ldots, K\}$ index the slots. Generally, a higher position slot has a smaller index. For the kth slot, β_k stands for its effective exposure, which means the probability that one user pays attention to this slot. Without lose of generality, assume that $\beta_1 > \beta_2 > \ldots > \beta_K > 0$.

Each (ad or organic) item i has a quality (weight) factor w_i to reflect its relative popularity compared to other items. In addition, each item i's estimate merchandise volume per click is g_i that is well known by the platform and advertisers. In detail, the estimate merchandise volume is an attribute that represents the expected sale amount per click. For ease of representation, we call g_i volume, instead. Each ad item i still has an extra private value v_i per click. More specifically, v_i means that the advertisers want to charge extra v_i to gain one click. Suppose that v_i is independently (not necessarily identical) drawn from $[0, u_i]$ according to a publicly known distribution $F_i(v_i)$ whose respective pdf is $f_i(v_i)$. Given this, the virtual value of ad item i can be represented as $\phi_i(v_i) = v_i - (1 - F_i(v_i))/f_i(v_i)$. We assume that the distribution $F_i(v_i)$ satisfies the regular condition, which implies that $\phi_i(v_i)$ is monotone non-decreasing. For the organic items, we have the following assumption:

Assumption 1. *For any organic item i, we assume its valuation is always 0, i.e., v_i is drawn from the degenerate distribution with one support point 0.*[3]

In the IAS, we consider the separable click-through-rate (CTR) model. That is, if the item i is allocated to slot k, the item i's CTR is $w_i \beta_k$. It means that if we put an item into a slot, the probability of being clicked is affected not only by the position but also by itself. Similarly, the corresponding merchandise volume is $g_i w_i \beta_k$.

In summary, the whole process of IAS can be described as: 1. The platform releases the information of slots; 2. The advertisers submit a bid price based on the value; 3. The platform ranks all ad items and organic items simultaneously and decides how much the advertisers should charge.

2.2   Mechanism Design

Let $v = (v_1, v_2, \ldots, v_{n_1})$ and $b = (b_1, b_2, \ldots, b_{n_1})$ be the value profile and the bid profile of all ad items. We may use v_{-i} and b_{-i} to represent the value profile and bid profile of all ad items except ad item i.

In the IAS, after receiving the bids from advertisers, a *mechanism* $\mathcal{M} = (x(b), p(b))$ consists of two rules, allocation rule $x(b)$ and payment rule $p(b)$.

---

[3] This assumption can be understood as that any organic items have no intention to charge extra money for click numbers. They do not submit any price to platform and make no contribution to the instant revenue.

More specifically, $x(b) = (x_1(b), x_2(b), \ldots, x_{n_1+n_2}(b))^4$ and $p(b) = (p_1(b), p_2(b), \ldots, p_{n_1}(b))$, where $x_i(b) = \sum_{k=1}^{K} x_{ik}(b)\beta_k$ and $x_{ik}(b)$ is the indicator function to imply whether item i is assigned to slot k. Note that the CTR of item i is $w_i x_i(b)$.

Since one slot should be allocated to one item and one item should be assigned to at most one slot, $x_{ik}(b)$ must satisfy the following constraints, denoted by $\mathcal{X}$:

$$\sum_k x_{ik}(b) \leq 1, \qquad\qquad \forall i \in A \text{ or } O,$$

$$\sum_{i \in A} x_{ik}(b) + \sum_{i \in O} x_{ik}(b) = 1, \qquad\qquad \forall k,$$

$$x_{ik}(b) \in \{0, 1\}, \qquad\qquad \forall i, k.$$

Given allocation rule and payment rule, the expected utility of ad item i with bid b_i, can be expressed as:

$$U_i(x, p, b_i) = \int_{V_{-i}} (v_i - p_i(b_i, v_{-i})) w_i x_i(b_i, v_{-i}) f_{-i}(v_{-i}) dv_{-i} \qquad (1)$$

In the IAS, to guarantee that advertisers have a desire to participate in the auction, we require that the utility of ad item should not be less than zero.

Definition 1 (Individual Rationality)

$$U_i(x, p, b_i) \geq 0, \qquad \forall b_i \in [0, u_i], \quad \forall i \in A.$$

In this paper, we also hope to design mechanisms that avoid advertisers false reporting. Since our model is in Bayesian setting, we consider the mechanisms with Bayesian Incentive Compatibility (BIC).

Definition 2 (Bayesian Incentive Compatibility)

$$U_i(x, p, v_i) \geq U_i(x, p, b_i), \qquad \forall b_i \in [0, u_i], \quad \forall i \in A.$$

We call a mechanism *feasible* iff it is IR, BIC and satisfies the condition $\mathcal{X}$. Fortunately, [20] has already given the equivalent characterization of IR, BIC mechanisms. With a little refinement, we can design mechanisms with IR and BIC in IAS as the following.

Lemma 1. (*[20]*). *A mechanism is IR, BIC if and only if, for any ad item i and bids of other items b_{-i} fixed,*
1. *$x_i(b_i, b_{-i})$ is monotone non-decreasing on b_i.*
2. *$p_i(b) = b_i - (\int_0^{b_i} x_i(s_i, b_{-i}) ds_i)/x_i(b)$, when $x_i(b) \neq 0$; Otherwise, $p_i(b) = 0$.*

In the following of this paper, we only concentrate on designing feasible mechanisms and use v_i to represent the bid price b_i directly for convenience.

---

[4] In the IAS, since we allow that ad items and organic items display in a mixture configuration, the outcome of an mechanism decides how to allocate both two types of items, simultaneously.

2.3 Core Problems

As mentioned before, e-commerce platforms, unlike conventional search engines, are concerned about instant revenue from ad items and GMV from both ad items and organic items. Recall the notations in Subsects. 2.1 and 2.2, the revenue and GMV are respectively given by

$$\text{Revenue} = \int_V \sum_{i \in A} p_i(v) w_i x_i(v) f(v) dv \tag{2}$$

and

$$\text{GMV} = \int_V \Big[\sum_{i \in A} g_i w_i x_i(v) + \sum_{i \in O} g_i w_i x_i(v) \Big] f(v) dv. \tag{3}$$

In this subsection, we put forward two different approaches to trade off the two objectives. The first approach is to think about the linear convex combination of the revenue and the GMV directly. Specifically, given a coefficient $\alpha \in [0,1]$ in advance, our problem is to find an optimal mechanism to maximize their convex combination with α. We call this problem as *Unconstrained Problem*.

Unconstrained Problem. $UCST(\alpha)$ Given the weighted coefficient $\alpha \in [0,1]$, the unconstrained problem $UCST(\alpha)$ can be written as

$$\max_{x \in \mathcal{X}} \quad \alpha \cdot \text{Revenue} + (1 - \alpha) \cdot \text{Volume}. \tag{4}$$

The second approach is that we optimize one metric while restricting the other metric. This problem is called *Constrained Problem*. In this paper, we always put GMV into constraints and optimize the revenue. Hence, in this problem, a threshold of GMV, denoted by V_0, is always given in advance.

Constrained Problem. $CST(V_0)$ Given a threshold V_0, the constrained problem $CST(V_0)$ can be written as

$$\max \quad \text{Revenue} \tag{P1}$$

$$\text{s.t.} \quad \text{Volume} \geq V_0 \tag{C1.1}$$

$$x \in \mathcal{X} \tag{C1.2}$$

3 The Optimal Mechanisms for the IAS

In this section, we mainly explore the optimal feasible mechanisms for both unconstrained problem and constrained problem. We start with the unconstrained problem and present the elegant optimal mechanism.

3.1   The Unconstrained Problem

From our intuition, if there is no ad item, i.e., nobody submits a bid, the platform will rank organic items by their volumes naturally. Because the organic item will not influence the revenue, even if ad items appear, the order among organic items does not change, as well. Therefore, the problem of designing the optimal mechanism is, in fact, how to pick up appropriate ad items and insert them properly into the list of organic items to achieve the optimum of goals. More specifically, we need to design the optimal ranking criterion to incorporate the ad items without disturbing the order of organic items, and the corresponding payment function. For simplicity, we can define $I = A \cup O$ as the new set to represent all items. Formally, we also extend $v, b, p(b)$ to $n_1 + n_2$ dimensions.

For the unconstrained problem with fully mixed allocation of the ads and organic items, we aim to find the criterion of rank to insert ad items into the sequence of organic items optimally. Because there is no other constraint except for the feasibility, we can simply the objective function to an easily observed form which can help us find the optimal ranking rule, shown in the Lemma 2.

Lemma 2. *To maximize the objective function (4), the mechanism needs to maximize the objective*

$$\int_V \sum_{i \in I} (\alpha \phi_i(v_i) + (1-\alpha)g_i)w_i x_i(v)f(v)dv - \alpha \sum_{i \in I} U_i(x, p, 0). \tag{5}$$

Because $p(b)$ only appear in $U_i(x, p, 0)$ and we need to guarantee the property of IR, by Lemma 1, if we choose

$$p_i(v) = v_i - \frac{\int_0^{v_i} x_i(s_i, v_{-i})ds_i}{x_i(v)} \tag{6}$$

as payment rule, then $U_i(x, p, 0)$ will be 0 and we only need to consider to select appropriate $x(b)$ to maximize the first part of (5).

As for allocation rule, we define $\psi_i(v_i) \overset{def}{=} (\alpha \phi_i(v_i) + (1-\alpha)g_i)w_i$ as the revised virtual value of item i. It is not difficult to check that the following allocation rule can maximize the first part of formula (5).

$$x_{ik}(v) = \begin{cases} 1 & \text{if } \psi_i(v_i) \text{ is the } k\text{th-highest revised virtual value}, k \in \{1, \dots, K\}, \\ 0 & \text{otherwise.} \end{cases} \tag{7}$$

Note that since $\phi_i(v_i)$ is regular, $\psi_i(v_i)$ is non-decreasing as well. Consequently, $x_i(v)$ satisfies the non-decreasing, when fix v_{-i}.

By the argument above, we can obtain the optimal mechanism for unconstrained problem by Theorem 1.

Theorem 1. *Given $\alpha \in [0, 1]$, the mechanism $\mathcal{M} = (x(v), p(v))$, where $x(v)$ and $p(v)$ are described by (7) and (6), is the optimal feasible mechanism for the unconstrained problem $UCST(\alpha)$.*

The optimal mechanism of the unconstrained problem has been given above. We will explain more about the optimal mechanism. For the special case that all items are organic items, the platform should rank them by their volumes. Our mechanism exactly has the same rank as this, because our mechanism ranks items by $\psi_i(v_i) = (\alpha\phi_i(v_i) + (1 - \alpha)g_i)w_i = (1 - \alpha)g_iw_i$. It is equivalent to ranking them by their volume. Based on this point of view, we can regard the rank of general case in another perspective. For ad items, the term $\phi_i(v_i)$ is different from that of organic item. In the same ranking criterion, we can rank the organic items first, then insert ad items optimally by ranking score $\psi_i(v_i)$. Another fact is that there exists no redundant slots that no items are assigned into. This is because, for the organic items, their revised virtual value is always greater than zero, which makes that, at least, the organic items can be shown in the result list. This point is also the difference with the classic Myerson auction which may not allocate items. In another words, our mechanism is based on the structure of Myerson auction, but allocation result is different. Finally, for the organic items, they also have the payment rule $p(b)$, but we can find that this value will be always equal to 0, which meets the practical requirement.

3.2 The Constrained Problem

In this subsection, we mainly concentrate on the constrained problem. Since the feasible region is discrete, i.e., $x_{ik}(v)$ is not a continuous function about bids v, it is difficult to optimize the objective directly. We first relax constraints $x_{ik}(v) \in \{0, 1\}$ in (C1.2) to $x_{ik}(v) \in [0, 1]$ and denote the relaxed feasible region as $\bar{\mathcal{X}}$. The relaxed constrained problem is

$$\max \quad R(x) \overset{def}{=} \int_V \sum_{i \in I} w_i p_i(v) x_i(v) f(v) dv \tag{P2}$$

$$\text{s.t.} \quad V(x) \overset{def}{=} \int_V \sum_{i \in I} g_i w_i x_i(v) f(v) dv \geq V_0 \tag{C2.1}$$

$$x \in \bar{\mathcal{X}}. \tag{C2.2}$$

Given this, we can prove that Program (P2) is a convex optimization and satisfies the strong duality, given an accessible V_0. For an accessible V_0, strong duality holding implies that

$$\underbrace{\max_{x: x \in \bar{\mathcal{X}}, V(x) \geq V_0} R(x)}_{\text{Primal}} = \underbrace{\min_{\lambda \geq 0} \max_{x \in \bar{\mathcal{X}}} \left(R(x) + \lambda(V(x) - V_0)\right)}_{\text{Dual}}$$

$$= \min_{\lambda \geq 0} \max_{x \in \bar{\mathcal{X}}} \left[\int_V \sum_{i \in I} (p_i(b_i) + \lambda g_i) w_i x_i(b) f(b) db - \lambda V_0 \right].$$

Observing the inner maximization problem of the dual problem, when given a λ, it is an unconstrained problem with relaxed feasible region $\bar{\mathcal{X}}$. Since the coefficient matrix of $\bar{\mathcal{X}}$ is totally unimodular matrix [22], the optimal solution of inner problem is integral form, i.e., 0–1 form. Therefore, the optimal solution of inner problem can be obtained by mechanism $\mathcal{M}$ with $\alpha = 1/(\lambda+1)$, that is,

$$x_{ik}^\lambda(v) = \begin{cases} 1 & \text{if } \psi_i^\lambda(v_i) \text{ is the } k\text{th-highest}, k \in \{1, 2, \ldots, K\}, \\ 0 & \text{otherwise,} \end{cases} \tag{8}$$

where $\psi_i^\lambda(v_i) \stackrel{def}{=} [(\phi_i(v_i) + \lambda g_i)w_i]/(\lambda+1)$. $p_i^\lambda(v)$ can be derived by replacing $x_{ik}(v)$ with $x_{ik}^\lambda(v)$ in (6).

As for the outer problem, according to the complementary-slackness, the optimal λ^* satisfies that $\lambda^*(V(x^{\lambda^*}) - V_0) = 0$. Hence, either $\lambda^* = 0$ or $V(x^{\lambda^*}) - V_0 = 0$ must hold. For the former case, we can testify whether $\lambda^* = 0$ by running the optimal mechanism $\mathcal{M}$ in Theorem 1 with $\alpha = 1$ and compare the output GMV with V_0. If the output GMV exceeds V_0, the optimal λ^* is 0. Otherwise, we turn to the latter case. Because $V(x^\lambda)$ is monotone on λ and it only contains one variable λ, we come up with a Dichotomy to output the optimal λ^*. Consequently, plugging the optimal λ^* into $x_{ik}^\lambda(v)$ and $p_{ik}^\lambda(v)$, we derive the optimal feasible mechanism, $\bar{\mathcal{M}} = (x_i^{\lambda^*}(v), p_i^{\lambda^*}(v))$, for the relaxed constrained problem. Since $x_{ik}^\lambda(v)$ is 0–1 form, $\bar{\mathcal{M}}$ is the optimal mechanism for the unrelaxed constrained problem.

Owing to the strong duality, the constrained problem is equivalent to an unconstrained problem in some sense, if we choose a proper parameter. Therefore, we have the following theorem to demonstrate their relationship.

Theorem 2. *The constrained problem with threshold V_0 is equivalent to one unconstrained problem with coefficient α^*, where $\alpha^* = 1/(\lambda^* + 1)$ and λ^* is the optimal Lagrangian multiplier in the constrained problem.*

4   Extensions

When allowing that advertisements and organic items can be arranged in a fully hybrid way, one inevitable problem is that sometimes there may be too many ad items displaying in the optimal allocations. Even though this type of allocations can achieve the optimal tradeoff, these scenarios will hurt the e-commerce platform in the long run. In this section, we generalize the integrated system to tailor for more practical requirements. We make two extensions: one is to restrict the number of ad items shown in one result page, the other is to add the constraint on the sparsity of ad items. In each extension, we find the optimal mechanism for the unconstrained problem (generalizing the Theorem 1), and build up the relationship between the constrained problem and the unconstrained problem (extending Theorem 2).

4.1 The Integrated Ad System with Number Budget on Ad Items

In this subsection, we still study designing the optimal mechanisms with an extra requirement of bounding the total number of ad items. Formally, if we require the number of ad items within c, we only need to add the constraint, $\sum_{i \in A} \sum_k x_{ik}(v) \leq c$, into $\mathcal{X}$ and denote new feasible region as $\mathcal{X}'$.

We first extend the results of unconstrained problem to this setting. From Lemma 1, 2 and mechanism $\mathcal{M}$, we have known that items are ranked by their revised virtual values, $\psi_i(v_i)$. To abide by the new constraints, ad items appear at most c times in the final layout. It is reasonable for us to take the top c ad items (ordered by their highest revised virtual value) into account while ranking all items. In view of this point, we generalize the mechanism $\mathcal{M}$ to $\mathcal{M}'(c)$:

Mechanism $\mathcal{M}'(c)$:
1. Sort all ad items into non-increasing order by $\psi_i(v_i)$. Remain the top c ad items and delete the other ad items;
2. For the remaining ad items and all organic items, run the mechanism $\mathcal{M}$ with coefficient α.

Due to the optimality of mechanism $\mathcal{M}$ and satisfying the restriction on the number of ad items, we can claim that mechanism $\mathcal{M}'(c)$ is the optimal mechanism in current setting.

Theorem 3. *Given $\alpha \in [0,1]$ and the budget c, the mechanism $\mathcal{M}'(c)$ is the optimal mechanism for the unconstrained problem $UCST(\alpha)$ with the budget on the number of ad items.*

As for the constrained problem, we follow the similar logic used in Sect. 3.2. However, the main difference and difficulty is that, the coefficient matrix of $\mathcal{X}'$ is not totally unimodular anymore, compared with the previous constrained problem. Fortunately, even though the coefficient matrix lacks of the unimodularity, we can prove that the relaxed problem still has the 0–1 form optimal solutions by transferring it to a min-cost max-flow problem (shown in full version). Based on the analysis in Sect. 3.2, the constrained problem still can simplify to an unconstrained problem by choosing the proper parameter, which means that we generalize the Theorem 2 into the case with budget constraint.

4.2 The Sparse Integrated Ad System

In this subsection, we investigate another extension that can control the sparsity of ad items. It is motivated by the reality that some e-commerce platforms hope that ad items appear intermittently, rather than emerge together. We consider two kinds of constraints on sparsity of ad items: the sparsity on row and the sparsity on column. The latter one can be found into full version.

The Sparsity on Row. When we search the products by personal computer on e-commerce platforms, the search results are usually shown in several rows and each row contains several items. Inspired by this layout, we hope to constrain the number of ads in each row to comfort the users. In view of this point, we divide one search-result page into multiple rows and each row contains consecutive l slots. Then, we require that there are at most c ad items displayed in these l slots. We call this model as the sparse integrated ad system – row (SIASR). From the mathematical perspective, the above requirement can be described as a group of constraints: $\sum_{i \in A} \sum_{k=ml+1}^{(m+1)l} x_{ik}(v) \leq c, \quad m = 0, 1, \ldots, \lfloor K/l \rfloor$ and $\sum_{i \in A} \sum_{k=\lfloor K/l \rfloor l+1}^{K} x_{ik}(v) \leq c$. Add them to $\mathcal{X}$ and denote the new constraint feasible set as $\tilde{\mathcal{X}}$.

We begin with the unconstrained problem in this case. Observing these new constraints, we can find that each one is actually a constraint on the number of ad items for l slots. In another word, if we divide one search-result page into $\lfloor K/l \rfloor + 1$ parts, every part is an IAS with number budget on ad items (introduced in Sect. 4.1). Consequently, We can run mechanism $\mathcal{M}'(c)$ for each part to output a feasible arrangement for SIASR. Since the optimality of $\mathcal{M}'(c)$, every part achieves the optimal arrangement, which induces the global optimal arrangement for all slots. Therefore, we derive the optimal mechanism in this setting, defined as $\tilde{\mathcal{M}}(c, l)$.

Mechanism $\tilde{\mathcal{M}}(c, l)$:
1. Run mechanism $\mathcal{M}'(c)$ to allocate the first l slots on all items;
2. Run mechanism $\mathcal{M}'(c)$ to allocate the next l slots on all remaining items;
3. Repeat step 2 until all slots are allocated. (In the last round, the number of slots may be less than l).

Theorem 4. *Given $\alpha \in [0, 1]$, the number of items in a row, l and the budget c, the mechanism $\tilde{\mathcal{M}}(c, l)$ is the optimal mechanism for the unconstrained problem $UCST(\alpha)$ of SIASR.*

For the constrained problem, guiding from the method in Sect. 3.2, the remaining question is to show the 0–1 form of the optimal solution of the relaxed problem. Compared to Subsect. 4.1, the problem in this subsection is trickier, since multiple parts enhance the difficulty on construction of the equivalent min-cost max-flow problem. However, we overcome it (shown in full version) and solve the constrained problem of SIASR, which extends the Theorem 2 to this setting.

5   Conclusion

To sum up, for the new features of e-commerce online platforms that combine advertisement and organic items for display purposes, we offer a thorough and exact study on designing mechanisms to balance both areas of "volume" and revenue. For different ways of tradeoff and practical requirements, optimal feasible mechanisms can be designed in order to realize them. We also empirically evaluate our mechanism and demonstrate its advantage.

As far as we know, our research is the first through study on mixed arrangement of advertisement items and organic items. There are still many problems remain open, to name a few:

- In this paper, we only focus on feasible mechanisms. There may exist mechanisms which are not limed to BIC or IR that can realize a higher trade-off.
- The valuation distribution is publicly known to each advertiser. A further area to explore would be how to design mechanisms that avoid cheating on distribution.

References

1. Abrams, Z., Schwarz, M.: Ad auction design and user experience. In: Deng, X., Graham, F.C. (eds.) Internet and Network Economics, pp. 529–534. Springer, Berlin (2007)
2. Aggarwal, G., Goel, A., Motwani, R.: Truthful auctions for pricing search keywords. In: Proceedings of the 7th ACM Conference on Electronic Commerce, pp. 1–7. ACM (2006)
3. Athey, S., Ellison, G.: Position auctions with consumer search. Q. J. Econ. **126**(3), 1213–1270 (2011)
4. Austin, D.: How google finds your needle in the web's haystack. Am. Math. Soc. Feat. Col. **10**(12) (2006)
5. Avrachenkov, K., Litvak, N.: Decomposition of the Google PageRank and optimal linking strategy (2004)
6. Bachrach, Y., Ceppi, S., Kash, I.A., Key, P., Kurokawa, D.: Optimising trade-offs among stakeholders in ad auctions. In: Proceedings of the Ffteenth ACM Conference on Economics and Computation, pp. 75–92. ACM (2014)
7. Chu, L.Y., Nazerzadeh, H., Zhang, H.: Position ranking and auctions for online marketplaces. Manag. Sci. **66**(8) (2017). SSRN 2926176
8. Clarke, E.H.: Multipart pricing of public goods. Public Choice **11**(1), 17–33 (1971)
9. Crowcroft, J.: Net neutrality: the technical side of the debate: a white paper. ACM SIGCOMM Comput. Commun. Rev. **37**(1), 49–56 (2007)
10. Edelman, B., Lockwood, B.: Measuring bias in 'organic' web search. Unpublished manuscript (2011)
11. Edelman, B., Ostrovsky, M., Schwarz, M.: Internet advertising and the generalized second-price auction: selling billions of dollars worth of keywords. Am. Econ. Rev. **97**(1), 242–259 (2007)
12. Groves, T., et al.: Incentives in teams. Econometrica **41**(4), 617–631 (1973)
13. Lahaie, S., Pennock, D.M.: Revenue analysis of a family of ranking rules for keyword auctions. In: Proceedings of the 8th ACM Conference on Electronic Commerce, pp. 50–56. ACM (2007)
14. Li, J., Liu, D., Liu, S.: Optimal keyword auctions for optimal user experiences. Decis. Support Syst. **56**, 450–461 (2013)
15. Li, W., Qi, Q., Wang, C.: Optimally integrating ad auction into e-commerce platforms. CoRR abs/2007.09359 (2020). https://arxiv.org/abs/2007.09359
16. Likhodedov, A., Sandholm, T.: Auction mechanism for optimally trading off revenue and efficiency. In: Proceedings of the 4th ACM Conference on Electronic Commerce, pp. 212–213. ACM (2003)
17. L'Ecuyer, P., Maillé, P., Stier-Moses, N.E., Tuffin, B.: Revenue-maximizing rankings for online platforms with quality-sensitive consumers. Oper. Res. **65**(2), 408–423 (2017)
18. Maillé, P., Tuffin, B.: Telecommunication Network Economics: From Theory to Applications. Cambridge University Press, Cambridge (2014)

19. Maskin, E., Riley, J.: Optimal multi-unit auctions'. Int. Lib. Criit. Writings Econ. **113**, 5–29 (2000)
20. Myerson, R.B.: Optimal auction design. Math. Oper. Res. **6**(1), 58–73 (1981)
21. Ostrovsky, M., Schwarz, M.: Reserve prices in internet advertising auctions: a field experiment. In: Proceedings of the 12th ACM Conference on Electronic Commerce, pp. 59–60. ACM (2011)
22. Papadimitriou, C.H., Steiglitz, K.: Combinatorial Optimization: Algorithms and Complexity. Courier Corporation, North Chelmsford (1998)
23. Roberts, B., Gunawardena, D., Kash, I.A., Key, P.: Ranking and tradeoffs in sponsored search auctions. ACM Trans. Econ. Comput. **4**(3), 17 (2016)
24. Shen, W., Tang, P.: Practical versus optimal mechanisms. In: Proceedings of the 16th Conference on Autonomous Agents and Multi-agent Systems, pp. 78–86. International Foundation for Autonomous Agents and Multiagent Systems (2017)
25. Sundararajan, M., Talgam-Cohen, I.: Prediction and welfare in ad auctions. Theory Comput. Syst. **59**(4), 664–682 (2016)
26. Thompson, D.R., Leyton-Brown, K.: Revenue optimization in the generalized second-price auction. In: Proceedings of the Fourteenth ACM Conference on Electronic Commerce, pp. 837–852. ACM (2013)
27. Varian, H.R.: Position auctions. Int. J. Ind. Organ. **25**(6), 1163–1178 (2007)
28. Vickrey, W.: Counterspeculation, auctions, and competitive sealed tenders. J. Financ. **16**(1), 8–37 (1961)
29. Williams, H.: Measuring search relevance (2010)
30. Wright, J.D.: Defining and Measuring Research Bias: Some Preliminary Evidence, pp. 12–14. International Center for Law & Economics, November 2011

Verifiable Crowd Computing: Coping with Bounded Rationality

Lu Dong[1], Miguel A. Mosteiro[1(✉)], and Shikha Singh[2]

[1] Pace University, New York, NY 10038, USA
{ld41349n,mmosteiro}@pace.edu
[2] Williams College, Williamstown, MA 01267, USA
shikha@cs.williams.edu

Abstract. In this paper we use the repeated-games framework to design and analyze a master-worker (MW) mechanism, where a master repeatedly outsources computational tasks to workers in exchange for payment. Previous work on MW models assume that all workers are *perfectly rational* and aim to maximize their expected utility. Perfect rationality is a strong behavioral assumption because it requires that all workers follow the prescribed equilibrium. Such a model may be unrealistic in a practical setting, where some agents are not perfectly rational and may deviate from the equilibrium for short-term gains. Since the correctness of these MW protocols relies on workers following the equilibrium strategies, they are not robust against such deviations.

We augment the game-theoretical MW model with the presence of such bounded-rational players or *deviators*. In particular, we show how to design a repeated-games based MW Verifiable Crowd Computing mechanism that incentivizes the rational workers (or *followers*) to effectively punish such deviators through the use of terminal payments. We show that the master can use terminal payments to obtain correct answers to computational problems even in this more general model. We supplement our theoretical results with simulations that show that our mechanism outperforms related approaches.

Keywords: Crowd computing · Master-worker computing · Internet computing · Verifiable computation outsourcing · Repeated games · Algorithmic game theory

1 Introduction

Due to the escalation of large data sets and the high cost of supercomputers, most computation today is not being performed locally, but rather outsourced either as Cloud Computing, Grid Computing, or to Internet-connected personal computers (which we refer to as Crowd Computing[1]). In a Crowd Computing platform, a client outsources computational tasks to several untrusted workers,

---

[1] The denomination Crowd Computing has been used for different systems. We define our model in Sect. 3.

© The Author(s), under exclusive license to Springer Nature Switzerland AG 2022
M. Li and X. Sun (Eds.): IJTCS-FAW 2022, LNCS 13461, pp. 59–78, 2022.
https://doi.org/10.1007/978-3-031-20796-9_5

often in exchange for money. Crowd Computing has proven to be a powerful and cost-effective alternative to setting up an expensive computational infrastructure locally. However, it also brings up several implementation challenges, the most prominent one being: how can a client ensure that the computational tasks have been performed correctly by the untrusted workers?

In this work, we study Crowd Computing in the master-worker (MW) model where computational tasks are assigned by a centralized entity called *master* to agents called *workers*. Workers are expected to compute the task and return the result. All communication between master and workers occurs in parallel, that is, tasks are assigned and the results are returned simultaneously by all workers.

Practical examples of such Crowd Computing systems include SETI@home [42], Foldit [32,45], and Mechanical Turk [3]. Reliable participation in these systems is induced through various incentives. For example, in Foldit the motivation is to participate in a game, in Mechanical Turk workers participate for profit compelled by a reputation system, whereas in SETI@home workers are altruistic volunteers.

Game-theoretical MW Model. Various models of worker-behavior have been studied in the MW literature in distributed computing, e.g., the workers have been assumed to be either *malicious* or *honest* [20,21,23,34–36,44].

The game-theoretic MW model assumes that all workers are rational and choose their strategy to maximize their expected utility [13,14,21,22,47]. While such a model is ideal for capturing the economic incentives that are present in these systems, it also imposes strong behavioral assumptions on the workers. In particular, the players are assumed to be perfectly rational. That is, all players play the prescribed equilibrium strategy. In practice, some strategic agents in a Crowd Computing system may not follow the equilibrium strategy for any number of reasons: carelessness, limited rationality (they may not believe in the expected utility guarantees of a probabilistic mechanism), they may be skeptical of the empty threats sustainining the equilbrium, or they may distrust the master. We call such players who are not perfectly-rational and may not follow the prescribed equilibrium as *deviators*.

There is evidence that even in systems that provide incentives for computing correctly, such deviators exist [4,26,30]. Since the master obtains the correct answer to the computational tasks only when all workers follow the equilibrium prescribed by the chosen incentives, the presence of such bounded-rational players, or *deviators* jeopardizes the correctness guarantees of such systems.

In this paper, we provide a principled approach to designing MW mechanisms for Crowd Computing in the presence of such "bounded rational" deviators. We say these deviators have bounded rationality to distinguish them from purely malicious players because while they initially deviate from the equilibrium, once they are punished by their rational peers, they realize that the threats are credible, and their utility will be maximized by following the equilibrium. That is, we assume that after being punished deviators behave as followers.

Master-Worker Model with Deviators. In this work, we augment the repeated-games-based MW model of Fernández Anta et al. [24] in two ways.

First, we consider pools of workers that include *deviators*.[2] In particular, we assume that some players in the system are perfectly-rational in the game-theoretic sense and, hence, will follow the prescribed equilibrium (we call them *followers*) and the rest are *deviators*, who initially may not follow the equilibrium strategy. Second, in contrast to the *infinitely* repeated game studied in [24], we consider *finitely* repeated games. That is, the MW interaction lasts for predetermined number of rounds (with a possible extension as explained in Sect. 3).

The presence of deviators in the MW model poses several challenges. First, we need a mechanism to identify and appropriately punish such deviators. Second, in a repeated-games framework such a punishment must be levied by the followers, who incur a short-term utility loss in doing so. Thus, we need a mechanism to incentivize the followers to punish deviators on behalf of the master. We achieve this through the use of **terminal payments**, which are additional payments provided to the followers at the end of a finitely-repeated game.

Terminal payments in finitely-repeated games has been studied by Gossner [27], who proved a Folk Theorem for such games when players use mixed-strategies. Gossner introduced these payments precisely to handle the presence of deviators. We follow Gossner's model of terminal payments and use them to compensate followers at the end of the MW protocol.

Finitely-repeated games and terminal payments are particularly appealing from a practical standpoint in Crowd Computing systems as they better capture the practice of (a) hiring workers for limited periods, and (b) keeping track of their behavior and rewarding compliance through payments, respectively. Extensions such as modeling different types of deviators, or collusion among workers are also interesting, and are left to future work.

1.1 Main Contributions

We summarize the main contributions of our paper below.

- *Modeling Deviators.* We generalize the repeated-games MW model to allow for the presence of bounded-rational workers or deviators. Bounded-rationality is a concept that has gained popularity in the computational applications of game theory (e.g. see. [5,10–12,16,28,29,43]) because it bridges the gap between game theory and rational behavior in practice. Our bounded-rationality models intentional deviations, as opposed to other models where deviations are accidental (e.g. trembling-hand equilibria [9]).
- *Terminal Payments.* We show how to implement terminal payments in a constructive way in the repeated-games model. This notion was previously only studied in an existential context. In particular, Gossner [27] showed that a mixed-strategy equilibrium of finitely-repeated games exists in the presence of terminal payments. In this paper, we construct a mechanism that admits such an equilibrium in the presence of deviators.

---

[2] The notion of deviators is only considered as part of the simulations in [24]; see Sects. 2 and 6 for further comparison.

- *Verifiable MW Computing in the Presence of Deviators.* As our main result (Theorem 1) we provide a MW mechanism that is robust against deviating workers—that is, the master obtains the correct answer in expectation and asymptotically almost surely under certain mild conditions. We achieve this through a finitely-repeated MW game and the use of terminal payments. Thus, our mechanism improves over the work of Fernández Anta et al. [24] in achieving correctness guarantees in a weaker behavioral model that includes bounded-rationality.
- *Simulation Results.* We also compare experimentally the performance of our mechanism to the previous work of Christoforou et al. [15] and Fernández-Anta et al. [24]. We compare these approaches on the following metrics: correctness of the answers obtained by the master and the cost incurred by the master. On the range of parameters tested, our simulations show that, even for a large number of tasks, our mechanism outperforms [15] in correctness at a similar cost. Moreover, our mechanism outperforms [24] in terms of the cost while providing similar correctness guarantees in a weaker model. We note that the repeated application of a one-shot verification mechanism was shown to be worse than [15] in [24]. Consequently, our mechanism outperforms such an approach and we omit including it in simulations for clarity.

2 Additional Related Work

We discuss the various MW models that have been studied in the literature.

MW with Malicious Workers. Distributed computations in presence of malice have been well studied (e.g. [20,34,35,44]). In [44], the workers are assumed to be malicious with some probability. To counter their effect, tasks with known outcomes are executed to detect malicious workers, which is argued to work better than voting techniques. However, later experimental work [34] showed that it takes a long time to achieve low error rates in practice. In [20], the model considers *malicious* workers who may deliberately return an incorrect result in an effort to harm the master. Workers have a predefined behavior: either they are malicious or honest. Majority voting is used to cope with malice in one-shot computations. This work was later extended [35] to many tasks. A distributed verification mechanism was introduced in [36], however, the model limits malice to 20% of the workers. None of these models study selfish (rational) behavior.

MW Model with Rational Workers. Distributed computations when workers are selfish (rational in the game-theoretic sense) have also been studied (e.g. [13,14,21–23]). In [47], all workers are assumed to be rational. Auditing and majority voting measures are used. If a non-audited computation does not yield an absolute majority, the task is recomputed to achieve reliability. Several works study coexistence of rational and malicious players [1,2,6–8,13,14,18,19,22,25,40]. For example, protocols in [19] tolerate up to k rational players that deviate from a NE, follow-up work [1] tolerates up to k colluding rational players and t Byzantine ones. The BAR model (Byzantine, Altruistic, and Rational) was introduced in [2] and later used in [37,38].

The above work applies to one-shot interactions between master and workers. That is, the system design is oriented to the reliable computation of one task.

Repeated Games and MW. In the repeated-games MW framework, the master sequentially assigns tasks to the same pool of workers. Christoforou et al. [15] (and follow-up [39]) use this framework to design a mechanism based on reinforcement-learning. In their approach, master and workers adjust their strategies for the next interaction according to the outcome of the previous one. The mechanism is shown to eventually converge to a state where the master always obtains the correct results with minimal verification. Depending on system parameters such as rewards, penalties and the workers' aspiration for profit, the time required for convergence may be long.

Our work builds upon the MW model of Fernández Anta et al. [24] and differs from it in two ways.

First, the MW model in [24] is based on infinitely repeated games [41], while we design a finitely-repeated MW game. The infinitely-repeated games approach is shown to improve upon repeated application of one-shot and reinforcement-learning mechanisms in terms of cost and reliability. However, the correctness of the model relies heavily on the assumption that the workers are uncertain about when the game ends. In particular, if the workers in [24] were to know when the interaction ends, they would stop following the equilibrium (and computing tasks) and default to cheating. Letting workers know the (possibly extended) length of the interaction a priori makes the model more realistic.

Second, the model in [24] assumes all players are perfectly rational, while we allow the presence of deviators. The term "deviators" is used in [24] to analyze deviations from the equilibrium strategy, but these deviations are never beneficial for the workers. Thus, since all workers are perfectly rational, no worker will ever deviate. In this paper, we define deviators as players who may not follow the prescribed equilibrium despite being aware of the consequence to their utility, perhaps because they do not trust the mechanism or the threats of punishment. To ensure that the punishment by followers is not just an empty threat, we incentivize followers to levy the punishment using terminal payoffs. In contrast, the punishment by peers in [24] is effectively an empty threat because imposing the punishment causes a utility loss to players, and they incur such a loss indefinitely in the infinitely repeated game.

Repeated Games and Crowdsourcing. More recently, a repeated-games approach to collect crowd sensor data was presented in [33]. Their mechanism uses a worker-reputation system, reinforcement learning to update strategies, and master verification in the presence of both rational and irrational workers. Their game model is finite but fails to address the deviation from equilibrium of rational workers when the sequence of task assignments is reaching the end.

Several crowdsourcing models focus on collusion reporting or cheating detection when there are only two workers [17,46], or considering master-worker interaction as a two-player game [31]. Such approaches are often case-by-case and do not model repeatedly outsourcing many computational tasks.

3 Preliminaries

We say that a stochastic event holds *with high probability* (*w.h.p.*), if it holds with probability at least $1 - 1/n^c$, for some constant $c > 0$, and that it holds *asymptotically almost surely* (or *a.a.s.*, for short), if it holds with probability at least $1 - 1/f(n)$, where $f(n) \in \omega(1)$.

Master-Worker Model. We consider a computing platform with a set N of processing entities called **workers**, where $|N| = n$, and a coordinator entity called **master**. We call this platform a master-work (MW) system.

The master assigns computing tasks to workers sequentially in **rounds**. We assume that the master has an unbounded source of tasks that need to be computed. We consider computation of tasks that have a unique correct answer. In each round, the master assigns the same computational task to all workers. Upon receiving the results, the master decides stochastically whether to **verify** them (e.g. checking the solutions received or computing the result itself if needed), or accept the majority. The **probability of verification** by the master is denoted as p_v. In order to avoid ties, we let n be odd. We assume that the cost of verification is negligible with respect to computing the task, e.g. NP-hard problems. If the master verifies the results of the computation, it pays a **reward** ρ to the workers that returned the correct result, and charges a **fine** ϕ to those that returned and incorrect result. If the master does not verify, workers in the majority receive the reward ρ and no worker is fined.

The **expected cost of the master** is defined as the sum of the expected verification cost and expected payments, minus the expected fines received (since they reduce the cost).

As a design decision, our mechanism specifies a **reward threshold** γ on the number of workers that will be rewarded. That is, if the number of workers to be rewarded exceeds the reward threshold, the master does not reward any worker (regardless of verification). This reward threshold reflects the natural assumption that the master has a limited budget. Because receiving fines from the workers is not the goal of the system, this parameter ensures that the cost reduction due to fines does not increase the budget of the master for rewards. Note the prescribed equilibrium will ensure that the number of workers following it (the followers) stay below the threshold γ. In other words, in absence of deviations, the workers strategic decision stochastically guarantees that the number of correct answers will be at most γ.

Computing a task involves a **cost** for the workers, which is assumed to be the same value c for all. For each round of task assignments, the strategy of the workers is defined by their probabilistic choice to either compute the result of the task or to return a default incorrect answer to avoid the cost of computing, albeit risking to be fined. For each worker $i \in N$, let the **probability of computing** the task be denoted as p_{ζ_i}, and let $p_{\not\zeta_i} = 1 - p_{\zeta_i}$ be the probability of not computing.

The above protocol is followed for each round of task assignments. We summarize the notation in Table 1.

Table 1. System parameters

ρ	: reward	γ	: reward threshold
ϕ	: fine	T	: number of rounds
c	: worker computing cost	N	: set of workers ($|N| = n$)
p_{ζ_i}	: worker i probability of computing	n_f	: number of followers
p_v	: master probability of verifying	n_d	: number of deviators ($n_d = n - n_f$)

Repeated Game Model. We model the interactions in the MW system as a finitely repeated game with mixed strategies and terminal payoffs [27]. That is, the master assigns T computing tasks sequentially, each to the same set of n workers. The strategic choice for each worker $i \in N$ is the value of the probability of computing p_{ζ_i}.

In each round of task assignments, the master and workers proceed as in the MW model above. The master computes an equilibrium where the workers expected utility is maximized and sends the equilibrium parameters and its computation to the workers. We assume that among n workers, there are n_f **followers** and $n_d = n - n_f$ **deviators**. The followers are rational in the game-theoretic sense and choose their strategies according to the prescribed equilibrium. The deviators, on the other hand, may not trust the mechanism and may choose a different strategy. The deviators motivation to steer away from equilibrium is to go unnoticed while reducing costs or increasing utilities. That is, they are not Byzantine malicious.

Upon detecting deviations, the followers change to a minmax strategy (defined below) that yields the lowest payoff that followers can force upon a deviator. The purpose of changing to a minmax strategy is to penalize the deviators for not following the equilibrium. This change of strategy lasts for a period of P rounds, which is called a **peer-punishment** period. P is large enough to counter any extra utility that may be obtained by the deviators. Followers compute the value P based on the detected deviation.

Without loss of generality, we assume that deviations are detected on or before the $(T - P)^{th}$ round, and that only one period of peer-punishment is needed throughout the whole computation. Otherwise, should a deviation occur after some round T_{late} such that $T_{late}+P > T$, or more than one peer-punishment period be needed, the master could assign more tasks[3] to extend the interaction period to $T_{late}+P$, or to make T much larger than the sum of all punishment periods needed respectively. Intuitively, one would expect that deviators "learn the lesson" after being punished—they confirm that deviating is indeed not worthwhile as any utility gain from deviating is nullified by the punishments. Also, we limit the number of deviations as an arbitrary number of deviations indicates malicious intent rather than bounded rationality. Furthermore, the deviators do not know whether subsequent deviations will cause punishments, and a single punishment is sufficient to establish the credibility of such a threat.

---

[3] Recall the assumption that the master has access to an unbounded number of computational tasks.

After the T rounds of task assignments are completed, the master pays a ***terminal payoff*** to followers to compensate them for the utility loss during the peer-punishment period, and the interaction between the master and the workers terminates.

Even though the master-worker interaction lasts a finite number of rounds, the above framework allows us to model the worker behavior as in an infinitely repeated game. That is, workers choose strategies taking into account long-term interaction. Indeed, the player behavior in infinitely repeated games is well-motivated even for finitely repeated games except in the last few rounds [41], and by introducing terminal payoffs (and extending the interaction beyond T rounds if needed) the long-term choice of strategies is not affected.

We now define the game formally. Let $G_1\langle N, (A_i), (\succsim_i)\rangle$ be a normal-form (one-round) game with set of players N, where $|N| = n$. For each player $i \in N$, $A_i = \{\zeta, \cancel{\zeta}\}$ is the action space (compute or not compute) of player $i \in N$, and $\succsim_i$ is the preference relation on the space of all workers actions $A = \times_{i \in N} A_i = \{\zeta, \cancel{\zeta}\}^N$. The preference relation of player i is represented by a ***utility function*** u_i , in the sense that $u_i(a) \geq u_i(b)$ whenever $a \succsim_i b$, for $a, b \in A$. The utility function u_i is based on the rewards and fines scheme defined in the MW model. That is, for $\#\zeta = \sum_{j \in N: A_j = \zeta} 1$ and $\#\cancel{\zeta} = \sum_{j \in N: A_j = \cancel{\zeta}} 1$,

$$
u_i = \begin{cases}
-c, & \text{if } A_i = \zeta \text{ and } \#\zeta > \gamma, \\
\rho - c, & \text{if } A_i = \zeta \text{ and } n/2 < \#\zeta \leq \gamma, \\
p_v\rho - c, & \text{if } A_i = \zeta \text{ and } \#\zeta < n/2, \\
-p_v\phi, & \text{if } A_i = \cancel{\zeta} \text{ and } \#\cancel{\zeta} > \gamma, \\
(1 - p_v)\rho - p_v\phi, & \text{if } A_i = \cancel{\zeta} \text{ and } n/2 < \#\cancel{\zeta} \leq \gamma, \\
-p_v\phi, & \text{if } A_i = \cancel{\zeta} \text{ and } \#\cancel{\zeta} < n/2.
\end{cases}
\tag{1}
$$

The expected utility of the workers in the normal-form game depends on the random choices of the master and other players. Following [41], we define the extended game $G_2\langle N, (S_i), (u_i)\rangle$ that differs from G_1 in that S_i is the mixed strategies space of player $i \in N$. Let $u_i : S \to \mathbb{R}$ be the ***expected utility function*** of player i, where the expected value represents player i's preferences over the space $S = \times_{i \in N} S_i$.

The overall expected utility is based on the expected utility in each of the T rounds and the terminal payoffs. We define the repeated game with terminal payoffs implemented by our mechanism as $G(T, \mathcal{W}, \omega)\langle N, (S_i), (u_i)\rangle$. We redefine the expected utility function $u_i : S^T \to \mathbb{R}$, that is, a function whose expected value represents player i's preferences over the space S^T. $\mathcal{W} \subseteq \mathbb{R}^N$ is the terminal payoffs set, and $\omega : H_T \to \mathcal{W}$ is the terminal payoffs function applied to H_T, where H_t is the set of t-tuples of elements of S, for $t \in [1, T]$. That is, H_T is the history of mixed-strategy actions at round T. We refer to the model as a strategic game for simplicity.

The ***minmax strategy*** is a follower mixed strategy (i.e., a choice of probability of computing) such that the payoff of the deviators is minimized, regardless of their strategy. The ***minmax payoff*** v_i is the lowest expected payoff of deviator i that the followers can force upon i. It is well-known (refer to Proposition 144.3 in [23]) that, for any enforceable payoff profile, that is, any payoff profile

---

Algorithm 1: Master algorithm with paramters $\gamma = \left\lceil n/2 + 2\sqrt{n \ln \ln n} \right\rceil$, $q = \frac{1}{2^n} \left(\sum_{j=n-\gamma}^{\gamma-1} \binom{n-1}{j} + (1 - p_v)\binom{n-1}{(n-1)/2} \right)$, and $p_v = c/\phi$. Messages are sent to (and received from) all workers.

---

1 send p_ζ, q, and certificate
2 **for** T *times* **do**
3   send a computational task
4   **upon** *receiving all answers* **do**
5   verified $\leftarrow$ *true*, with probability p_v or *false*, with probability $1 - p_v$
6   **if** verified **then**
7   verify the answers
8   fine the incorrect workers
9   **if** *correct count* $\leq \gamma$ **then**
10   reward workers that were correct
11   **else**
12   accept the answer of the majority
13   **if** *majority count* $\leq \gamma$ **then**
14   reward workers in the majority
15   send list $\langle$answer, count$\rangle$ and verified
16 verify complete-information certificate received
17 send payoffs according to $\omega(H_T)$

---

where the expected utility for each worker is at least the minmax payoff, there exists a Nash equilibrium payoff profile that all workers will follow due to long-term rationality. We assume that followers aim for **Pareto efficiency** (refer to Sect. 1.7 in [41]), that is, an equilibrium mixed strategy corresponding to an enforceable payoff profile that maximizes the workers expected utility.

4 Algorithmic Mechanism

In this section we describe the mechanism that implements the MW computing platform. The algorithm for the master (refer to Algorithm 1) follows the model defined in Sect. 3 in Lines 2 to 14. After computing the equilibrium p_ζ based on the parameters defined for the system (reward, fine, etc.), in Line 1 the master sends the value of p_ζ to workers together with a certificate showing the calculation of such equilibrium, and a parameter q that is a function of the number of workers n, the cost c, and the fine ϕ. In Line 15, the master sends a list of the answers received and the number of each, which the workers will use to detect deviations. Based on the complete-information certificate received from workers (i.e. H_T, rounds of punishment, correct answers, etc.), the master computes the terminal payoffs and sends to followers in Lines 16 and 17.

Algorithm 2: Algorithm for each follower i. Messages are sent to (and received from) the master.

---

1 receive p_ζ, q, and verify certificate
2 $sum_\delta \leftarrow 0$, $sum_u \leftarrow 0$, $r \leftarrow 0$, $t \leftarrow 0$, $p \leftarrow p_\zeta$
3 **while** $t \leq T$ **do**
4      $t++$
       // computation phase
5   receive computing task
6   action $\leftarrow \zeta$, with probability p or $\not\zeta$, with probability $1 - p$
7   **if** action $= \zeta$ **then** result $\leftarrow$ computed task **else** result $\leftarrow$ fabricated bogus result
8   send result
       // deviation-detection phase
9   receive list⟨answer, count⟩ and verified
10   **for** *each* (answer, count) *in* list **do**
11   **if** action $= \not\zeta$ **then** verify answer
12   **if** answer *is correct and* count $\neq np_\zeta$ **then**
13              $sum_\delta \leftarrow sum_\delta + (np_\zeta - \text{count})^2$
14              $sum_u \leftarrow sum_u + u_{deviator}(\text{count, verified})$
15              $r++$

       // peer-punishment phase
16   **if** $sum_\delta \geq np_\zeta(1.6r + \ln n)$ **then**   // Lemma 2
17          $P \leftarrow (sum_u - \rho q + c)/(\rho q)$ // Lemma 4
18          $sum_\delta \leftarrow 0$, $sum_u \leftarrow 0$, $r \leftarrow 0$, $p \leftarrow 1$
19   execute computation phase P times
20          $p \leftarrow p_\zeta$, $t \leftarrow t + P$

21 send complete-information certificate
22 receive terminal payoff

---

5 Analysis

In our analysis, we relate the system parameters to the utility of workers and the correctness of the master. First, we show bounds on the mixed strategy equilibrium, that is, a probability of computing, in which workers maximize their expected utility (refer to Lemma 1). Then, we proceed to analyze our mechanism to handle deviations. Specifically, we prove that the deviation detection method is correct w.h.p. (refer to Lemma 2), and we provide bounds on the length of the peer-punishment period and on the terminal payoffs that compensate followers for their loss of profit during such period (refer to Lemma 4). In our final theorem, we connect all the aspects of our mechanism showing additionally that the master achieves correctness in expectation and a.a.s. under some conditions.

Due to space restrictions, we defer some proofs to the full version of this paper.

To simplify exposition, we define the following probability functions and use them throughout the analysis.

$$p_1' = \sum_{j=\frac{n+1}{2}}^{\gamma} \frac{j}{n-j} \binom{n-1}{j} p^{j-1}(1-p)^{n-j-1}(j-np),$$

$$p_2' = \sum_{j=1}^{\frac{n-1}{2}} \frac{j}{n-j} \binom{n-1}{j} p^{j-1}(1-p)^{n-j-1}(j-np), \text{ and}$$

$$p_3' = \sum_{j=n-\gamma}^{\frac{n-1}{2}} \binom{n-1}{j} p^{j-1}(1-p)^{n-j-1}(j-np).$$

5.1   A Pareto-Efficient Repeated Game Equilibrium

Recall that, for any enforceable payoff profile, that is, any payoff profile where the expected utility for each worker is at least the minmax payoff, there exists a Nash equilibrium payoff profile that all workers will follow due to long-term rationality (refer to Proposition 144.3 in [23]). To ensure Pareto efficiency, we identify a strategy profile that maximizes the workers expected utility (refer to Sect. 1.7 in [41]).

Lemma 1. *Consider any MW system with $n > 1$ workers where $n_f = n$, $n_d = 0$, $p_v = c/\phi$, $\gamma = \lceil n/2 + 2\sqrt{n \ln \ln n} \rceil$ such that $\gamma < n$, and $\phi \geq c(|p_2'| - |p_3'|)/(p_1' - |p_3'|)$ for any $0 \leq p \leq 1/2$. Then, the following holds.*

- *The mixed strategy equilibrium of G is such that all workers use the same probability of computing p_ζ, and such probability is in the range $1/2 < p_\zeta \leq \gamma/n$.*
- *The maximum expected utility that any worker $i \in N$ can obtain with a mixed strategy equilibrium of G is $E(u_i) = \rho q - c$, where q is in the range*

$$\frac{1}{2^n} \left(\sum_{j=n-\gamma}^{\gamma-1} \binom{n-1}{j} + (1-p_v)\binom{n-1}{(n-1)/2} \right) \leq q \leq 1.$$

5.2   Deviation-Detection Method

In this section, we establish a method for the followers to detect deviations from the prescribed equilibrium, and if so move to a peer-punishment phase. Using Chernoff bounds, the following lemma bounds the number of correct answers that should be obtained w.h.p. when workers follow an equilibrium. Based on those bounds, we provide a test that keeps track of such deviations over possibly many rounds. Thus, either if strong deviations occur over a few rounds, or slight deviations occur over many rounds, or both, the deviation is detected w.h.p.

Lemma 2. *For any MW system where all workers use the same p_ζ such that $p_\zeta > 1/2$, let X_t be the number of correct answers in round t. For any sequence of rounds (not necessarily contiguous) of task assignments $t_1, \ldots, t_r$, such that $X_t \neq np_\zeta$ for all $t \in \{t_1, \ldots, t_r\}$. For any constant $\xi > 0$, the following holds with probability at least $1 - 1/n^\xi$: $\sum_{t=t_1}^{t_r}(X_t - np_\zeta)^2 < np_\zeta(1.6r + \xi \ln n)$.*

Lemma 2 gives a method of detection for the followers. Namely, for any sequence of r rounds, possibly not contiguous, where the number of correct answers is not np_ζ, keep track of what is the difference with respect to np_ζ. If $p_\zeta < 1$, some workers do not compute the task, but they need to know how many correct answers were received to decide on the peer punishment. They can do so by verifying which, if any, of the answers sent by the master are correct, and use the corresponding count to compute the total number of correct answers. Then, whenever the inequality proved in Lemma 2 is not true, followers move to a peer-punishment phase.

5.3 Deviation Punishment and Terminal Payoffs

In this section we analyze conditions on the system parameters that enable deviation peer-punishment. That is, what is the minmax strategy that followers should use, and what is the duration of the peer-punishment phase. Recall that the minmax strategy is a follower mixed strategy, i.e. a choice of probability of computing, such that the payoff of deviators is minimized, independently of what the deviators do, for each round of peer-punishment. We define v_i as the lowest expected payoff that can be forced upon worker i by other workers. That is, for mixed strategies σ_i and σ_{-i} of worker i and workers in $N \setminus \{i\}$ respectively, we have $v_i = \min_{\sigma_{-i}} \max_{\sigma_i} E(u_i)$.

Lemma 3. *For any MW system with $n_f = n - 1$ followers and a deviator i, if all followers use $p_\zeta = 1$, it is $v_i = \max(-c, -p_v\phi)$.*

Lemma 3 states that the minmax strategy for MW systems where $n_f = n-1$ is $p_\zeta = 1$. Indeed, as long as the number of followers exceeds the reward threshold, if followers use the same minmax strategy, deviators receive the same minmax payoff. We establish this observation in Corollary 1.

Corollary 1. *For any MW system with $n_f > \gamma$ followers and $n_d = n - n_f$ deviators, if all followers follow the strategy $p_\zeta = 1$, then the minmax payoff $v_i = \max(-c, -p_v\phi)$.*

Lemma 4 relates the length of the peer-punishment phase to the extra payoff attained by a deviator until it is detected, and the utility loss from being punished.

Lemma 4. *For any MW system with $n_f > \gamma$ followers and $n_d = n - n_f$ deviators, $p_v = c/\phi$, $\gamma = \lceil n/2 + 2\sqrt{n \ln \ln n} \rceil$ such that $\gamma < n$, and $\phi \geq$*

$c(|p_2'| - |p_3'|)/(p_1' - |p_3'|)$, or any $0 \leq p \leq 1/2$. If deviations from the equilibrium are detected in rounds $t_1, \ldots, t_r$, then the following holds.

(i) If followers use the minmax strategy for $P \geq \frac{\sum_{t=t_1}^{t_r} u_j(t) - \rho q + c}{\rho q - c + \min(c, \ p_v \phi)}$, rounds of peer-punishment, where $q = \frac{1}{2^n} \left(\sum_{j=n-\gamma}^{\gamma-1} \binom{n-1}{j} + (1 - p_v)\binom{n-1}{(n-1)/2} \right)$, and $u_j(t)$ is the utility obtained by any deviator j in each round t in $t_1, \ldots, t_r$, then any utility gain by deviators during the $t_1, \ldots, t_r$ rounds of deviation from the prescribed equilibrium, is lost in those P rounds of peer-punishment.

(ii) To compensate the followers for their loss while using the minmax strategy, it is enough for the master to pay terminal payoffs of $\omega \geq P(\rho - c + \min(c, p_v \phi))$.

5.4 Mechanism Properties

We now establish our main result in the following theorem.

Theorem 1. Consider the mechanism specified in Algorithms 1 and 2 under the models of Sect. 3, where the reward threshold is $\gamma = \left\lceil n/2 + 2\sqrt{n \ln \ln n} \right\rceil$ such that $\gamma < n$, the set of workers has at least $n_f > \gamma$ followers, the probability of verification is $p_v = c/\phi$, and the fine is $\phi \geq c(|p_2'| - |p_3'|)/(p_1' - |p_3'|)$, for any $0 \leq p \leq 1/2$. Then, the following holds.

1. The mixed-strategy equilibrium is some probability p_ζ where $1/2 < p_\zeta \leq \gamma/n$.
2. Deviators are detected w.h.p.
3. The expected utility of each follower for T rounds is at least

$$T \left(\frac{\rho}{2^n} \left(\sum_{j=n-\gamma}^{\gamma-1} \binom{n-1}{j} + (1 - p_v)\binom{n-1}{(n-1)/2} \right) - c \right).$$

4. The expected cost of the master is at most $T(np_\zeta \rho - p_v(1 - p_\zeta)\phi)$.
5. In each round of peer-punishment, the master obtains the correct answer, and in each round of task assignments without deviations or peer-punishment,

   (a) in expectation the master obtains the correct answer,
   (b) if $p_\zeta \geq 1/2 + \sqrt{n \ln \ln n}$ the master obtains the correct answer a.a.s., and
   (c) if $p_\zeta \leq 1/2 + \sqrt{n \ln \ln n}$ the master pays less than $\gamma \rho$ in rewards a.a.s.

Proof. Claims 1 to 3 follow from Lemmas 1 and 2, and the correctness of the mechanism shown in Lemma 4.

For Claim 4, given that during punishment periods followers use $p_\zeta = 1$ and $n_f > \gamma$, no worker is rewarded, the worst case cost is incurred when there is no deviation. Thus, the claimed upper bound on the expected cost is a straightforward application of the payment model specified in Sect. 3 for T rounds of task assignments without deviations.

We prove Claim 5 as follows. From Corollary 1, we know that during rounds of peer punishment followers use $p_\zeta = 1$. Given that $n_f > \gamma > n/2$, the master

obtains the correct answer during those rounds. For any round without deviations or peer-punishment, let the random variable X_ζ be the number of workers
that compute. Then, the following holds.

a) It is $E(X_\zeta) = np_\zeta$. Hence, given that $p_\zeta > 1/2$ as shown in Lemma 1, it
follows that the master obtains the correct answer in expectation.
b) Let $\delta = \sqrt{n \ln \ln n}/E(X_\zeta)$. Then,

$$1 - \delta = 1 - \frac{\sqrt{n \ln \ln n}}{np_\zeta} = 1 - \frac{\sqrt{n \ln \ln n}}{n/2 + \sqrt{n \ln \ln n}} = \frac{n}{2E(X_\zeta)}.$$

Given that $0 < \delta < 1$, by Chernoff bound:

$$\Pr(X_\zeta \leq n/2) = \Pr(X_\zeta \leq E(X_\zeta)(1 - \delta))$$

$$\leq \exp(-E(X_\zeta)\delta^2/2) = \exp\left(-\frac{\ln \ln n}{2p_\zeta}\right)$$

$$= \left(\frac{1}{\ln n}\right)^{\frac{1}{2p_\zeta}} \leq \frac{1}{\sqrt{\ln n}}.$$

Thus, $\Pr(X_\zeta \leq (n-1)/2) \leq 1/\sqrt{\ln n}$. Therefore, the master obtains the
correct answer a.a.s.
c) Let $\delta = \gamma/E(X_\zeta) - 1$. Given that $0 < \delta < 1$, by Chernoff bound $\Pr(X_\zeta \geq \gamma) \leq \exp(-E(X_\zeta)\delta^2/3)$ and

$$\exp(-E(X_\zeta)\delta^2/3) = \exp\left(-\frac{\ln \ln n}{3p_\zeta}\right)$$

$$= \left(\frac{1}{\ln n}\right)^{\frac{1}{3p_\zeta}} \leq \frac{1}{\sqrt[3]{\ln n}}.$$

Thus, $\Pr(X_\zeta \geq \gamma) \leq 1/\sqrt[3]{\ln n}$. Therefore, the master pays less than $\gamma\rho$ in
rewards a.a.s.

6   Simulations

In this section, we present our simulations of the master and worker algorithms
(Algorithms 1 and 2) in the presence of deviators. To compare the performance
of our approach, we also simulated the reinforcement-learning-based mechanism
in [15], and the mechanism based on infinitely-repeated games in [24].

Throughout this section, we refer to our finitely-repeated game with terminal payoffs mechanism as FRG. We refer to the *evolutionary dynamics* based
approach of the mechanism in [15] as ED, and to the infinitely-repeated games
approach in [24] as IRG.

Simulation Design. To the best of our knowledge, the impact of deviators on
the gaurantees of ED and IRG has not been theoretically analyzed. In ED, workers update their probability of computing p_ζ in each round based on the previous

round's p_ζ, payment received, and some measure of their profit aspiration. It is assumed that all workers comply with the mechanism. On the other hand, in IRG it is assumed that, all rational workers will follow the equilibrium because of the threat of punishment.

In our experimental evaluation, we assume that 40% of workers are deviators in all three mechanisms. In FRG, the deviators and followers proceed as described in our model. In ED, the deviators and followers start with a different p_ζ, and they converge over time. In IRG, we follow the simulations in [24] and assume that the deviation is readily detected after one round (optimal detection), and that the deviators become followers after one round of punishment (optimal effect of peer punishment). For consistency we assume that deviations occur only once in all three mechanisms.

In the IRG mechanism, the expected utility of the workers is computed at equilibrium of the infinitely-repeated game. Hence, for this comparison, we assume that workers in the IRG mechanism do not know that the interaction is finite (once the number of task assignments desired by the master is completed, the interaction stops). Note that if the workers in the IRG mechanism do know the length of interaction, its correctness collapses. Comparing our approach against a stronger but less realistic execution of IRG [24] only strengthens the results of our evaluation.

We provide preliminary simulations based on the following system parameters. We evaluate all mechanisms for $n = 9$, 99, and 999 workers and the number of rounds of task assignments in the range $T \in [20, 1000]$. The payoffs scheme is evaluated for $\rho = 10$, $\phi = 10$, and $c = 2$. The master's probability of verification is set to $p_v = c/\phi = 0.2$.

For FRG, we approximate the probability of computing of the followers at equilibrium as $p_\zeta - 0.55$. For the deviators, we use $p_\zeta = 0.9$, to model their motivation to compute to avoid being fined by the master.

For ED, we set the worker aspiration for profit $a = 0.1$, learning rate $\alpha = 0.01$, the master tolerance to error $\tau = 0.5$. We set initial probabilities $p_\zeta = 0.5$ for the followers and $p_\zeta = 0.9$ for deviators, and the probability of verification $p_v = 0.2$.

We configured the parameters of IRG as follows (using the notation in [24]). The reward $WB_A = \rho = 10$, the fine $WP_C = \phi = 10$, and the cost of computing $WC_T = c = 2$. The additional master parameters in [24] are set as follows. Cost of verification MC_V, profit from being correct MB_R and cost of being wrong MP_W are all set to 0; whereas the cost of accepting an answer is $MC_A = \rho = 10$. In [24], the followers use $p_\zeta = 0.9$ and the deviators $p_\zeta \leq 0.5$. We simulate the same $p_\zeta = 0.9$ for followers; for the deviators, we set $p_\zeta = 0.5$ which is the most favorable choice for IRG.

These parameters choices satisfy the analysis of our work and [24]. A broader range of simulation parameters is left to the full version.

The simulation code was written in Python 3.6 and executed on a PC. The results presented are the average of 10 executions of each mechanism.

Discussion of Results. The results of our simulations can be seen in Figs. 1, 2, and 3. In Fig. 1, we compare the mechanisms with respect to the ratio of the

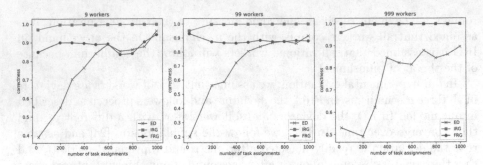

Fig. 1. Comparison of FRG, ED and IRG: number of correct answers obtained divided by the total number of task assignments vs. number of task assignments.

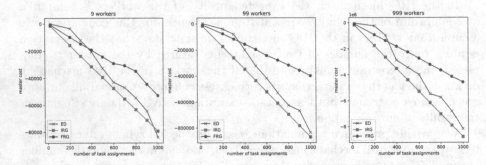

Fig. 2. Comparison of the cost incurred by the Master in FRG, ED and IRG.

number of tasks for which the master obtains the correct answer to the total number of task assignments. In Fig. 2, we compare the cost incurred by the master for each mechanism. In Fig. 3, we plot the utilities of each worker in our FRG mechanism and show that they obtain positive utilities from participating.

We observe in Figs. 1 and 2 that for any T up to 400 task assignments FRG performs much better than ED in correctness, at similar costs incurred by the master. In other words, FRG outperforms ED, unless the number of task assignments is very large. The number of task assignments T is a design choice in the mechanism. That is, the master may configure the platform to run for $T = 400$ task assignments, and hire a new pool of workers for each batch of computations.

Compared to IRG, the cost incurred by the master in FRG is much smaller, at a slightly reduced correctness rate. Note that the correctness guarantees achieved by IRG in our simulations is optimistic, because we assume that the players are unaware of the number of rounds. Thus, we conclude that for a wide range of parameters, FRG outperforms ED, and achieves similar correctness with much lower cost incurred by the master compared to IRG (assuming that the workers are unaware of the finite nature of the game).

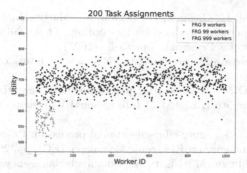

Fig. 3. Workers utility in the FRG mechanism. The x axis is the worker IDs.

Finally, in Fig. 3 we show that the expected utility of each worker is positive for each of the n values tested and for $T = 200$ (other values of T gave similar results). This confirms the feasibility of the mechanism—by participating, the workers receive non-negative utility.

7 Conclusion

In this paper, we design a finitely-repeated MW mechanism to perform verifiable crowd-computing in the presence of deviators, workers who are not perfectly rational and may not follow the prescribed equilibrium. This model better reflects real-world crowd-computing applications, where non-compliant workers exist [4, 26,30], and contracts often have a prespecified length of interaction.

We use the notion of terminal payments to incentivize the followers (rational workers who follow the equilibrium) to punish such deviators. We prove that the master is able to obtain correct answers from such a mechanism always in expectation, and asymptotically almost surely under certain mild conditions.

Finally, we simulate our mechanism and compare it with previous approaches: ED [15] and IRG [24]. Our simulations show that our mechanism outperforms ED in terms of correctness, and IRG in terms of the cost incurred by the master.

Acknowledgement. The work presented in this manuscript was partially supported by NSF CCF 1947789, and the Pace University SR Grant and Kenan Fund.

References

1. Abraham, I., Dolev, D., Goden, R., Halpern, J.: Distributed computing meets game theory: robust mechanisms for rational secret sharing and multiparty computation. In: Proceedings of the 25th Annual ACM Symposium on Principles of Distributed Computing, pp. 53–62 (2006)

2. Aiyer, A.S., Alvisi, L., Clement, A., Dahlin, M., Martin, J., Porth, C.: Bar fault tolerance for cooperative services. In: Proceedings. of the 20th ACM Symposium on Operating Systems Principles, pp. 45–58 (2005)

3. Amazon.com.: Amazon Mechanical Turk. http://www.mturk.com. Accessed 2 Oct 2017

4. Anderson, D.: BOINC: a system for public-resource computing and storage. In: Proceedings of the 5th IEEE/ACM International Workshop on Grid Computing, pp. 4–10 (2004)

5. Azar, P.D., Micali, S.: Super-efficient rational proofs. In: Proceedings of the 14th Annual ACM conference on Electronic Commerce (EC), pp. 29–30 (2013)

6. Babaioff, M., Feldman, M., Nisan, N.: Combinatorial agency. In: Proceedings of the 7th ACM Conference on Electronic Commerce, pp. 18–28 (2006)

7. Babaioff, M., Feldman, M., Nisan, N.: Mixed strategies in combinatorial agency. In: Proceedings of the 2nd international Workshop on Internet & Network Economics, pp. 353–364 (2006)

8. Babaioff, M., Feldman, M., Nisan, N.: Free-riding and free-labor in combinatorial agency. In: Mavronicolas, M., Papadopoulou, V.G. (eds.) SAGT 2009. LNCS, vol. 5814, pp. 109–121. Springer, Heidelberg (2009). https://doi.org/10.1007/978-3-642-04645-2_11

9. Bielefeld, R.S.: Reexamination of the perfectness concept for equilibrium points in extensive games. In: Models of Strategic Rationality. Theory and Decision Library C, vol. 2, pp. 1–31. Springer, Dordrecht (1988). https://doi.org/10.1007/978-94-015-7774-8_1

10. Chen, J., McCauley, S., Singh, S.: Rational proofs with multiple provers. In: Proceedings of the 7th Innovations in Theoretical Computer Science Conference (ITCS), pp. 237–248 (2016)

11. Chen, J., McCauley, S., Singh, S.: Efficient rational proofs with strong utility-gap guarantees. In: Deng, X. (ed.) SAGT 2018. LNCS, vol. 11059, pp. 150–162. Springer, Cham (2018). https://doi.org/10.1007/978-3-319-99660-8_14

12. Chen, J., McCauley, S., Singh, S.: Non-cooperative rational interactive proofs. In: 27th Annual European Symposium on Algorithms (ESA 2019). Schloss Dagstuhl-Leibniz-Zentrum fuer Informatik (2019)

13. Christoforou, E., Fernández Anta, A., Georgiou, C., Mosteiro, M.A.: Algorithmic mechanisms for Internet supercomputing under unreliable communication. In: Proceedings of the 10th IEEE International Symposium on Network Computing and Applications, pp. 275–280 (2011)

14. Christoforou, E., Fernández Anta, A., Georgiou, C., Mosteiro, M.A.: Algorithmic mechanisms for reliable master-worker Internet-based computing. IEEE Trans. Comput. 63(1), 179–195 (2014)

15. Christoforou, E., Fernández Anta, A., Georgiou, C., Mosteiro, M.A., Sánchez, A.: Applying the dynamics of evolution to achieve reliability in master-worker computing. Concurr. Comput. Pract. Exp. 25(17), 2363–2380 (2013)

16. Conlisk, J.: Why bounded rationality? J. Econ. Lit. 34(2), 669–700 (1996)

17. Dong, C., Wang, Y., Aldweesh, A., McCorry, P., van Moorsel, A.: Betrayal, distrust, and rationality: Smart counter-collusion contracts for verifiable cloud computing. In: Proceedings of the 2017 ACM SIGSAC Conference on Computer and Communications Security, pp. 211–227 (2017)

18. Eidenbenz, R., Schmid, S.: Combinatorial agency with audits. In: Proceedings of the International Conference on Game Theory for Networks, pp. 374–383 (2009)

19. Eliaz, K.: Fault tolerant implementation. Rev. Econ. Stud. 69, 589–610 (2002)

20. Fernández, A., Georgiou, C., Lopez, L., Santos, A.: Reliable Internet-based computing in the presence of malicious workers. Parall. Process. Lett. **22**(1) (2012)
21. Fernández Anta, A., Georgiou, C., Mosteiro, M.A.: Designing mechanisms for reliable Internet-based computing. In: Proceedings of the 7th IEEE International Symposium on Network Computing and Applications, pp. 315–324 (2008)
22. Fernández Anta, A., Georgiou, C., Mosteiro, M.A.: Algorithmic mechanisms for Internet-based master-worker computing with untrusted and selfish workers. In: Proceedings of the 24th IEEE International Parallel and Distributed Processing Symposium, pp. 1–11 (2010)
23. Fernández Anta, A., Georgiou, C., Mosteiro, M.A., Pareja, D.: Algorithmic mechanisms for reliable crowdsourcing computation under collusion. Public Lib. Sci. One **10**(3) (2015)
24. Fernández Anta, A., Georgiou, C., Mosteiro, M.A., Pareja, D.: Multi-round master-worker computing: a repeated game approach. In: Proceedings of the IEEE 35th Symposium on Reliable Distributed Systems, pp. 31–40. IEEE (2016)
25. Gairing, M.: Malicious Bayesian congestion games. In: Proceedings of the 6th Workshop on Approximation and Online Algorithms, pp. 119–132 (2008)
26. Golle, P., Mironov, I.: Uncheatable distributed computations. In: Proceedings of the Cryptographer's Track at RSA Conference 2001, pp. 425–440 (2001)
27. Gossner, O.: The folk theorem for finitely repeated games with mixed strategies. Internat. J. Game Theory **24**(1), 95–107 (1995)
28. Guo, S., Hubáček, P., Rosen, A., Vald, M.: Rational arguments: single round delegation with sublinear verification. In: Proceedings of the 5th Annual Conference on Innovations in Theoretical Computer Science (ITCS), pp. 523–540 (2014)
29. Guo, S., Hubáček, P., Rosen, A., Vald, M.: Rational sumchecks. In: Theory of Cryptography Conference, pp. 319–351 (2016)
30. Heien, E., Anderson, D., Hagihara, K.: Computing low latency batches with unreliable workers in volunteer computing environments. J. Grid Comput. **7**, 501–518 (2009)
31. Hu, Q., Wang, S., Cheng, X., Ma, L., Bie, R.: Solving the crowdsourcing dilemma using the zero-determinant strategies. IEEE Trans. Inf. Forensics Secur. **15**, 1778–1789 (2019)
32. It, F.: http://fold.it/portal/.Accessed 11 June 2016
33. Jin, X., Li, M., Sun, X., Guo, C., Liu, J.: Reputation-based multi-auditing algorithmic mechanism for reliable mobile crowdsensing. Pervasive Mob. Comput. **51**, 73–87 (2018)
34. Kondo, D., et al.: Characterizing result errors in internet desktop grids. In: Proceedings of the 13th International European Conference on Parallel and Distributed Computing, pp. 361–371 (2007)
35. Konwar, K., Rajasekaran, S., Shvartsman, A.: Robust network supercomputing with malicious processes. In: Proceedings of the 20th International Symposium on Distributed Computing, pp. 474–488 (2006)
36. Kuhn, M., Schmid, S., Wattenhofer, R.: Distributed asymmetric verification in computational grids. In: Proceedings of the 22nd IEEE International Parallel & Distributed Processing Symposium, pp. 1–10 (2008)
37. Li, H.C., et al.: Flightpath: obedience vs choice in cooperative services. In: Proceedings of the 8th USENIX Symposium on Operating Systems Design and Implementation, pp. 355–368 (2008)
38. Li, H.C., et al.: Bar gossip. In: Proceedings of the 6th USENIX Symposium on Operating Systems Design and Implementation, pp. 191–204 (2006)

39. Lu, K., Yang, J., Gong, H., Li, M.: Classification-based reputation mechanism for master-worker computing system. In: Wang, L., Qiu, T., Zhao, W. (eds.) QShine 2017. LNICST, vol. 234, pp. 238–247. Springer, Cham (2018). https://doi.org/10.1007/978-3-319-78078-8_24

40. Moscibroda, T., Schmid, S., Wattenhofer, R.: When selfish meets evil: byzantine players in a virus inoculation game. In: Proceedings of the 25th Annual ACM Symposium on Principles of Distributed Computing, pp. 35–44 (2006)

41. Osborne, M.J., Rubinstein, A.: A Course in Game Theory. The MIT Press (1994)

42. Project, T.S.: http://setiathome.berkeley.edu. Accessed 11 June 2016

43. Rubinstein, A.: Modeling Bounded Rationality. MIT Press, London (1998)

44. Sarmenta, L.: Sabotage-tolerance mechanisms for volunteer computing systems. Futur. Gener. Comput. Syst. **18**(4), 561–572 (2002)

45. Treuille, A., et al.: Predicting protein structures with a multiplayer online game. Nature **466** (2010)

46. Yu, J., Li, Y.: New methods of uncheatable grid computing. Comput. Inform. **37**(6), 1293–1312 (2019)

47. Yurkewych, M., Levine, B., Rosenberg, A.: On the cost-ineffectiveness of redundancy in commercial p2p computing. In: Proceedings of the 12th ACM Conference on Computer and Communications Security, pp. 280–288 (2005)

Game Theory in Block Chain

Equilibrium Analysis of Block Withholding Attack: An Evolutionary Game Perspective

Zhanghao Yao, Yukun Cheng(✉), and Zhiqi Xu

Suzhou University of Science and Technology, Suzhou 215009, China
zhyao@post.usts.edu.cn, ykcheng@amss.ac.cn

Abstract. With the advancement of blockchain technology, blockchain-based digital cryptocurrencies, like Bitcoin, have received broad interest. Due to the permissionless environment, the blockchain is vulnerable to different kinds of attacks, such as the block withholding (BWH) attack. BWH attack is one common selfish mining attack, by which the attacking pool infiltrates the attacked pool, and the infiltrating miners withhold all the blocks newly discovered in the attacked pool. Therefore, the attacking pool benefit by withholding blocks, damaging the benefits of victim pools. In this paper, we introduce the reward reallocation mechanism by paying additional rewards to the miners who successfully mine blocks, and propose an evolutionary game model for BWH attack among pools to study the strategy selection of pools. By constructing the replicator dynamic equations, the evolutionary stable strategies of pools are explored based on different levels of additional rewards. Our results provide enlightening significance to mitigate the negative influence from BWH attacks in practice.

Keywords: Block withholding attack · Evolutionary game · Mitigation measure · Blockchain

1 Introduction

Bitcoin, being one of the most popular blockchain applications, has risen in popularity in recent years. Satoshi Nakamoto [1] first proposed Bitcoin as a digital cryptocurrency. The bitcoin system uses Proof of Work (PoW) as the consensus protocol to secure transactions [2]. Under PoW protocol, the agents, named as miners, compete to mine blocks by expending their computation power and then obtain the corresponding rewards. As the computation power of the whole network grows, the Bitcoin system automatically increases the difficulty of block generation to preserve the block generation interval. This makes it more difficult for an single miner to successfully mine a block. Therefore, miners join in mining pools to mine blocks by aggregating their computation power [3].

This research is supported by the National Nature Science Foundation of China (No. 11871366).

The manager of a mining pool assigns the mining work to all miners in this pool. Miners need to complete the work and submit a partial proof of work (PPoW). Miners also have the probability of directly finding a hash that meets the bitcoin network's difficulty goal, namely full proof of work (FPoW). When a miner generates a FPoW, she submits it to the manager, who then sends it to the Bitcoin system. As a result, the pool receives the entire reward of a newly discovered block. This reward is then distributed by the manager, which depends on each miner's contribution, including the PPoW miners.

During the mining process in a mining pool, a miner may only submits PPoW and withholds FPoW. Such a malicious behavior is called the block withholding (BWH) attack. A mining pool is also susceptible to BWH attack from the opponent pools. Eyal [4] studied BWH attack among pools, in which the attacking pool sends some infiltrators to the victim pool, and these infiltrators play BWH attack by throwing away FPoW. Generally, the ratio of the infiltrator's computation power to the computation power of the mining pool is called the infiltration ratio. The infiltrators share the rewards of the victim pool, but do not make any contribution.

BWH attack was first proposed by Rosenfeld [5]. The attacker participates open mining pool, submits only PPoW to the manager and withholds FPoW, damaging both his own and the pool's revenue. Courtois and Bahack expanded the strategy choices of attackers who launch BWH attacks in [6], arguing that attackers can freely allocate their computation power to mine and launch BWH attack at the same time and demonstrating that higher rewards are achieved in this situation. Luu et al. [7] developed a computational power-splitting game model to analyze revenue under various computation power distributions. They demonstrated that attackers are encouraged to launch a BWH attack in an equilibrium state. Eyal [4] investigated a game between two mining pools to evaluate the Nash equilibrium of BWH attack, and he discovered that miners would confront the miner's dilemma. Alkalay-Houlihan and Shah proposed [8] a detailed analysis based for this game. They proved that this game always admits a pure Nash equilibrium, and the pure price of anarchy (PPoA) is at most 3. Wang and Chen [9] gave a tight bound of $(1, 2)$ for PPoA, and showed the tight bound holds in a more general setting, in which infiltrators may betray. Chen et al. [10] proposed a revised approach to reallocate the reward to the miners. Instead of proportionally allocating the reward to all miners, a pool manager deducts a fraction from the reward to award the miner who actually mined the block. Under this setting, the authors proved that for any number of mining pools, no-pool-attacks is always a Nash equilibrium. Recently, many research focus on how to mitigate BWH attack. Bag et al. [11] improved the existing Bitcoin mining protocol to make it difficult for miners to distinguish between PPoW and FPoW so as to resist BWH attack. Schrijver et al. [12] proposed a game-theoretic model, in which a novel reward mechanism to encourage miners to submit blocks as soon as possible is introduced. Kim et al. [13] explored the evolutionary game model of the miner's dilemma, and the change of the miner numbers in the pool over time.

In this article, we introduce an approach to mitigate the negative effect from BWH attack by giving additional rewards to honest miners, and propose an evolutionary game model for BWH attack between two mining pools to study the strategy selection of pools.

2   The Evolutionary Game Model for BWH Attack

2.1   Basic Evolutionary Game Model

We consider that there are two pools in the evolutionary game, pool 1 and pool 2, with the same computation power of m units. There may be other mining power outside the game, solo miners or other mining pools, but we assume they have no interaction with these two pools. Each of these two pools has two strategies. One is to play BWH attack (denoted by A) by sending some miners with an amount of computation power, called *infiltration computation power*, to another pool. The other (denoted by N) is not to infiltrate another pool. Thus the strategy profiles of two pools are: (N, N), (N, A), (A, N) and (A, A). To be specific, let the infiltration rate of a pool during BWH attack be α, $\alpha \in (0, 1)$, and the infiltration computation power of the pool be αm. When a pool behaves honestly, the mining cost per unit time is denoted by C_1. If a pool launches BWH attack, then its free-riding behavior could decrease the mining cost, which is denoted by C_2, $0 \le C_2 < C_1$. When a block is successfully appended on the blockchain by a pool, this pool would obtain an amount of revenue. Here let R, $R \ge 0$, be the mining revenue per unit of computation power and per unit time. If neither of these two pools plays BWH attack, then the favorable mining environment will bring more revenue to the pools. Thus we assume that the revenue per unit computation power and per unit time increases to be γR, $\gamma > 1$.

To mitigate the negative influence from BWH attack and to encourage the miners to behave honestly, we introduce an approach to reallocate the reward, that is to assign a portion of revenue βR, $\beta \in [0, 1)$, to the honest miners who submit the FPoW in advance. Here parameter β is named as the additional rewards ratio. The rest of reward is shared by all miners proportional to their computation power.

Based on the setting of the evolutionary game model, the payoff matrix of two mining pools under different strategy profiles is shown in Table 1.

Table 1. Payoff matrix of the model.

Pool 1	Pool 2	
	Not attack(N)	Attack(A)
Not attack(N)	$m\gamma R - C_1$	$\beta mR + (1-k)mR\frac{m}{m+\alpha m} - C_1$
	$m\gamma R - C_1$	$(m - \alpha m)R + (1 - \beta)mR\frac{\alpha m}{m+\alpha m} - C_2$
Attack(N)	$(m - \alpha m)R + (1 - \beta)mR\frac{\alpha m}{m+\alpha m} - C_2$	$(m - \alpha m)R - C_2$
	$\beta mR + (1-k)mR\frac{m}{m+\alpha m} - C_1$	$(m - \alpha m)R - C_2$

2.2   The Stable Solutions of the Evolutionary Game

Our evolutionary game model is symmetric because each mining pool has the same amount of computation power and the infiltration ratio of each pool is the same too. Thus we assume the probability that a pool does not attack by x ($0 \leq x \leq 1$), and thus the probability to launch BWH attack is $1 - x$. As each pool constantly adjusts its probability of strategy selection by learning and imitating the behavior of the opponent with higher revenue, we are interested in the convergence of the probability, which is then the evolutionary stable solution (ESS), denoted by x^*.

Since the evolutionary game is symmetric, the payoff of a pool is denoted to be U_1, when it plays honestly. Let U_2 be the payoff, when a pool plays BWH attack. Based on the payoff matrix Table 1, the expected payoffs U_1 and U_2 are proposed in the following.

$$U_1 = x(m\gamma R - C_1) + (1 - x)[\beta mR + (1 - \beta mR)\frac{m}{m + \alpha m} - C_1].$$

$$U_2 = x[(m - \alpha m)R + (1 - \beta)mR\frac{\alpha m}{m + \alpha m} - C_2] + (1 - x)[(m - \alpha m)R - C_2].$$

Hence, the average expected payoff U_3 of this pool is $U_3 = xU_1 + (1 - x)U_2$.

By [14], the difference between the payoff by adopting this strategy and the player's average expected payoff determines the growth rate of strategy selection probability. Hence,

$$F(x) = \frac{\mathrm{d}x}{\mathrm{d}t} = x(U_1 - U_3) = x(1 - x)\left[(\gamma - 1)mRx + \frac{(\beta\alpha + \alpha^2)mR}{1 + \alpha} - C_1 + C_2\right].$$

$$(1)$$

Clearly, when the growth rate of strategy selection probability is equal to 0, the participant will stably select this strategy at last, and thus this strategy is stable. So, by setting $F(x) = 0$, we have three fixed points:

$$x_1 = 0, \ x_2 = 1, \ x_3 = \frac{(C_1 - C_2)(1 + \alpha) - (\beta + \alpha)\alpha mR}{(1 + \alpha)(\gamma - 1)mR}. \qquad (2)$$

2.3   Evolutionary Equilibrium Analysis on the Decision of BWH Attack

In our discussion for BWH attack between two mining pools, each pool needs to decide the infiltration ratio α, besides whether to attack. In addition, to mitigate the influence from BWH attack, we impose the additional reward mechanism to the miners who submit the FPoW. So the additional rewards ratio β is also a variable we need to determine.

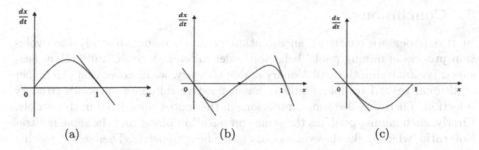

Fig. 1. The pool's dynamic evolution

According to (2), we have three distinguished cases: $x_3 \leq 0$, $0 < x_3 < 1$ and $x_3 \geq 1$.

- Case 1. $x_3 \leq 0$. Under this case, there are two fixed points $x_1 = 0$ and $x_2 = 1$. Furthermore, as $x_3 = \frac{(C_1-C_2)(1+\alpha)-(\beta+\alpha)\alpha mR}{(1+\alpha)(\gamma-1)mR} \leq 0$, we have $\beta \geq -\alpha + \frac{C_1-C_2}{\alpha mR} + \frac{C_1-C_2}{mR}$. Based on (1), the figure for this case is shown in Fig. 1(a). If the slope of the tangent line at the intersection of the curve and the horizontal coordinate is negative in the dynamic evolution diagram, then this intersection point is an evolutionary stable strategy [15]. From the Fig. 1(a), we can observe the evolutionary stable strategy is $x^* = 1$, indicating that when the additional rewards ratio β is higher, the mining pool gains fewer when it attacks, which motivates it to mine honestly.

- Case 2. $0 < x_3 < 1$. Under this case, there are three fixed points $x_1 = 0$, $x_2 = 1$ and $x_3 = \frac{(C_1-C_2)(1+\alpha)-(\beta+\alpha)\alpha mR}{(1+\alpha)(\gamma-1)mR}$. As $0 < x_3 < 1$, we have $-\alpha + \frac{[C_1-C_2-(\gamma-1)mR]}{\alpha mR} + \frac{[C_1-C_2-(\gamma-1)mR]}{mR} < \beta < -\alpha + \frac{(C_1-C_2)}{\alpha mR} + \frac{(C_1-C_2)}{mR}$. By (1), the figure for this case is shown in Fig. 1(b). Two points $x^* = 0$ or $x^* = 1$ may be the evolutionary stable strategies, which depends on the initial probability selected by the mining pool. To be specific, if the initial probability $x \in (0, x_3)$, then $F(x) \leq 0$, and the probability converges to $x^* = 0$ at last, meaning that the pool selects to attack eventually. When $x \in (x_3, 1)$, we have $F(x) \geq 0$ and the pool finally decides not to attack.

- Case 3. $x_3 \geq 1$. Under this case, there are two fixed points $x_1 = 0$ and $x_2 = 1$. Because $\frac{(C_1-C_2)(1+\alpha)-(\beta+\alpha)\alpha mR}{(1+\alpha)(\gamma-1)mR} \geq 1$, we have $0 < \beta \leq -\alpha + \frac{[C_1-C_2-(\gamma-1)mR]}{\alpha mR} + \frac{[C_1-C_2-(\gamma-1)mR]}{mR}$. The phase diagram under this case, shown in Fig. 1(c), and the evolutionary stable strategy is $x^* = 0$. It demonstrates that if reward ratio β is relatively small, then the infiltrators get more from the victim pool, increasing the revenue of the attacker. So for the case of $x_3 \geq 1$, launching BWH attack is the evolutionary stable strategy of each pool.

3   Conclusions

In this paper, we construct an evolutionary game model to study the evolution process of mining pools' behavior under different levels of mitigation measures. By analyzing the evolutionary stable strategy, we discover that the higher additional reward to honest miners has a positive influence on pool's strategy selection. There are also some limitations in the model considered in this article. Firstly, each mining pool has the same computation power and the same infiltration ratio, which make the evolutionary game be symmetric. Therefore exploring the asymmetric evolutionary game model is our future work. Secondly, we only consider the case that two mining pools may launch BWH attack in this paper. Thus how to extend the study to the scenarios with more mining pools is very interesting.

References

1. Nakamoto S. Bitcoin: A Peer-to-Peer Electronic Cash System (2008)
2. Tschorsch, F., Scheuermann, B. Bitcoin and beyond: a technical survey on decentralized digital currencies. IEEE Commun. Surv. Tutor. **18**(3), 2084–2123 (2016)
3. Liu, Y., Chen, X., Zhang, L., et al.: An intelligent strategy to gain profit for bitcoin mining pools. In: 2017 10th International Symposium on Computational Intelligence and Design (ISCID), vol. 2, pp. 427–430. IEEE (2017)
4. Eyal, I.: The miner's dilemma. In 2015 IEEE Symposium on Security and Privacy, pp. 89–103. IEEE (2015)
5. Rosenfeld, M. Analysis of bitcoin pooled mining reward systems. arXiv preprint arXiv:1112.4980 (2011)
6. Courtois, N. T., Bahack, L. On subversive miner strategies and block withholding attack in bitcoin digital currency. arXiv preprint arXiv:1402.1718 (2014)
7. Luu, L., Saha, R., Parameshwaran, I., et al.: On power splitting games in distributed computation: The case of bitcoin pooled mining. In 2015 IEEE 28th Computer Security Foundations Symposium, pp. 397–411. IEEE (2015)
8. Alkalay-Houlihan, C., Shah, N.: The pure price of anarchy of pool block withholding attacks in bitcoin mining. Proc. AAAI Conf. Artif. Intell. **33**(01), 1724–1731 (2019)
9. Wang, Q., Chen, Y. The tight bound for pure price of anarchy in an extended miner's dilemma game. arXiv preprint arXiv:2101.11855 (2021)
10. Chen, Z., Li, B., Shan, X. Discouraging pool block withholding attacks in Bitcoin. J. Combinat. Optim. **43**, 444–459 (2021)
11. Bag, S., Ruj, S., Sakurai, K.: Bitcoin block withholding attack: analysis and mitigation. IEEE Trans. Inf. Forensics Secur. **12**(8), 1967–1978 (2016)
12. Schrijvers, O., Bonneau, J., Boneh, D., Roughgarden, T.: Incentive compatibility of bitcoin mining pool reward functions. In: Grossklags, J., Preneel, B. (eds.) FC 2016. LNCS, vol. 9603, pp. 477–498. Springer, Heidelberg (2017). https://doi.org/10.1007/978-3-662-54970-4_28
13. Kim, S., Hahn, S.G.: Mining pool manipulation in blockchain network over evolutionary block withholding attack. IEEE Access **7**, 144230–144244 (2019)
14. Friedman, D.: On economic applications of evolutionary game theory. J. Evol. Econ. **8**(1), 15–43 (1998)
15. Gong, H., Jin, W. Analysis of urban car owners commute mode choice based on evolutionary game model. J. Control Sci. Eng. **2015**(6), 1–5 (2015)

Frontiers of Algorithmic Wisdom

An Approximation Algorithm for the H-Prize-Collecting Power Cover Problem

Han Dai[1], Weidong Li[1], and Xiaofei Liu[2]($\boxtimes$)

[1] School of Mathematics and Statistics, Yunnan University, Kunming, China
[2] School of Information Science and Engineering, Yunnan University, Kunming, China

lxfjl2016@163.com

Abstract. We are given a set U of user points, a set S of sensors in a d-dimensional space $\mathbb{R}^d$ and a lower bound H. Each user point $u \in U$ has a profit $h(u)$ and a penalty cost $\pi(u)$. Each sensor $s \in S$ can adjust its power, and the cover range of sensors with power $p(s)$ is a d-dimensional ball of radius $r(s)$, where $p(s) = r(s)^\alpha$ and $\alpha \geq 1$ is a constant. The goal of the H-prize-collecting power cover problem is to determine a power assignment such that the total profit of covered user points is at least H and the total power of sensors plus the total penalty cost of uncovered user points is minimized. First, we proved that this problem is NP-hard even when $\alpha = 1$, and $d = 1$ and $\pi(u) = 0$ for any $u \in U$. Then, by utilizing primal-dual and Lagrangian relaxation techniques, we present a $(4 \cdot 3^{\alpha-1} + \epsilon)$-approximation algorithm for any desired accuracy $\epsilon > 0$.

Keywords: Power cover problem · H-prize-collecting · Approximation algorithm · Lagrangian relaxation

1 Introduction

The goal of the minimum power partial coverage problem is to keep at least a specified number of user points of interest under monitoring, such that the total power of sensors is minimized, where the service area of a sensor is a disk centered at the sensor whose radius is determined by the power of the sensor [7]. However, the user points may have different priorities, which can be described as profits in the real world. In this paper, we consider a generalized version of the minimum power partial coverage problem [3], called the H-prize-collecting power cover problem, which is defined as follows. We are given a set U of n user points, a set S of m sensors on a d-dimensional space $\mathbb{R}^d$ and a lower bound H. Each user point $u \in U$ has a profit $h(u)$ and a penalty cost $\pi(u)$. Each sensor $s \in S$ can adjust its power, and the cover range of sensors with power $p(s)$ is a d-dimensional ball of radius $r(s)$, where $p(s) = r(s)^\alpha$. The aim of the H-prize-collecting power cover (H-PCPC, for short) problem is to determine a power assignment such that the total profit of covered user points is at least H and the total power of sensors plus the total penalty cost of uncovered user points is minimized.

M. Li and X. Sun (Eds.): IJTCS-FAW 2022, LNCS 13461, pp. 89–98, 2022.
https://doi.org/10.1007/978-3-031-20796-9_7

When $h(u) = 1$ for any $u \in U$, the H-PCPC problem is exactly the k-prize-collecting minimum power cover problem introduced by Liu et al. [13], where a feasible power assignment must cover at least k users and k is a given constant. They presented a 3^α-approximation algorithm. When the penalty cost of the uncovered user is generalized to a submodular function, Liu et al. [9] presented a $5 \cdot 2^\alpha + 1$-approximation algorithm. More related results can be found in [5,14]

When $h(u) = 1$ and $\pi(u) = 0$ for any $u \in U$, the H-PCPC problem is exactly the minimum power partial cover problem introduced by [3], who presented a polynomial-time $(12 + \epsilon)$-approximation algorithm when $\alpha = 2$ if the space $\mathbb{R}^d$ is a plane. When $\alpha \geq 1$, Li et al. [7] presented a 3^α-approximation algorithm, and Dai et al. [2] presented a $O(\alpha)$-approximation algorithm when the space $\mathbb{R}^d$ is a plane. Ran et al. [16] presented a polynomial time approximation scheme (PTAS) based on a plane subdivision technique. Moreover, Liang et al. [8] presented an exact algorithm when all user points and all sensors are on a line, i.e., $d = 1$.

When $H = \sum_u h(u)$, the H-PCPC problem is exactly the minimum power cover problem introduced by Biló et al. [1], who presented a PTAS based on a plane subdivision technique. Additionally, Biló et al. [1] presented an exact algorithm for when all user points and all sensors are on a line.

When $H = 0$, the H-PCPC problem is exactly the prize-collecting power cover problem, which is a special case of the k-prize-collecting minimum power cover problem. There is a 3^α-approximation algorithm based on the algorithm in [13]. Liu et al. [12] presented a PTAS based on a plane subdivision technique. In the same paper, they presented a $3^\alpha + 1$-approximation algorithm for the problem with submodular penalties, where submodular function has the property of decreasing marginal return and occur in many mathematical models in [10,11,15].

The rest of this paper is organized as follow. In the second part, we formally introduce the H-PCPC problem and prove that this problem is NP-hard, even when $\alpha = 1$, and $d = 1$ and $\pi(u) = 0$ for any $u \in U$. In the third section, we recall the prize-collecting power cover problem and introduce a 3^α-approximation algorithm from [13]. In the fourth section, we consider the Lagrangian relaxation of the H-PCPC problem and present a $(4 \cdot 3^{\alpha-1} + \epsilon)$-approximation algorithm. Finally, we give brief conclusions.

2   Preliminaries

Suppose U is a set of n user points, S is a set of m sensors on a d-dimensional space $\mathbb{R}^d$, and H is a nonnegative profit bound. A user point $u \in U$ is covered by a sensor $s \in S$ with power $p(s)$ if u belongs to the d-dimensional ball $B(s, r(s))$ supported by $p(s)$, that is, $u \in B(s, r(s))$, where $B(s, r(s))$ is a d-dimensional ball $B(s, r(s))$ centered at s whose radius $r(s)$ is determined by

$$p(s) = r(s)^\alpha, \text{ where } \alpha \geq 1,$$

For convenience, we abbreviate 'd-dimensional' as 'ball'. A user point is covered by a power assignment $p : S \rightarrow R^+$ if it is covered by some ball supported by p. Each user point $u \in U$ has a profit $h(u)$ and a penalty $\pi(u)$. The H-prize-collecting power cover (H-PCPC) problem aims to determine a power assignment

p such that the total profit of covered user points is at least H and the total power of sensors plus the total penalties of uncovered user points is minimized. Here, we assume that there is no limit on the power at a sensor.

Theorem 1. *The H-PCPC problem is NP-hard, even when $\alpha = 1$, and $d = 1$ and $\pi(u) = 0$ for any $u \in U$.*

Proof. We use the NP-complete partition problem [4] for the reduction. Given t positives $a_1, a_2, \ldots, a_t$, is there a subset $A \subseteq \{a_1, a_2, \ldots, a_t\}$ such that $\sum_{a_i : a_i \in A} a_i = \frac{1}{2} \sum_{i=1}^{t} a_i$?

For any instance of the partition problem, we construct an instance $(U, \mathcal{S}; p; h; \pi)$ of the H-PCPC problem as follows: Given $U = \{u_1, u_2, \ldots, u_t\}$ and $\mathcal{S} = \{s_1, s_2, \ldots, s_t\}$ on a line, let L be a large constant, and

$$\begin{cases} d(s_i, s_{i+1}) = L, \forall i \in \{1, 2, \ldots, t-1\}; \\ d(u_i, s_i) = a_i, \forall i \in \{1, 2, \ldots, t\}, \end{cases}$$

where $d(a, b)$ is the Euclidean distance from point a to point b. $\alpha = 1$, $H = \frac{1}{2} \sum_{i=1}^{t} a_i$, $h(u_i) = a_i$ and $\pi(u_i) = 0$ for each $i = \{1, 2, \ldots, t\}$.

It is easy to prove that the partition problem has a solution if and only if the objective value of the optimal solution of instance $(U, \mathcal{S}; p; h; \pi)$ of the H-PCPC problem is $\frac{1}{2} \sum_{i=1}^{t} a_i$. $\square$

Note that in any optimal solution for an instance of the H-PCPC problem, for any sensor s with $r(s) > 0$, there is at least one user point on the boundary of ball $B(s, r(s))$ since otherwise the ball could be shrunk to result in less power. Hence, at most $m \cdot n$ balls, denoted as $\mathcal{B}$, must be considered in the H-PCPC. For each ball $B \in \mathcal{B}$, use $U(B)$, $r(B)$, $p(B)$ and $c(B)$ to represent the set of user points covered, the radius, the power, and the center of ball B, respectively. Then, we have

$$p(B) = r(B)^\alpha.$$

Let (F^*, R^*) be an optimal solution for the H-PCPC, and its objective value is OPT, where $F \subseteq \mathcal{B}$ and $R^* \subseteq U$. Similar to the preprocessing step in [6], for any $\epsilon > 0$, we can assume that each ball $B \in \mathcal{B}$ satisfies

$$p(B) \leq \epsilon \cdot OPT. \tag{1}$$

3   The Prize-Collecting Power Cover Problem

In this section, we begin with a brief description of the primal-dual algorithm in [13], denoted by Algorithm $\mathbb{PCPC}$, and present the structural properties of its output solution.

As in [13], for each ball $B \in \mathcal{B}$, a variable $x_B \in \{0, 1\}$ indicates whether ball B is picked; for each user point $u \subseteq U$, a variable $z_u \in \{0, 1\}$ indicates whether

u is a penalty user point. The PCPC problem can be formulated as the following integer programming problem:

$$\min \sum_{B:B\in\mathcal{B}} p(B)x_B + \sum_{u:u\in U} \pi(u)z_u$$

$$s.t. \quad \sum_{B:u\in U(B)} x_B + z_u \geq 1, \ \forall u \in U \tag{2}$$

$$x_B \in \{0,1\}, \ \forall B \in \mathcal{B},$$

$$z_u \in \{0,1\}, \ \forall u \in U,$$

where $U(B)$ is the set of user points covered by ball B. The first constraint states that any user point $u \in U$ is either covered by some disk or penalized. By relaxing the integrality constraints, a linear program is obtained and its dual program is

$$\max \sum_{u:u\in U} y_u$$

$$s.t. \quad \sum_{u:u\in U(D)} y_u \leq p(B), \ \forall B \in \mathcal{B}, \tag{3}$$

$$y_u \leq \pi(u), \ \forall u \in U,$$

$$y_u \geq 0, \ \forall u \in U.$$

Let $(\{y_u\}_{u\in U(D)})$ be a feasible solution of dual program (3). For any ball $B \in \mathcal{B}$, B is called *tight* if $\sum_{u\in U(D)} y_u = c \cdot r(D)^\alpha$; for any user point $u \in U$, u is called *tight* if $y_u = \pi(u)$.

Algorithm $\mathbb{PCPC}$ consists of three steps:

(1) The primal-dual scheme is employed to find a tight ball set $\mathcal{B}^{tight}$ and a temporarily rejected user point set R^{temp}, where $R^{temp}\cup\bigcup_{B:B\in\mathcal{B}^{tight}} U(B) = U$ and $U(B)$ is the set of user points covered by B.
(2) A maximally independent set $\mathcal{I} \subseteq \mathcal{B}^{tight}$ is found based on the greedy method.
(3) Each ball in $\mathcal{I}$ has its radius enlarged three times. This set F of balls and uncovered user points R are the output solutions of the algorithm.

Algorithm $\mathbb{PCPC}$ in [13] is described as follows.

Let $\{y_u\}_{u\in U}$ be the dual value generated by Algorithm 1; then, it is not difficult to obtain that $\{y_u\}_{u\in U}$ is a feasible solution of dual program (3), *i.e.*,

$$\sum_{u:u\in U} y_u \leq OPT,$$

where OPT is the optimal value of the PCPC problem. Based on Lemma 2 and Theorem 2 in [13], it is not hard to obtain the following theorem.

Theorem 2. *(F, R) generated by Algorithm $\mathbb{PCPC}$ is a feasible solution of the PCPC problem, and*

$$\sum_{B:B\in F} p(B) + 3^\alpha \cdot \sum_{u:u\in R} \pi(u) \leq 3^\alpha \cdot OPT.$$

Algorithm 1: [13] Algorithm $\mathbb{PCPC}$

Input: A user point set U; a ball set $\mathcal{B}$.

Output: A feasible solution (F, R).

1 Initially, set the dual variable $y_u = 0$ and $\mathcal{B}^{tight} = \mathcal{I} = R^{temp} = \emptyset$.

2 **while** $U \neq \emptyset$ **do**

3    $\quad$ Increase $\{y_u\}_{u \in U}$ simultaneously until either some ball B becomes tight or some use point u' becomes tight.

4    $\quad$ If ball B becomes tight, $\mathcal{B}^{tight} := \mathcal{B}^{tight} \cup \{B\}$ and $U := U \setminus U(B)$; otherwise, use point u' becomes tight, $R^{temp} := R^{temp} \cup \{u'\}$ and $U := U \setminus \{u'\}$.

5 **while** $\mathcal{B}^{tight} \neq \emptyset$ **do**

6    $\quad$ $B' := \arg\max_{B \in \mathcal{B}^{tight}} r(B)$, $\mathcal{I} := \mathcal{I} \cup \{B'\}$.

7    $\quad$ **for** $B \in \mathcal{B}^{tight}$ **do**

8    $\quad\quad$ **if** $U(B) \cap U(B') \neq \emptyset$ **then**

9    $\quad\quad\quad$ $\mathcal{B}^{tight} := \mathcal{B}^{tight} \setminus \{B\}$

10 Set $F := \{B(c(B), 3 \cdot r(B)) | B \in \mathcal{I}\}$ and $R := U \setminus \bigcup_{B:B \in \mathcal{I}} U(B)$. Output (F, R).

---

4 The H-Prize-Collecting Power Cover Problem

As with the definitions of variables $\{x_B\}_{B \in \mathcal{B}}$ and $\{z_u\}_{u \in U}$ in integer program (2), the H-PCPC problem can be formulated as the following integer program:

$$\min \sum_{B:B \in \mathcal{B}} p(B)x_B + \sum_{u:u \in U} \pi(u)z_u$$

$$s.t. \quad \sum_{B:u \in U(B)} x_B + z_u \geq 1, \ \forall u \in U \tag{4}$$

$$\sum_{u:u \subseteq U} h(u)z_u \leq H_U - H,$$

$$x_B \in \{0, 1\}, \ \forall B \in \mathcal{B},$$

$$z_u \in \{0, 1\}, \ \forall u \in U.$$

where the second constraint of (4) guarantees that the total profit of uncovered user points of any feasible solution is at most $H_U - H$ and $H_U = \sum_{u:u \in U} h(u)$.

Then, the resulting Lagrangian relaxation of integer program (2) is

$$\lambda - \text{HLP}: \quad \min \sum_{B:B \in \mathcal{B}} p(B)x_B + \sum_{u:u \in U} \pi(u)z_u + \lambda \cdot \left(\sum_{u:u \subseteq U} h(u)z_u - (H_U - H) \right)$$

$$s.t. \quad \sum_{B:u \in U(B)} x_B + z_u \geq 1, \ \forall u \in U$$

$$x_B \in \{0, 1\}, \ \forall B \in \mathcal{B},$$

$$z_u \in \{0, 1\}, \ \forall u \in U.$$

Given any $\lambda \geq 0$, we remark that, excluding the constant term of $-\lambda(H_U - H)$ in the objective function, λ−HLP is exactly an integer program of the prize-collecting power cover problem on instance $\mathfrak{I}_\lambda$, in which each user point $u \in U$ has a penalty $\pi(u) + \lambda \cdot h(u)$, where

$$\mathfrak{I}_\lambda : \quad \min \sum_{B:B \in \mathcal{B}} p(B)x_B + \sum_{u:u \in U} (\pi(u) + \lambda \cdot h(u))z_u$$

$$s.t. \quad \sum_{B:u \in U(B)} x_B + z_u \geq 1, \ \forall u \in U$$

$$x_B \in \{0,1\}, \ \forall B \in \mathcal{B},$$

$$z_u \in \{0,1\}, \ \forall u \in U.$$

For any $\lambda \geq 0$, let OPT_λ and $OPT(\lambda - \text{HLP})$ be the optimal values of $\mathfrak{I}_\lambda$ and λ−HLP, respectively; then, we have $OPT_\lambda - \lambda \cdot (H_U - H) = OPT(\lambda - \text{HLP})$. Furthermore, any optimal solution of integer program (2) is also a feasible solution of λ−HLP. Let OPT be the optimal value of integer program (2); then, we have

$$OPT_\lambda - \lambda \cdot (H_U - H) = OPT(\lambda - \text{HLP}) \leq OPT, \ \forall \ \lambda \geq 0. \tag{5}$$

For any $\lambda \geq 0$, let (F_λ, R_λ) be the output solution generated by Algorithm $\mathbb{PCPC}$ and let

$$H_\lambda = \sum_{u:u \in U \setminus R_\lambda} h(u)$$

be the total profit of the user points covered by (F_λ, R_λ). Thus, (F_λ, R_λ) is a feasible solution of integer program (2) if $H_\lambda \geq H$. In particular, if $H_0 \geq H$, we can output (F_λ, R_λ), and its objective value is no more than $3^\alpha \cdot OPT$. Without loss of generality, we assume that

$$H_0 < H.$$

Note that when $\lambda > \frac{\sum_{B \in \mathcal{B}} p(B)}{\min_{u \in U} h(u)}$, (F_λ, R_λ) generated by Algorithm $\mathbb{PCPC}$ covers all user points in U. This statement and $H_0 < H$ imply that we can find λ_1 and λ_2 in polynomial time based on a binary search over the interval $[0, \frac{\sum_{B \in \mathcal{B}} p(B)}{\min_{u \in U} h(u)}]$ and Algorithm $\mathbb{PCPC}$, where

$$\lambda_1 - \lambda_2 \leq \frac{\epsilon p_{\min}}{H_U}$$

and

$$H_{\lambda_1} \geq H \geq H_{\lambda_2},$$

where $p_{\min} = \min_{B:B \in \mathcal{B}} p(B)$.

Lemma 1. *For any $\lambda \geq 0$, (F_λ, R_λ) generated by Algorithm $\mathbb{PCPC}$ on instance $\mathfrak{I}_\lambda$ satisfies*

$$OUT(F_\lambda, R_\lambda) \leq 3^\alpha \cdot (OPT + \lambda \cdot (H_\lambda - H)).$$

Proof. For any $\lambda \geq 0$, the objective value of (F_λ, R_λ) is

$$OUT(F_\lambda, R_\lambda)$$
$$= \sum_{B:B\in F_\lambda} p(B) + \sum_{u:u\in U\backslash R_\lambda} \pi(u)$$
$$= \sum_{B:B\in F_\lambda} p(B) + \sum_{u:u\in U\backslash R_\lambda} (\pi(u) + \lambda \cdot h(u)) - \sum_{u:u\in U\backslash R_\lambda} \lambda \cdot h(u)$$
$$\leq 3^\alpha \cdot OPT_\lambda - (3^\alpha - 1) \sum_{u:u\in U\backslash R_\lambda} (\pi(u) + \lambda \cdot h(u)) - \sum_{u:u\in U\backslash R_\lambda} \lambda \cdot h(u)$$
$$\leq 3^\alpha \cdot (OPT + \lambda \cdot (H_U - H)) - 3^\alpha \cdot (H_U - H_\lambda)$$
$$= 3^\alpha \cdot (OPT + \lambda \cdot (H_\lambda - H)),$$

where the first inequality follows from Theorem 1 and the second inequality follows from inequality (5). □

Lemma 2. *If $H_{\lambda_1} = H$ or $H_{\lambda_2} = H$, then $(F_{\lambda_1}, R_{\lambda_1})$ or $(F_{\lambda_2}, R_{\lambda_2})$ is a feasible solution of the H-PCPC problem, and its objective value is no more than $3^\alpha \cdot OPT$.*

Proof. If $H_{\lambda_1} = H$, $(F_{\lambda_1}, R_{\lambda_1})$ is a feasible solution, and its objective value by Lemma 1 is

$$OUT(F_{\lambda_1}, R_{\lambda_1}) \leq 3^\alpha \cdot (OPT + \lambda \cdot (H_{\lambda_1} - H)) = 3^\alpha \cdot OPT.$$

If $H_{\lambda_2} = H$, similarly, $(F_{\lambda_2}, R_{\lambda_2})$ is a feasible solution satisfying

$$OUT(F_{\lambda_2}, R_{\lambda_2}) \leq 3^\alpha \cdot OPT.$$

□

Then, we consider the case with

$$H_{\lambda_1} > H > H_{\lambda_2}.$$

Thus, $(F_{\lambda_1}, R_{\lambda_1})$ is a feasible solution of the H-PCPC problem, and $(F_{\lambda_2}, R_{\lambda_2})$ is not a feasible solution of the H-PCPC problem. Then, by adding some ball in F_{λ_1} to F_{λ_2}, we construct an augmenting feasible solution (F_a, R_a), where $F_a = F_{\lambda_2} \cup F'_{\lambda_1}$, $F'_{\lambda_1} \subseteq F_{\lambda_1} \backslash F_{\lambda_2}$ and R_a is the set of user points not covered by F_a. We select the minimum solution between $(F_{\lambda_1}, R_{\lambda_1})$ and (F_a, R_a) as the output.

Then, we illustrate how to construct F'_{λ_1}. First, each user point $u \in (U \backslash R_{\lambda_1}) \cap R_{\lambda_2}$ is assigned to an arbitrary ball in $F_{\lambda_1} \backslash F_{\lambda_2}$ that covers it, and $\varphi(B)$ represents the total profit of the covered user points assigned to B. Then, the balls in $F_{\lambda_1} \backslash F_{\lambda_2}$ are sorted such that $\frac{p(B_1)}{\varphi(B_1)} \leq \frac{p(B_2)}{\varphi(B_2)} \leq \cdots$. Finally, let $F'_{\lambda_1} = \{B_1, \ldots, B_q\}$, where q is the minimal index for which $\sum_{i=1}^q \varphi(B_i) \geq H - H_{\lambda_2}$.

Let

$$\gamma = \frac{H - H_{\lambda_2}}{H_{\lambda_1} - H_{\lambda_2}};$$

we then have the following lemma.

Lemma 3. $\sum_{B:B\in F'_{\lambda_1}} P(B) \leq \gamma \cdot \sum_{B:B\in F_{\lambda_1}\setminus F_{\lambda_2}} p(B) + 3^\alpha \cdot \epsilon \cdot OPT.$

Theorem 3. $\min\{OUT(F_{\lambda_1}, R_{\lambda_1}), OUT(F_a, R_a)\} \leq (4 \cdot 3^{\alpha-1} + 2 \cdot 3^\alpha \cdot \epsilon) \cdot OPT.$

Proof. By $\gamma = \frac{H - H_{\lambda_2}}{H_{\lambda_1} - H_{\lambda_2}}$, we have

$$\gamma \cdot OUT(F_{\lambda_1}, R_{\lambda_1}) + (1 - \gamma) \cdot OUT(F_{\lambda_2}, R_{\lambda_2})$$
$$\leq 3^\alpha \gamma \cdot (OPT + \lambda_1(H_{\lambda_1} - H)) + 3^\alpha(1 - \gamma) \cdot (OPT + \lambda_2(H_{\lambda_2} - H))$$
$$+ 3^\alpha \gamma \cdot \frac{\epsilon p_{\min}}{H_U} \cdot (H_{\lambda_1} - H)$$
$$\leq 3^\alpha \cdot OPT + 3^\alpha \cdot \epsilon p_{\min} \leq 3^\alpha(1 + \epsilon) \cdot OPT, \tag{6}$$

where the first inequality follows from Lemma 1, the second inequality follows from $p_{\min} \leq p(B) \leq \epsilon \cdot OPT$ for any $B \in \mathcal{B}$.

Let

$$\theta = \frac{OUT(F_{\lambda_2}, R_{\lambda_2})}{OPT}.$$

By inequality (6), the objective value of $(F_{\lambda_1}, R_{\lambda_1})$ is

$$OUT(F_{\lambda_1}, R_{\lambda_1}) \leq \frac{3^\alpha(1 + \epsilon) \cdot OPT - (1 - \gamma) \cdot OUT(F_{\lambda_2}, R_{\lambda_2})}{\gamma}$$
$$= \frac{3^\alpha(1 + \epsilon) - (1 - \gamma) \cdot \theta}{\gamma} OPT.$$

Since $F_a = F'_{\lambda_1} \cup F_{\lambda_2}$, we have $R_a \subseteq F_{\lambda_2}$, and the objective value of $(F_{\lambda_2}, R_{\lambda_2})$ is

$$OUT(F_a, R_a)$$
$$= \sum_{D:D\in F_{\lambda_2}} p(D) + \sum_{D:D\in F'_{\lambda_1}} p(D) + \sum_{u:u\in R_{\lambda_2}} \pi(u)$$
$$\leq (3^\alpha \cdot (1 + \epsilon) + \gamma \cdot \theta + 3^\alpha \epsilon) \cdot OPT,$$

where the inequality follows from $R_a \subseteq R_{\lambda_2}$, $\pi(u) \geq 0$ for any $u \in U$, and inequality (6).

Therefore,

$$\min\{OUT(F_{\lambda_1}, R_{\lambda_1}), OUT(F_a, R_a)\}$$
$$= \min\{\frac{3^\alpha \cdot (1 + \epsilon) - (1 - \gamma) \cdot \theta}{\gamma}, 3^\alpha \cdot (1 + \epsilon) + \gamma \cdot \theta + 3^\alpha \epsilon\} \cdot OPT$$
$$\leq 4 \cdot 3^{\alpha-1} \cdot OPT + 2 \cdot 3^\alpha \epsilon \cdot OPT$$

where the last inequality follows from $\gamma = \frac{1}{2}$ and $\theta = 2 \cdot 3^{\alpha-1}$. □

5   Conclusion

In this paper, we consider the H-prize-collecting power cover (H-PCPC) problem, which is a space case of the H-prize-collecting set cover problem. We proved that this problem is NP-hard even when $\alpha = 1$, and $d = 1$ and $\pi(u) = 0$ for any $u \in U$. Additionally, we present a $(4 \cdot 3^{\alpha-1} + \epsilon)$-approximation algorithm utilizing the primal-dual and Lagrangian relaxation techniques for any desired accuracy $\epsilon > 0$.

It would be interesting to design a better algorithm for the H-PCPC problem. Moreover, the generalization of this problem to metric space is worthy of further study. A submodular function has the property of decreasing marginal return and there has been considerable work on submodular penalty optimization, in which the penalty is determined by a submodular function. The generalization of this problem to submodular penalties is worth considering.

Acknowledgement. The work is supported in part by the National Natural Science Foundation of China [No. 12071417].

References

1. Biló, V., Caragiannis, I., Kaklamanis, C., Kanellopoulos, P.: Geometric clustering to minimize the sum of cluster sizes. In: Brodal, G.S., Leonardi, S. (eds.) Algorithms - ESA 2005, LNCS, vol. 3669, pp. 460–471. Springer, Heidelberg (2005) https://doi.org/10.1007/11561071_42
2. Dai, H., Deng, B., Li, W., Liu, X.: A note on the minimum power partial cover problem on the plane. J. Combinat. Optim. **44**(2), 1–9 (2022). https://doi.org/10.1007/s10878-022-00869-8
3. Freund, A., Rawitz, D.: Combinatorial Interpretations of Dual Fitting and Primal Fitting. In: Solis-Oba, R., Jansen, K. (eds.) WAOA 2003. LNCS, vol. 2909, pp. 137–150. Springer, Heidelberg (2004). https://doi.org/10.1007/978-3-540-24592-6_11
4. Garey, M.R., Johnson, D.S.: Computers and Intractability: A Guide to the Theory of NP-Completeness. W.H Freeman and Company, New York (1990)
5. Guo, J., Liu, W., Hou, B.: An approximation algorithm for P-prize-collecting set cover problem. J. Operat. Res. Soc. China (2021). https://doi.org/10.1007/s40305-021-00364-7
6. Könemann, J., Parekh, O., Segev, D.: A unified approach to approximating partial covering problems. Algorithmica **59**(4), 489–509 (2011)
7. Li, M., Ran, Y., Zhang, Z.: A primal-dual algorithm for the minimum power partial cover problem. J. Combinat. Optim. **39**, 725–746 (2020)
8. Liang, W., Li, M., Zhang, Z., Huang, X.: Minimum power partial multi-cover on a line. Theoret. Comput. Sci. **864**, 118–128 (2021)
9. Liu, X., Dai, H., Li, S., Li, W.: The k-prize-collecting minimum power cover problem with submodular penalties on a plane . Sci. Sin. Inform. **52**, 947–959 (2022). (in Chinese). https://doi.org/10.1360/SSI-2021-0445
10. Liu, X., Li, W.: Approximation algorithms for the multiprocessor scheduling with submodular penalties. Optim. Lett. **15**(6), 2165–2180 (2021). https://doi.org/10.1007/s11590-021-01724-1

11. Liu, X., Li, W.: Combinatorial approximation algorithms for the submodular multicut problem in trees with submodular penalties. J. Combinat. Optim. **44**,1964–1976 (2020). https://doi.org/10.1007/s10878-020-00568-2

12. Liu, X., Li, W., Dai, H.: Approximation algorithms for the minimum power cover problem with submodular/linear penalties. Theor. Comput. Sci. **923**, 256–270 (2022). https://doi.org/10.1016/j.tcs.2022.05.012 (2022)

13. Liu, X., Li, W., Xie, R.: A primal-dual approximation algorithm for the k-prize-collecting minimum power cover problem. Optim. Lett. **16**, 2373–2385 (2021). https://doi.org/10.1007/s11590-021-01831-z

14. Liu, X., Li, W., Yang, J.: A primal-dual approximation algorithm for the k-prize-collecting minimum vertex cover problem with submodular penalties. Front. Comput. Sci. **17** (2022). https://doi.org/10.1007/s11704-022-1665-9

15. Liu, X., Li, W., Zhu, Y.: Single machine vector scheduling with general penalties. Mathematics **9**, 1965 (2021)

16. Ran, Y., Huang, X., Zhang, Z., Du, D.-Z.: Approximation algorithm for minimum power partial multi-coverage in wireless sensor networks. J. Global Optim. **80**(3), 661–677 (2021). https://doi.org/10.1007/s10898-021-01033-y

Online Early Work Maximization on Three Hierarchical Machines with a Common Due Date

Man Xiao and Weidong Li[✉]

School of Mathematics and Statistics, Yunnan University, Kunming 650504, China
weidongmath@126.com

Abstract. In this paper, we consider the online early work problems on three hierarchical machines with a common due date. When there is one machine of hierarchy 1, we propose an optimal online algorithm with a competitive ratio of 1.302. When there are two machines of hierarchy 1, we give a lower bound 1.276, and propose an online algorithm with a competitive ratio of 1.302.

Keywords: Online · Early work · Hierarchy · Competitive ratio

1 Introduction

Since Chen et al. [5] proposed the online early work maximization problem, more and more online and semi-online related problems are considered [6,16]. For a maximization (minimization) problem, the competitive ratio of an online algorithm A is defined as the minimum value of ρ such that $C^{OPT}(I) \leq \rho C^A(I)(C^A(I) \leq \rho C^{OPT}(I))$ for any instance I, where $C^A(I)(C^A$, for short) denotes the output value of the online algorithm A, and $C^{OPT}(I)(C^{OPT}$, for short) denotes the off-line optimal value. For an online problem, if there is no algorithm with competitive ratio less than ρ, then ρ is a lower bound of the problem. If there is an algorithm whose competitive ratio matches the lower bound, this algorithm is called an optimal online algorithm.

In the online scheduling problem, we are given n independent jobs and m identical machines, each job must be assigned irrevocably to a machine before the next job can be revealed. For the online makespan minimization scheduling problem, if $m = 2$, Feigle et al. [7] proved that list scheduling is the optimal online algorithm. Kellerer et al. [13] gave an optimal online algorithm with a competitive ratio of $\frac{4}{3}$ when the total size of all jobs is known in advance. Azar and Regev [3] designed an optimal online algorithm with a competitive ratio of $\frac{4}{3}$ when the optimal value is known in advance. He and Zhang [10] proposed an optimal online algorithm with a competitive ratio of $\frac{4}{3}$ when the largest processing time is known in advance. If $m = 3$, Feigle et al. [7] proved that list scheduling is an optimal online algorithm with a competitive ratio of $\frac{5}{3}$. When the total size of all jobs is known in advance, Angelelli et al. [1] gave a lower bound $\frac{\sqrt{129}-3}{6}$, and Lee and Lim [14] proposed an online algorithm with

M. Li and X. Sun (Eds.): IJTCS-FAW 2022, LNCS 13461, pp. 99–109, 2022.
https://doi.org/10.1007/978-3-031-20796-9_8

a competitive a ratio of $\frac{7}{5}$. When the optimal value is known in advance, Bohm et al. [4] proposed a lower bound $\frac{15}{11}$ and an online algorithm with a competitive ratio of $\frac{11}{8}$. If m is arbitrary, at present, the best lower bound is 1.85 [9], and the best upper bound is 1.92 [8]. When the total size of all jobs is known in advance, Albers and Hellwig [2] gave a lower bound 1.585, and Kellerer et al. [12] designed an optimal online algorithm.

Assume that there are m machines with two hierarchies and the goal is to minimize makespan, where the machines of hierarchy 1 can process all the jobs, and the machines of hierarchy 2 can only process the jobs of hierarchy 2. If $m = 2$, Park et al. [15] and Jiang et al. [11] independently gave an optimal online algorithm with a competitive ratio of $\frac{5}{3}$. Park et al. [15] also proposed an optimal online algorithm with a competitive ratio of $\frac{3}{2}$ when the total size of all jobs is known in advance. Wu and Yang [17] proposed two online algorithms with competitive ratios of 1.618 and 1.5 respectively when the largest processing time or the optimal value is known in advance. If there are one machine of hierarchy 1 and two machines of hierarchy 2, Zhang et al. [19] gave a lower bound 1.824, and an online algorithm with a competitive ratio of 1.857. If there are two machines of hierarchy 1 and one machine of hierarchy 2, Zhang et al. [19] gave a lower bound 1.801, and an online algorithm with a competitive ratio of 1.857. If there are k machines of hierarchy 1 and $m-k$ machines of hierarchy 2, Zhang et al. [19] presented an online algorithm with a competitive ratio of $1 + \frac{m^2-m}{m^2-km+k^2} < \frac{7}{3}$. Xiao et al. [20] considered the buffer model on three machines.

For the online early work maximization problem on two identical machines, Chen et al. [5] gave an optimal online algorithm with a competitive ratio of $\sqrt{5}-1$. When the total size of all jobs is known in advance, Chen et al. [6] designed an optimal online algorithm with a competitive ratio of $\frac{6}{5}$. If two machines have different hierarchies, Xiao et al. [18] proposed three optimal online algorithms when the total size of low-hierarchy jobs, the total size of high-hierarchy jobs, and both the total size of low-hierarchy and high-hierarchy jobs are known in advance, respectively.

In this paper, we consider online early work maximization problem on three machines with two hierarchies as in [19], and the corresponding results are shown in Table 1. The rest of this paper is organized as follows. In Sect. 2, we describe some preliminaries. In Sect. 3, we consider the model that has one machine of hierarchy 1, and design an optimal online algorithm. In Sect. 4, we consider the model that has two machine of hierarchy 1, give a lower bound 1.276, and propose an online algorithm with a competitive ratio of 1.302. Finally, we make a summary.

Table 1. The related results on the three machines with two hierarchies

	Makespan minimization		Early work maximization	
	Lower bound	Upper bound	Lower bound	Upper bound
One machine of hierarchy 1	1.801 [19]	1.857 [19]	1.302	1.302
Two machines of hierarchy 1	1.824 [19]	1.857 [19]	1.276	1.302

2 Preliminaries

We are given three machines M_1, M_2 and M_3, and a job set $\mathcal{J} = \{J_1, J_2, \ldots, J_n\}$. Denote the j-th job as $J_j = (p_j, g_j)$, $j \in \{1, 2, \ldots, n\}$, where p_j indicates the processing time of job J_j (also called the size of J_j), and g_j indicates the hierarchy of job J_j. We know the information of J_j only when J_j arrives, and the next job J_{j+1} arrives only when J_j is scheduled irrevocably on some machine. If $g_j = 1$, we call J_j as a job of hierarchy 1 or low-hierarchy job, and call J_j as a job of hierarchy 2 or high-hierarchy job, otherwise. Only if $g(M_i) \leq g_j$ for $i \in \{1, 2, 3\}$ and $j \in \{1, 2, \ldots, n\}$ where $g(M_i)$ is the hierarchy of machine M_i, machine M_i can process the job J_j, and each job can only be processed on one machine. As in Chen et al. [5,6], assume that each job has a common due date $d > 0$, and the size is no more than d, i.e.,

$$p_j \leq d, \text{ for } j = 1, 2, \ldots, n.$$

The early work of job J_j is $X_j \in [0, p_j]$. If the job J_j is completed before the due date d, then the job is called totally early, and $X_j = p_j$. If the job J_j starts at the time of $S_j < d$, but finishes after the d, J_j is called partially early, and $X_j = d - S_j$. If the job J_j starts at the time of $S_j \geq d$, J_j is called totally late, and $X_j = 0$.

A schedule is actually a partition (S_1, S_2, S_3) of the job set J, such that $S_1 \cup S_2 \cup S_3 = J$ and $S_i \cap S_j = \emptyset$ for any $i \neq j$. Let $L_i = \sum_{J_j \in S_i} p_j$ be the load of M_i, for $i \in \{1, 2, 3\}$. The objective is to find a schedule such that total early work

$$X = \sum_{j=1}^{n} X_j = \sum_{i=1}^{3} \min\{L_i, d\}$$

is maximized.

In the next sections, for convenience, let T_k ($k \in \{1, 2\}$) be the total size of the jobs with hierarchy k, and L_i^j be the load of M_i after job J_j is scheduled for $i \in \{1, 2, 3\}$ and $j \in \{1, 2, \ldots, n\}$. Obviously, we have $L_i^n = L_i$.

3 One Machine of Hierarchy 1

In this section, we consider the model with one machine of hierarchy 1 and two machines of hierarchy 2. For convenience, denote this problem as $P3(1, 2, 2)|online, d_j = d|max(X)$. Without loss of generality, let $g(M_1) = 1$, and $g(M_2) = g(M_3) = 2$, which means that M_2 and M_3 can only process the high-hierarchy jobs, and M_1 can process all the jobs. Meanwhile, based on the definition of total early work X, we have

Lemma 1. *The optimal value C^{OPT} is at most* $\min\{3d, T_1 + T_2, d + T_2\}$.

Proof. By the definition of X, $C^{OPT} \leq 3d$ and $C^{OPT} \leq T_1 + T_2$. Since M_2 and M_3 can only process the high-hierarchy jobs, the total early work of jobs processed on machine M_2 and M_3 is no more than T_2, implying that $C^{OPT} \leq d + T_2$. Thus, the lemma holds. ∎

Theorem 1. *Any online algorithm A for $P3(1,2,2)|online, d_j = d|max(X)$ has a competitive ratio at least $\frac{\sqrt{13}-1}{2}$.*

Proof. For convenience, assume than $d = 1$. The first two jobs are $J_1 = (\frac{\sqrt{13}-1}{4}, 2)$ and $J_2 = (\frac{\sqrt{13}-1}{4}, 2)$.

If at least one job is assigned to M_1, the last job $J_3 = (1,1)$ arrives, implying that $L_1 \geq \frac{\sqrt{13}+3}{4}$, and $L_2 + L_3 \leq \frac{\sqrt{13}-1}{4}$. Thus, we have $C^{OPT} = 1 + \frac{\sqrt{13}-1}{4} + \frac{\sqrt{13}-1}{4} = \frac{\sqrt{13}+1}{2}$ and $C^A \leq 1 + \frac{\sqrt{13}-1}{4} = \frac{\sqrt{13}+3}{4}$, implying that

$$\frac{C^{OPT}}{C^A} \geq \frac{\frac{\sqrt{13}+1}{2}}{\frac{\sqrt{13}+3}{4}} = 5 - \sqrt{13} > \frac{\sqrt{13}-1}{2}.$$

If both J_1 and J_2 are assigned to M_2 or M_3, no more jobs arrive. Since $\frac{\sqrt{13}-1}{2} > 1$, we have $C^{OPT} = \frac{\sqrt{13}-1}{2}$ and $C^A = 1$. Therefore, $\frac{C^{OPT}}{C^A} \geq \frac{\sqrt{13}-1}{2}$.

Else, the next job $J_3 = (1,2)$ arrives. If J_3 is assigned to M_1, the last job $J_4 = (1,1)$ arrives. We have $L_1 = 2$, $L_2 = L_3 = \frac{\sqrt{13}-1}{4}$, and $C^A = 1 + \frac{\sqrt{13}-1}{4} + \frac{\sqrt{13}-1}{4} = \frac{\sqrt{13}+1}{2}$. Since $C^{OPT} = 3$, we have

$$\frac{C^{OPT}}{C^A} \geq \frac{3}{\frac{\sqrt{13}+1}{2}} = \frac{\sqrt{13}-1}{2}.$$

If J_3 is assigned to M_2 or M_3, then no more job arrives, implying that $C^{OPT} = 1 + \frac{\sqrt{13}-1}{4} + \frac{\sqrt{13}-1}{4} = \frac{\sqrt{13}+1}{2}$ and $C^A = 1 + \frac{\sqrt{13}-1}{4} = \frac{\sqrt{13}+3}{4}$. Therefore, we have $\frac{C^{OPT}}{C^A} = \frac{\frac{\sqrt{13}+1}{2}}{\frac{\sqrt{13}+3}{4}} = 5 - \sqrt{13} > \frac{\sqrt{13}-1}{2}$. Thus, the theorem holds. ∎

Our algorithm is described as follows, where we assign the high-hierarchy jobs preferentially to the machines of hierarchy 2.

Theorem 2. *The competitive ratio of Algorithm A1 is at most $\frac{\sqrt{13}-1}{2}$.*

Proof. If $\min\{L_1, L_2, L_3\} \geq d$, we have $C^{A2} = 3d \geq C^{OPT}$. If $\max\{L_1, L_2, L_3\} \leq d$, we have $C^{A1} = L_1 + L_2 + L_3 = T_1 + T_2 \geq C^{OPT}$. Therefore, we only need to consider the case where $\min\{L_1, L_2, L_3\} < d < \max\{L_1, L_2, L_3\}$. We distinguish the following four cases.

Case 1. $L_1 > d$, $L_2 < d$ and $L_3 < d$.

In this case, we have $C^{A1} = d + L_2 + L_3$. If no high-hierarchy job is assigned to M_1, Algorithm A1 reaches optimality.

If there exist some high-hierarchy jobs assigned to M_1, by the choice of Algorithm A1, we have $L_2 + L_3 > \frac{\sqrt{13}-1}{2}d$. Therefore, by Lemma 1, we have

$$\frac{C^{OPT}}{C^{A1}} \leq \frac{3d}{d + L_2 + L_3} \leq \frac{3d}{d + \frac{\sqrt{13}-1}{2}d} = \frac{\sqrt{13}-1}{2}.$$

Algorithm 1: A1

---

1 Initially, let $L_1^0 = L_2^0 = L_3^0 = 0$.

2 When a new job $J_j = (p_j, g_j)$ arrives,

3 **if** $g_j = 1$ **then**

4 $\quad$ Assign the job J_j to the M_1.

5 **else**

6 $\quad$ **if** $L_2^{j-1} + p_j \leq \frac{\sqrt{13}-1}{2}d$ **then**

7 $\quad\quad$ Assign the job J_j to M_2.

8 $\quad$ **else**

9 $\quad\quad$ **if** $L_3^{j-1} + p_j \leq \frac{\sqrt{13}-1}{2}d$ **then**

10 $\quad\quad\quad$ Assign the job J_j to M_3.

11 $\quad\quad$ **else**

12 $\quad\quad\quad$ Assign the job J_j to M_1.

13 If there is another job, $j \leftarrow j + 1$, go to **step 2**. Otherwise, stop.

Case 2. $L_1 > d$, $L_2 > d$ and $L_3 < d$.

In this case, we have $C^{A1} = 2d + L_3$. If no high-hierarchy job assigned to M_1, we have $T_2 = L_2 + L_3$. According to the Algorithm A1, we have $L_2 \leq \frac{\sqrt{13}-1}{2}d$. By Lemma 1, we have

$$\frac{C^{OPT}}{C^{A1}} \leq \frac{d + T_2}{L_3 + 2d} = \frac{d + L_2 + L_3}{2d + L_3} \leq \frac{d + L_3 + \frac{\sqrt{13}-1}{2}d}{L_3 + 2d} = \frac{L_3 + \frac{\sqrt{13}+1}{2}d}{L_3 + 2d}$$

$$= 1 + \frac{(\frac{\sqrt{13}+1}{2} - 2)d}{L_3 + 2d} \leq \frac{\sqrt{13}+1}{4} < \frac{\sqrt{13}-1}{2}.$$

If there exist some high-hierarchy jobs assigned to M_1, by the choice of Algorithm A1, we have $L_3 \geq \frac{\sqrt{13}-3}{2}d$. Thus, by Lemma 1, we have

$$\frac{C^{OPT}}{C^{A1}} \leq \frac{3d}{L_3 + 2d} \leq \frac{3d}{\frac{\sqrt{13}-3}{2}d + 2d} = \frac{\sqrt{13}-1}{2}.$$

Case 3. $L_1 < d$, $L_2 > d$ and $L_3 > d$.

In this case, we have $C^{A1} = 2d + L_1$. According to the choice of Algorithm A1, we have $L_2 \leq \frac{\sqrt{13}-1}{2}d$ and $L_3 \leq \frac{\sqrt{13}-1}{2}d$. By Lemma 1, we have

$$\frac{C^{OPT}}{C^{A1}} \leq \frac{T_1 + T_2}{L_1 + 2d} = \frac{L_1 + L_2 + L_3}{L_1 + 2d} \leq \frac{L_1 + \frac{\sqrt{13}-1}{2}d + \frac{\sqrt{13}-1}{2}d}{L_1 + 2d}$$

$$= \frac{L_1 + (\sqrt{13} - 1)d}{L_1 + 2d} = 1 + \frac{(\sqrt{13} - 3)d}{L_1 + 2d} \leq \frac{\sqrt{13} - 1}{2}.$$

Case 4. $L_1 < d$, $L_2 > d$ and $L_3 < d$.

In this case, we have $C^{A1} = d + L_1 + L_3$. According to the choice of Algorithm A1, we have $L_2 \leq \frac{\sqrt{13}-1}{2}d$. By Lemma 1, we have

$$\frac{C^{OPT}}{C^{A1}} \leq \frac{T_1 + T_2}{L_1 + L_3 + d} = \frac{L_1 + L_2 + L_3}{L_1 + L_3 + d} \leq \frac{L_1 + L_3 + \frac{\sqrt{13}-1}{2}d}{L_1 + L_3 + d} \leq \frac{\sqrt{13}-1}{2}.$$

Thus, the theorem holds. ∎

4   Two Machines of Hierarchy 1

In this section, we consider the model with two machines of hierarchy 1 and one machine of hierarchy 2. For convenience, denote this problem as $P3(1,1,2)|online, d_j = d|max(X)$. Without loss of generality, let $g(M_1) = g(M_2) = 1$, and $g(M_3) = 2$, which means that M_1 and M_2 can process all the jobs, and M_3 can only process the high-hierarchy jobs. Based on the definition of total early work X, we have

Lemma 2. *The optimal value C^{OPT} is at most* $\min \{3d, T_1 + T_2, 2d + T_2\}$.

Proof. By the definition of X, $C^{OPT} \leq 3d$ and $C^{OPT} \leq T_1 + T_2$. Since M_3 can only process the high-hierarchy jobs, the total early work of jobs processed on machine M_3 is no more than T_2, implying that $C^{OPT} \leq 2d + T_2$. Thus, the lemma holds. ∎

For convenience, let $\beta \approx 0.352$ be the positive real root of equation $\beta^3 + 4\beta^2 + 7\beta - 3 = 0$ and $\alpha = \frac{3}{2+\beta} - 1 \approx 0.276$.

Theorem 3. *Any online algorithm A for $P3(1,1,2)|online, d_j = d|max(X)$ has a competitive ratio at least $1 + \alpha$.*

Proof. For convenience, assume $d = 1$, the first job is $J_1 = (\alpha, 2)$. We distinguish the following two cases.

Case 1. J_1 is assigned to M_3.

The next job $J_2 = (1, 2)$ arrives. If J_2 is assigned to M_3, no more job arrives. We have $C^{OPT} = 1 + \alpha$, $C^A = 1$, and $\frac{C^{OPT}}{C^A} \geq 1 + \alpha$. Else, the last two jobs $J_3 = (1, 1)$ and $J_4 = (1, 1)$ arrive. We have $C^{OPT} = 3$, $C^A \leq 2 + \alpha$, and $\frac{C^{OPT}}{C^A} \geq \frac{3}{2+\alpha} > 1 + \alpha$.

Case 2. J_1 is assigned to M_1 or M_2.

The next job $J_2 = (\beta, 2)$ arrives. If J_2 is assigned to M_3, the next job $J_3 = (1, 2)$ arrives. If J_3 is assigned to M_3, no more job arrives. We have $C^{OPT} = 1 + \alpha + \beta$, $C^A = 1 + \alpha$ and $\frac{C^{OPT}}{C^A} \geq \frac{1+\alpha+\beta}{1+\alpha} = 1 + \alpha$. If J_3 is assigned to M_1 or M_2, the last two jobs $J_4 = (1, 1)$ and $J_5 = (1, 1)$ arrive, we have $C^{OPT} = 3$ and

$C^A \leq 2 + \beta$. Therefore, $\frac{C^{OPT}}{C^A} \geq \frac{3}{2+\beta} = 1 + \alpha$. Else, J_2 is assigned to M_1 or M_2, the last two jobs $J_3 = (1,1)$ and $J_4 = (1,1)$ arrive. We have $C^{OPT} = 2 + \alpha + \beta$ and $C^A \leq 2$. Therefore, $\frac{C^{OPT}}{C^A} \geq \frac{2+\alpha+\beta}{2} > 1 + \alpha$.

Thus, the theorem holds. ∎

Our online algorithm is described as in Algorithm A2.

Algorithm 2: A2

---

1   Initially, let $L_1^0 = L_2^0 = L_3^0 = 0$.
2   When a new job $J_j = (p_j, g_j)$ arrives,
3   **if** $g_j = 1$ **then**
4   **if** $L_1^{j-1} + p_j \leq \frac{\sqrt{13}-1}{2} d$ **then**
5   ⌊ Assign the job J_j to M_1.
6   **else**
7   **if** $L_2^{j-1} + p_j \leq \frac{\sqrt{13}-1}{2} d$ **then**
8   ⌊ Assign the job J_j to M_2.
9   **else**
10   ⌊ Assign the job J_j to machine with the least current load among M_1 and M_2.
11 **else**
12   **if** $L_3^{j-1} + p_j \leq \frac{\sqrt{13}-1}{2} d$ **then**
13   ⌊ Assign the job J_j to M_3.
14   **else**
15   **if** $L_1^{j-1} + p_j \leq \frac{\sqrt{13}-1}{2} d$ **then**
16   ⌊ Assign the job J_j to M_1.
17   **else**
18   **if** $L_2^{j-1} + p_j \leq \frac{\sqrt{13}-1}{2} d$ **then**
19   ⌊ Assign the job J_j to M_2.
20   **else**
21   ⌊ Assign the job J_j to machine with the least current load among M_1, M_2 and M_3.
22 If there is another job, $j == j + 1$, go to **step 2**. Otherwise, stop.

Theorem 4. *The competitive ratio of Algorithm A2 is at most $1 + \alpha$*

Proof. If $\max\{L_1, L_2, L_3\} \leq d$, we have $C^{A2} = L_1 + L_2 + L_3 = T_1 + T_2 \geq C^{OPT}$. If $\min\{L_1, L_2, L_3\} \geq d$, we have $C^{A2} = L_1 + L_2 + L_3 = 3d \geq C^{OPT}$. Hence, we only to consider the $\min\{L_1, L_2, L_3\} < d < \max\{L_1, L_2, L_3\}$. Without loss of generality, we distinguish the following four cases.

Case 1. $L_1 > d$, $L_2 > d$ and $L_3 < d$.

In this case, we have $C^{A2} = 2d + L_3$. If there is no high-hierarchy job assigned to M_1 or M_2, Algorithm A2 reaches the optimality. Else, let $J_l = (p_l, 2)$ be the last high-hierarchy job assigned to M_1 or M_2. According to the choice of Algorithm A2, we have $L_3^{l-1} + p_l > \frac{\sqrt{13}-1}{2}d$. Since $p_l \leq d$, we have

$$L_3 \geq L_3^{l-1} > \frac{\sqrt{13}-1}{2}d - p_l \geq \frac{\sqrt{13}-1}{2}d - d = \frac{\sqrt{13}-3}{2}d.$$

By Lemma 2, we have

$$\frac{C^{OPT}}{C^{A2}} \leq \frac{3d}{2d + L_3} \leq \frac{3d}{2d + \frac{\sqrt{13}-3}{2}d} = \frac{\sqrt{13}-1}{2}.$$

Case 2. $L_1 > d$, $L_2 < d$ and $L_3 < d$.

In this case, we have $C^{A2} = d + L_2 + L_3$. If $L_1 \leq \frac{\sqrt{13}-1}{2}d$, by Lemma 2, we have

$$\frac{C^{OPT}}{C^{A2}} \leq \frac{T_1 + T_2}{d + L_2 + L_3} = \frac{L_1 + L_2 + L_3}{d + L_2 + L_3} \leq \frac{\frac{\sqrt{13}-1}{2}d + L_2 + L_3}{d + L_2 + L_3} \leq \frac{\sqrt{13}-1}{2}.$$

If $L_1 > \frac{\sqrt{13}-1}{2}d$, we distinguish the following two subcases.

Case 2.1. There is no high-hierarchy job assigned to M_1 or M_2. Let J_t be the last job assigned to M_1. Since $L_1 > \frac{\sqrt{13}-1}{2}d$ and $p_t \leq d$, by the choice of Algorithm A2, we have $0 < L_1 - p_t \leq L_2$, implying that $L_1 - p_t + L_2 > \frac{\sqrt{13}-1}{2}d$ and $L_2 > \frac{\sqrt{13}-1}{4}d$. By Lemma 2, we have

$$\frac{C^{OPT}}{C^{A2}} \leq \frac{2d + T_2}{d + L_2 + L_3} = \frac{2d + L_3}{d + L_2 + L_3} \leq \frac{2d + L_3}{d + \frac{\sqrt{13}-1}{4}d + L_3}$$

$$= 1 + \frac{2d - \frac{\sqrt{13}+3}{4}d}{\frac{\sqrt{13}+3}{4}d + L_3} \leq 1 + \frac{\frac{5-\sqrt{13}}{4}d}{\frac{\sqrt{13}+3}{4}d} = 2\sqrt{13} - 6 < \frac{\sqrt{13}-1}{2}.$$

Case 2.2. There exist some high-hierarchy jobs assigned to M_1 or M_2. If $L_2 + L_3 \geq \frac{\sqrt{13}-1}{2}d$, by Lemma 2, we have

$$\frac{C^{OPT}}{C^{A2}} \leq \frac{3d}{d + L_2 + L_3} \leq \frac{3d}{d + \frac{\sqrt{13}-1}{2}d} = \frac{\sqrt{13}-1}{2}.$$

If $L_2 + L_3 < \frac{\sqrt{13}-1}{2}d$, by the choice of Algorithm A2, there is no high-hierarchy job assigned to M_2. Let J_t be the last job assigned to M_1. Since $L_1 > \frac{\sqrt{13}-1}{2}d$, by the choice of Algorithm A2, we have $L_1 - p_t \leq L_2$. If there exists one high-hierarchy job assigned to M_1 before J_t, by the choice of Algorithm A2, we have

$$L_3 + L_2 \geq L_3 + L_1 - p_t > \frac{\sqrt{13}-1}{2}d,$$

which contradicts the assumption $L_2 + L_3 < \frac{\sqrt{13}-1}{2}d$. Else, there is no high-hierarchy job assigned to M_1 before J_t, which implies that J_t is a high-hierarchy job. By the choice of Algorithm A2, we have $L_1 - p_t \le L_2$ and $L_1 - p_t \le L_3$. Let $L_2 + L_3 = x$. Thus, $L_1 \le \frac{L_2+L_3}{2} + p_t = \frac{x}{2} + p_t \le \frac{x}{2} + d$. From Lemma 2 and $x < \frac{\sqrt{13}-1}{2}d$, we have

$$\frac{C^{OPT}}{C^{A2}} \le \frac{T_1 + T_2}{d + L_2 + L_3} = \frac{L_1 + x}{d + x} \le \frac{d + \frac{x}{2} + x}{d + x} = 1 + \frac{\frac{x}{2}}{d + x}$$

$$\le 1 + \frac{\frac{\sqrt{13}-1}{4}d}{d + \frac{\sqrt{13}-1}{2}d} = 1 + \frac{7 - \sqrt{13}}{12} < \frac{\sqrt{13}-1}{2}.$$

Case 3. $L_1 < d$, $L_2 < d$ and $L_3 > d$.

In this case, we have $C^{A2} = L_1 + L_2 + d$. If $L_3 \le \frac{\sqrt{13}-1}{2}d$, by Lemma 2, we have

$$\frac{C^{OPT}}{C^{A2}} \le \frac{T_1 + T_2}{L_1 + L_2 + d} = \frac{L_1 + L_2 + L_3}{L_1 + L_2 + d} \le \frac{\sqrt{13}-1}{2}.$$

If $L_3 > \frac{\sqrt{13}-1}{2}d$, let J_k be the last job assigned to M_3. By the choice of Algorithm A2, and $p_k \le d$, we have $0 < L_3 - p_k \le \min\{L_1, L_2\}$, implying that $L_1 + L_2 > \frac{\sqrt{13}-1}{2}d$. Thus,

$$\frac{C^{OPT}}{C^{A2}} \le \frac{3d}{L_1 + L_2 + d} \le \frac{3d}{\frac{\sqrt{13}-1}{2}d + d} = \frac{\sqrt{13}-1}{2}.$$

Case 4. $L_1 > d$, $L_2 < d$ and $L_3 > d$.

In this case, we have $C^{A2} = 2d + L_2$. We distinguish the following three subcases.

Case 4.1. $L_1 \le \frac{\sqrt{13}-1}{2}d$ and $L_3 \le \frac{\sqrt{13}-1}{2}d$. By Lemma 2, we have

$$\frac{C^{OPT}}{C^{A2}} \le \frac{T_1 + T_2}{2d + L_2} = \frac{L_1 + L_2 + L_3}{2d + L_2} \le \frac{(\sqrt{13}-1)d + L_2}{2d + L_2} = 1 + \frac{(\sqrt{13}-3)d}{2d + L_2} \le \frac{\sqrt{13}-1}{2}.$$

Case 4.2. $L_1 > \frac{\sqrt{13}-1}{2}d$ and $L_3 \le \frac{\sqrt{13}-1}{2}d$. Let J_t be the last job assigned to M_1. By the choice of Algorithm A2, we have $L_2 \ge L_1 - p_t > 0$ and $L_1 - p_t + L_2 > \frac{\sqrt{13}-1}{2}d$, implying that $L_2 > \frac{\sqrt{13}-1}{4}d$. By Lemma 2, we have

$$\frac{C^{OPT}}{C^{A2}} \le \frac{3d}{2d + L_2} \le \frac{3d}{2d + \frac{\sqrt{13}-1}{4}d} = \frac{12}{7 + \sqrt{13}} = \frac{7 - \sqrt{13}}{3} < \frac{\sqrt{13}-1}{2}.$$

Case 4.3. $L_3 > \frac{\sqrt{13}-1}{2}d$. Let J_k be the last job assigned to M_3. By the choice of Algorithm A2, we have $\min\{L_1, L_2\} \ge L_3 - p_k > 0$. Since $p_k \le d$, we have $L_2 \ge L_3 - p_k \ge (\frac{\sqrt{13}-1}{2} - 1)d = \frac{\sqrt{13}-3}{2}d$. By Lemma 2, we have

$$\frac{C^{OPT}}{C^{A2}} \le \frac{3d}{2d + L_2} \le \frac{3d}{2d + \frac{\sqrt{13}-3}{2}d} = \frac{\sqrt{13}-1}{2}.$$

Thus, the theorem holds. ∎

5   Discussion

In this paper, we considered the online early work problems on three hierarchical machines. When only one machine of hierarchy 1, we design an optimal online algorithm. When only one machine of hierarchy 2, we give a lower bound 1.276 and an upper bound 1.302. In the future, it is interesting to study the online and semi-online early work maximization problems on m hierarchical machines.

Acknowledgement. The work is supported in part by the National Natural Science Foundation of China [No. 12071417].

References

1. Angelelli, E., Speranza, M., Tuza, Z.: Semi on-line scheduling on three processors with known sum of the tasks. J. Sched. **10**, 263–269 (2007). https://doi.org/10.1007/s10951-007-0023-y
2. Albers, S., Hellwig, M.: Semi-online scheduling revisited. Theor. Comput. Sci. **443**, 1–9 (2012)
3. Azar, Y., Regev, O.: On-line bin-stretching. Theor. Comput. Sci. **168**, 17–41 (2001)
4. Böhm, M., Sgall, J., van Stee, R., Veselý, P.: Online bin stretching with three bins. J. Sched. **20**(6), 601–621 (2017). https://doi.org/10.1007/s10951-016-0504-y
5. Chen, X., Sterna, M., Han, X., Blazewicz, J.: Scheduling on parallel identical machines with late work criterion: offline and online cases. J. Sched. **19**, 729–736 (2016). https://doi.org/10.1007/s10951-015-0464-7
6. Chen, X., Kovalev, S., Liu, Y., Sterna, M., Chalamon, I., Blazewicz, J.: Semi-online scheduling on two identical machines with a common due date to maximize total early work. Discrete Appl. Math. **290**, 71–78 (2021)
7. Faigle, U., Kern, W., Turan, G.: On the performance of on-line algorithms for partition problems. Acta Cybernet. **9**(2), 107–119 (1989)
8. Fleischer, R., Wahl, M.: On-line scheduling revisited. J. Sched. **3**, 343–353 (2000)
9. Gormley, T., Reingold, N., Torng, E., Westbrook, J.: Generating adversaries for request-answer games. In: Proceedings of the 11th Annual ACM-SIAM Symposium on Discrete Algorithms, pp. 564–565. SIAM, Philadelphia (2000)
10. He, Y., Zhang, G.: Semi on-line scheduling on two identical machines. Computing **62**, 179–187 (1999). https://doi.org/10.1007/s006070050020
11. Jiang, Y., He, Y., Tang, C.: Optimal online algorithms for scheduling on two identical machines under a grade of service. J. Zhejiang Univ., Sci., A **7**, 309–314 (2006). https://doi.org/10.1631/jzus.2006.A0309
12. Kellerer, H., Kotov, V., Gabay, M.: An efficient algorithm for semi-online multiprocessor scheduling with given total processing time. J. Sched. **18**(6), 623–630 (2015). https://doi.org/10.1007/s10951-015-0430-4
13. Kellerer, H., Kotov, V., Speranza, M., Tuza, Z.: Semi on-line algorithms for the partition problem. Oper. Res. Lett. **21**(5), 235–242 (1997)
14. Lee, K., Lim, K.: Semi-online scheduling problems on a small number of machines. J. Sched. **16**, 461–477 (2013). https://doi.org/10.1007/s10951-013-0329-x
15. Park, J., Chang, S., Lee, K.: Online and semi-online scheduling of two machines under a grade of service provision. Oper. Res. Lett. **34**(6), 692–696 (2006)
16. Sterna, M.: Late and early work scheduling: a survey. Omega **104**(15–16), 102453 (2021)

17. Wu, Y., Yang, Q.: Optimal semi-online scheduling algorithms on two parallel identical machines under a grade of service provision. In: Chen, B. (ed.) AAIM 2010. LNCS, vol. 6124, pp. 261–270. Springer, Heidelberg (2010). https://doi.org/10.1007/978-3-642-14355-7_27
18. Xiao, M., Liu, X., Li, W.: Semi-online early work maximization problem on two hierarchical machines with partial information of processing time. In: Wu, W., Du, H. (eds.) AAIM 2021. LNCS, vol. 13153, pp. 146–156. Springer, Cham (2021). https://doi.org/10.1007/978-3-030-93176-6_13
19. Zhang, A., Jiang, Y., Tan, Z.: Online parallel machines scheduling with two hierarchies. Theor. Comput. Sci. **410**, 3597–3605 (2009)
20. Xiao, M., Ding, L., Zhao, S., Li, W.: Semi-online algorithms for hierarchical scheduling on three parallel machines with a buffer size of 1. In: He, K., Zhong, C., Cai, Z., Yin, Y. (eds.) NCTCS 2020. CCIS, vol. 1352, pp. 47–56. Springer, Singapore (2021). https://doi.org/10.1007/978-981-16-1877-2_4

Secure Computations Through Checking
Suits of Playing Cards

Daiki Miyahara[1,2]([✉]) [iD] and Takaaki Mizuki[2,3] [iD]

[1] The University of Electro-Communications, Tokyo, Japan
miyahara@uec.ac.jp
[2] National Institute of Advanced Industrial Science and Technology, Tokyo, Japan
[3] Tohoku University, Sendai, Japan

Abstract. Card-based cryptography started with the "five-card trick"
designed by Den Boer (EUROCRYPT 1989); it enables Alice and Bob
to securely evaluate the AND value of their private bits using a physical
deck of five cards. It was then shown that the same task can be done
with only four cards, i.e., Mizuki et al. proposed a four-card AND protocol (ASIACRYPT 2012). These two AND protocols are simple and easy
even for non-experts, such as high school students, to execute. Their only
common drawback is the need to prepare a customized deck consisting of
red and black cards such that all cards of the same color must be identical. Fortunately, several existing protocols are based on a standard deck
of playing cards (commercially available). Among them, the state-of-the-
art AND protocol was constructed by Koch et al. (ASIACRYPT 2019);
it uses four playing cards (such as 'A, J, Q, K') to securely evaluate the
AND value. The protocol is elaborate, while its possible drawback is the
need to repeat a shuffling operation six times (in expectation), which
makes it less practical.

This paper aims to provide the first practical protocol working on a
standard deck of playing cards. We present an extremely simple AND
protocol that terminates after only one shuffle using only four cards;
our proposed protocol relies on a new operation, called the "half-open"
action, whereby players can check only the suit of a face-down card without revealing the number on it. We believe that this new operation is
easy-to-implement, and hence, our four-card AND protocol working on a
standard deck is practical. We formalize the half-open action to present a
formal description of our proposed protocol. Moreover, we discuss what
is theoretically implied by introducing the half-open action and show
that it can be applied to efficiently solving Yao's Millionaires' problem
with a standard deck of cards.

Keywords: Card-based cryptography · Secure computation · Real-life
hands-on cryptography

1 Introduction

Card-based cryptography enables people including non-specialists to easily
conduct cryptographic tasks, such as secure multiparty computations and

© The Author(s) 2022
M. Li and X. Sun (Eds.): IJTCS-FAW 2022, LNCS 13461, pp. 110–128, 2022.
https://doi.org/10.1007/978-3-031-20796-9_9

zero-knowledge proofs, in daily activities using a deck of physical cards. Typically, we use a *two-colored deck* of cards, i.e., a deck consisting of black ♣ and red cards ♥ whose backs are all identical ?. In history, the first card-based protocol called the *five-card trick* was presented by Den Boer [2] at EUROCRYPT 1989; it enables Alice and Bob holding private bits $a \in \{0, 1\}$ and $b \in \{0, 1\}$, respectively, to securely evaluate the AND value $a \wedge b$ using five cards ♣ ♣ ♥ ♥ ♥, as described below.

1.1   The Five-Card Trick

Assume that, based on a pair of cards of different colors, Alice and Bob agree upon the following encoding rule:

$$♣ ♥ = 0, \quad ♥ ♣ = 1. \tag{1}$$

If two face-down cards ? ? represent a bit $x \in \{0, 1\}$ according to the above encoding (1), then we call them a *commitment* to x and denote it by

$$\underbrace{? \ ?}_{x}.$$

The five-card trick [2] proceeds as follows.

1. Alice and Bob privately create commitments to $\overline{a}$ and b, respectively, and between them, place one helping red card ♥; then, turn it face down:

$$\underbrace{? \ ?}_{\overline{a}} \ ♥ \ \underbrace{? \ ?}_{b} \ \rightarrow \ \underbrace{? \ ?}_{\overline{a}} \ ? \ \underbrace{? \ ?}_{b}.$$

   Note that the three cards in the middle would be ♥ ♥ ♥ if and only if $a = b = 1$.

2. Apply a *random cut*, denoted by $\langle \cdot \rangle$, to the sequence of five cards:

$$\underbrace{? \ ?}_{\overline{a}} \ ? \ \underbrace{? \ ?}_{b} \ \rightarrow \ \left\langle ? \ ? \ ? \ ? \ ? \right\rangle \ \rightarrow \ ? \ ? \ ? \ ? \ ?.$$

   A random cut is a cyclic shuffling operation such that the resulting sequence is randomly shifted. Note that a secure implementation of a random cut called the Hindu cut has been known [49].

3. Reveal all the five cards; then, we learn the value of $a \wedge b$, which depends on whether or not the three red cards ♥ ♥ ♥ are consecutive (apart from cyclic rotation):

$a \wedge b = 0$ or $a \wedge b = 1$.

Thus, the five-card trick can elegantly evaluate the AND value securely.

1.2 Protocols with a Standard Deck of Cards

After twenty-three years since the invention of the five-card trick, it was reported at ASIACRYPT 2012 that the same task can be conducted without any helping card [30]. These two AND protocols [2,30] are simple and easy even for non-experts, such as high school students, to understand. Actually, both the protocols are practical and used for introducing the notion of secure computations in university classes [21,39]. On the other hand, their only common drawback is the need to prepare a customized deck consisting of red and black cards (♣ ♣ ♥ ♥ ⋯) such that all cards of the same color must be indistinguishable.

Fortunately, there are several existing protocols that work on a *standard deck* of playing cards (which is commercially available), such as:

$$A_\clubsuit\;A_\heartsuit\;A_\diamondsuit\;A_\spadesuit\;2_\clubsuit\;2_\heartsuit\;2_\diamondsuit\;2_\spadesuit\;\cdots\;K_\clubsuit\;K_\heartsuit\;K_\diamondsuit\;K_\spadesuit.$$

Table 1 enumerates the existing AND protocols working on a standard deck. In these protocols, a standard deck is regarded as a total order on $\{1, 2, \ldots, 52\}$ (because of 13 numbers × 4 suits): that is, a protocol is supposed to work on cards like $\boxed{1}\,\boxed{2}\,\boxed{3}\,\boxed{4}\cdots$ whose backs are all $\boxed{?}$. In this standard deck setting, a Boolean value can be also represented by a pair of cards: a bit $x \in \{0, 1\}$ is encoded with the order of two cards $\boxed{i}$ and $\boxed{j}$, $1 \le i < j \le 52$, according to

$$\boxed{i}\,\boxed{j} = 0, \quad \boxed{j}\,\boxed{i} = 1. \tag{2}$$

Therefore, such two face-down cards serve a *commitment* to $x \in \{0, 1\}$, which is denoted by

$$\underbrace{\boxed{?}\,\boxed{?}}_{[x]^{\{i,j\}}},$$

where the set $\{i, j\}$ is called its *base*.

Among the existing AND protocols (shown in Table 1), the state-of-the-art one was presented by Koch et al. at ASIACRYPT 2019 [10]; it is a card-minimal AND protocol, i.e., it uses only four cards (such as $\boxed{1}\,\boxed{2}\,\boxed{3}\,\boxed{4}$). Given two input commitments to $a, b \in \{0, 1\}$, the protocol produces a commitment to $a \wedge b$ after applying a random cut six times in expectation:

$$\underbrace{\boxed{?}\,\boxed{?}}_{[a]^{\{1,2\}}}\underbrace{\boxed{?}\,\boxed{?}}_{[b]^{\{3,4\}}} \;\rightarrow\; \text{Random cut 6 times (exp.)} \;\rightarrow\; \underbrace{\boxed{?}\,\boxed{?}}_{[a\wedge b]^B},$$

where the base B will be one of $\{1, 2\}, \{1, 3\}, \{1, 4\}, \{2, 4\}, \{3, 4\}$.

The description of this protocol [10] will be presented in Sect. 2. Although the protocol is elaborate as will be seen, its possible drawback is the need to repeat a random cut six times (in expectation), which makes it less practical.

Table 1. The existing AND protocols with a standard deck

	# of cards	# of shuffles	not Las Vegas?
Niemi and Renvall [37]	5	9.5 (exp.)	
Mizuki [26]	8	4	✓
Koch and Schrempp and Kirsten [10,11]	4	6 (exp.)	

1.3   Contribution

As mentioned above and implied by Table 1, the existing AND protocols working on a standard deck are somewhat impractical due to their numbers of required shuffles and/or cards. Thus, our aim is to provide the first practical AND protocol working on a standard deck of playing cards.

In this study, we present an extremely simple AND protocol that terminates after only *one* random cut using only *four* cards. The key idea behind our construction is to make use of the simple fact that every card in a standard deck has a suit in addition to its number. In other words, we do not regard a standard deck just as a total order, but we directly utilize the suits (♣, ♠, ♥, ♦) to perform an efficient secure AND computation. More precisely, our AND protocol works on a deck of four cards

$$\boxed{3_\clubsuit}\,\boxed{3_\heartsuit}\,\boxed{9_\clubsuit}\,\boxed{9_\heartsuit}\,,$$

whose suits (♣ or ♥) will play an important role. Our four-card AND protocol relies on a new operation, called the *half-open* action, whereby players are able to check only the suit of a face-down card without revealing the number on it. Thus, briefly, given two input commitments to $a, b \in \{0, 1\}$, our protocol securely evaluates the value of $a \wedge b$ by using one random cut and the half-open action:

$$\boxed{?}\,\boxed{?}\,\boxed{?}\,\boxed{?} \quad \rightarrow \quad \text{Random cut once + Half-open}$$
$$\underset{[a]^{\{a\}}\ \ [b]^{\{b\}}}{}$$

$$\rightarrow \quad \begin{cases} a \wedge b = 1 & \text{if } \boxed{3_\clubsuit}\boxed{\clubsuit},\ \boxed{9_\clubsuit}\boxed{\clubsuit},\ \boxed{3_\heartsuit}\boxed{\heartsuit},\ \text{or } \boxed{9_\heartsuit}\boxed{\heartsuit} \text{ appears,} \\ a \wedge b = 0 & \text{otherwise.} \end{cases}$$

We present the details of our protocol in Sect. 4. As will be seen in Sect. 3, our new operation, the "half-open" action, is easy-to-implement, and hence, we believe that our four-card AND protocol working on a standard deck is practical.

In Sect. 5, we construct, by extending the computational model of card-based protocols, which has been developed in [8,13,31,32,48], a formal computation model that admits the operation mentioned above, i.e., the half-open action. Based on this model, we present a formal description of our proposed protocol in Sect. 6.

Moreover, we discuss what is theoretically implied by introducing the half-open action and show that it can be applied to efficiently solving Yao's Millionaires' problem [50] with a standard deck of cards in Sect. 7. Somewhat surprisingly, our solution requires only one more half-open action compared to the existing solutions [23] that work on a two-colored deck of cards.

2 The Existing Card-Minimal AND Protocol

In this section, we introduce the existing card-minimal AND protocol constructed by Koch et al. [10,11] using four cards $\boxed{1}\,\boxed{2}\,\boxed{3}\,\boxed{4}$.

1. Given input commitments to $a, b \in \{0, 1\}$, apply a random cut:

$$\underbrace{\boxed{?}\,\boxed{?}}_{[a]^{\{1,2\}}}\,\underbrace{\boxed{?}\,\boxed{?}}_{[b]^{\{3,4\}}} \rightarrow \left\langle \boxed{?}\,\boxed{?}\,\boxed{?}\,\boxed{?} \right\rangle \rightarrow \boxed{?}\,\boxed{?}\,\boxed{?}\,\boxed{?}.$$

2. Turn over the first card; assume that the revealed card is $\boxed{1}$ (the other cases are similar). Then, turn it face down:

$$\boxed{1}\,\boxed{?}\,\boxed{?}\,\boxed{?} \rightarrow \boxed{?}\,\boxed{?}\,\boxed{?}\,\boxed{?}.$$

   (a) Swap the third and fourth cards.

$$\overset{1\ \ 2\ \ 3\ \ 4}{\boxed{?}\,\boxed{?}\,\boxed{?}\,\boxed{?}} \rightarrow \overset{1\ \ 2\ \ 4\ \ 3}{\boxed{?}\,\boxed{?}\,\boxed{?}\,\boxed{?}}.$$

   (b) Apply a random cut:

$$\left\langle \boxed{?}\,\boxed{?}\,\boxed{?}\,\boxed{?} \right\rangle \rightarrow \boxed{?}\,\boxed{?}\,\boxed{?}\,\boxed{?}.$$

   (c) Turn over the first card. If $\boxed{3}$ appears, proceed to Step 3; otherwise, go back to (b) after turning over the face-up card.

3. Apply a random cut to the second, third, and fourth cards:

$$\boxed{3}\left\langle \boxed{?}\,\boxed{?}\,\boxed{?} \right\rangle \rightarrow \boxed{3}\,\boxed{?}\,\boxed{?}\,\boxed{?}.$$

4. Reveal the second card.

   (a) If either $\boxed{1}$ or $\boxed{4}$ is revealed, then we obtain a commitment to $a \wedge b$:

$$\boxed{3}\,\boxed{1}\,\underbrace{\boxed{?}\,\boxed{?}}_{[a\wedge b]^{\{2,4\}}} \quad \text{or} \quad \boxed{3}\,\boxed{4}\,\underbrace{\boxed{?}\,\boxed{?}}_{[a\wedge b]^{\{1,2\}}}.$$

   (b) If $\boxed{2}$ is revealed, we obtain a commitment to $\overline{a \wedge b}$ (which can be easily changed into a commitment to $a \wedge b$ just by swapping the two cards):

$$\boxed{3}\,\boxed{2}\,\underbrace{\boxed{?}\,\boxed{?}}_{[a\wedge b]^{\{1,4\}}}.$$

Note that in Steps 2(b) and (c), the protocol searches for $\boxed{3}$ among the four cards by repeating the application of a random cut; the expected number of shuffles is four (because of the four cards). Therefore, in total, this protocol needs six random cuts (in expectation), and hence, it is a Las Vegas algorithm.

In the next sections, we aim to reduce the number of shuffles required for a secure AND computation.[1]

Fig. 1. Typical playing cards   **Fig. 2.** Half-open with fingers

3   New Action: Half-Open of Playing Cards

In this section, we propose a novel action in card-based cryptography: the *half-open* action reveals only the suit of a given face-down card (of a standard deck) without leaking any information about its number:

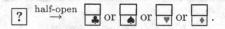

We present a couple of implementations of this action.

Because implementing the half-open action depends on a physical design of cards, let us consider typical playing cards as shown in Fig. 1. Here, we mention the general policy of implementing the half-open action, which would be helpful when considering non-typical playing cards. Because the half-open action reveals only a suit, we should find a specific area on the front where a suit is "identically" placed for all cards. The half-open action can be achieved by revealing such an area while hiding the others.

Figure 2 describes a playing card such that nothing but its suit (♣) is visible. One possible way to have such a situation is as follows. Note that a typical

---

[1] Recently, a zero-knowledge proof protocol for Sudoku has been developed using standard decks of cards [42], and protocols based on private operations have been constructed using a standard deck [19]. In addition, efficient copy and XOR protocols [15], a card-minimal three-input AND protocol [14], and other three-input protocols [5] have been devised.

playing card has a number and suit in the upper left and lower right corners. Given a face-down card (to which apply the half-open), another (face-up) card is inserted below it, the two cards are stacked, and they are turned over. Then, slide out one card while keeping the number hidden with fingers so that only the suit becomes visible.

Because mastering this method requires some effort, we consider an easier method for the half-open action. For this, we made covers that disclose only suits of cards without leaking their numbers, as shown in Fig. 3. Here, we used an A3-size notebook and scissors to make the covers, but any sheet of paper can be used, and it is effortless to make.

Next, we show how to use the covers specifically. We put a cover under a target card (Fig. 4), lift the card and cover together (Fig. 5), and turn them over to check the suit (Fig. 6). We describe this result as ♥. Thus, the half-open action can be easily implemented: it is an effortless task to create a cover, insert the cover under the card, and turn them over.

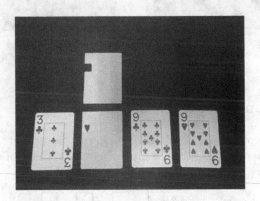

Fig. 3. The use of covers

Fig. 4. Put a cover under the card

Fig. 5. Lift up the card and cover together

Fig. 6. Turn them over. We describe this as ♥.

It should be noted that Marcedone et al. [21, Solution 4] first considered a similar idea of folding up a portion of a customized card such as a square card to obtain partial information. Shinagawa [46] used specialized cards with invisible ink to obtain partial information by illuminating a black light with a cover. Compared to their studies, our study uses a standard deck of commercially available cards, i.e., we do not need to prepare a specialized deck of cards.

4   Our Simple AND Protocol Based on Half-Open Action

In this section, we present our efficient AND protocol working on a standard deck with the help of the half-open action.

In the sequel, we use the following four playing cards:

$$\boxed{3\clubsuit}\,\boxed{3\heartsuit}\,\boxed{9\clubsuit}\,\boxed{9\heartsuit},$$

although any four cards can be chosen as long as two of them have the same suit and the others have another same suit.

Table 2. The principle behind our proposed protocol

(a,b)	Sequence	Right of $\boxed{3\clubsuit}$	Left of $\boxed{9\heartsuit}$	Right of $\boxed{3\heartsuit}$	Left of $\boxed{9\clubsuit}$
$(0,0)$	$\boxed{3\clubsuit}\boxed{9\heartsuit}\boxed{3\heartsuit}\boxed{9\clubsuit}$	$\boxed{9\heartsuit}$	$\boxed{3\clubsuit}$	$\boxed{9\clubsuit}$	$\boxed{3\heartsuit}$
$(0,1)$	$\boxed{3\clubsuit}\boxed{9\heartsuit}\boxed{9\clubsuit}\boxed{3\heartsuit}$	$\boxed{9\heartsuit}$	$\boxed{3\clubsuit}$	$\boxed{3\clubsuit}$	$\boxed{9\heartsuit}$
$(1,0)$	$\boxed{9\heartsuit}\boxed{3\clubsuit}\boxed{3\heartsuit}\boxed{9\clubsuit}$	$\boxed{3\heartsuit}$	$\boxed{9\clubsuit}$	$\boxed{9\clubsuit}$	$\boxed{3\heartsuit}$
$(1,1)$	$\boxed{9\heartsuit}\boxed{3\clubsuit}\boxed{9\clubsuit}\boxed{3\heartsuit}$	$\boxed{9\heartsuit}$	$\boxed{3\clubsuit}$	$\boxed{9\heartsuit}$	$\boxed{3\clubsuit}$

Alice and Bob hold their private bits $a \in \{0,1\}$ and $b \in \{0,1\}$, respectively. Our proposed protocol proceeds as follows.

1. Alice takes $\boxed{3\clubsuit}\boxed{9\heartsuit}$ and places a commitment to a based on the following encoding similar to (2):

$$\boxed{3\clubsuit}\boxed{9\heartsuit} = 0, \quad \boxed{9\heartsuit}\boxed{3\clubsuit} = 1.$$

Bob takes $\boxed{3\heartsuit}\boxed{9\clubsuit}$ and places a commitment to b in the same way as Alice, i.e., by focusing only on their numbers:

$$\boxed{3\heartsuit}\boxed{9\clubsuit} = 0, \quad \boxed{9\clubsuit}\boxed{3\heartsuit} = 1.$$

Thus, we have the following two commitments:

$$\underbrace{\boxed{?}\boxed{?}}_{[a]\{3\clubsuit,9\heartsuit\}}\quad\underbrace{\boxed{?}\boxed{?}}_{[b]\{3\heartsuit,9\clubsuit\}}.$$

Table 2 indicates the actual sequence of cards for each input. Let us focus on $\boxed{3_\clubsuit}$: observe that the suit of the card on the right of $\boxed{3_\clubsuit}$ has a suit $\clubsuit$ if and only if $a = b = 1$. Therefore, they can obtain only the value of $a \wedge b$ by checking the suit.

In the same way, the suits of the cards on the right of $\boxed{3_\heartsuit}$ and on the left of $\boxed{9_\clubsuit}$ and $\boxed{9_\heartsuit}$ determine the value of $a \wedge b$. See the third to sixth columns of Table 2. Our protocol uses this relationship to perform a secure computation of the logical AND function.

2. Apply a random cut to the sequence of four cards:

$$\underbrace{\boxed{?}\,\boxed{?}}_{[a]\{3\clubsuit,9\heartsuit\}}\ \underbrace{\boxed{?}\,\boxed{?}}_{[b]\{3\heartsuit,9\clubsuit\}}\ \rightarrow\ \left\langle\boxed{?}\boxed{?}\boxed{?}\boxed{?}\right\rangle\ \rightarrow\ \boxed{?}\boxed{?}\boxed{?}\boxed{?}.$$

Note that the relationship shown in Table 2 remains unchanged (despite the random cut).

3. Reveal the first card. Note that the revealed card should be one of $\boxed{3_\clubsuit}$, $\boxed{3_\heartsuit}$, $\boxed{9_\clubsuit}$, $\boxed{9_\heartsuit}$ with the equal probability (i.e., $1/4$), and hence, information about the input is never leaked.

4. If the revealed card is either $\boxed{3_\clubsuit}$ or $\boxed{3_\heartsuit}$, then apply the half-open action to its right card, namely the second card (because a '3' is placed on the right side of a clock or wristwatch). If it is either $\boxed{9_\clubsuit}$ or $\boxed{9_\heartsuit}$, then apply the half-open action to its left card, namely the fourth card (because a '9' is placed on the left side of a wristwatch). Alice and Bob obtain the value of $a \wedge b$ as follows:

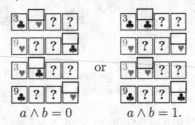

$$a \wedge b = 0 \qquad a \wedge b = 1.$$

This is our four-card AND protocol, which uses only one random cut and one half-open action. Although we believe that the correctness and security of our protocol are clear from the above description, we present their formal proofs in Sect. 6.2.

It should be noted that if one wants to use the first method shown in Fig. 2 to implement the half-open action in Step 4, the card revealed in Step 3 can be used as a cover (so that no additional card needs).

In the next sections, we formally define the half-open action and formally describe our protocol.

5   Formalizing Half-Open Action

In this section, we formalize a card-based protocol using a standard deck of playing cards such that the *half-open* action is allowed.

In the literature [13, 31], a deck was typically represented as a multiset over a symbol set, such as $[\clubsuit, \clubsuit, \heartsuit, \heartsuit, \heartsuit]$ and $\{1, 2, 3, 4\}$. Remember that our protocol proposed in Sect. 4 employs the fact that every card in a standard deck has a suit and its number. Therefore, we call a pair of a number and a suit, such as $(1, \clubsuit)$, $(1, \spadesuit)$, $(2, \heartsuit)$, and $(2, \diamondsuit)$, an *atomic card*. Following this, we denote a *standard deck* $\mathcal{D}$ by a multiset of atomic cards[2]. For an atomic card $c = (i, s)$, we denote its suit symbol by $\mathsf{ss}(c) := s$. For example, $\mathsf{ss}(1, \clubsuit) = \clubsuit$ and $\mathsf{ss}(2, \diamondsuit) = \diamondsuit$.

5.1 Notations

For a deck $\mathcal{D}$, a *face-up* card and a *face-down* card are represented as $\frac{c}{?}$ and $\frac{?}{c}$ for $c \in \mathcal{D}$, respectively. In addition, a face-down card to which the half-open action was applied is represented as $\frac{\mathsf{ss}(c)}{c}$. Given such a face-up, face-down, or "half-open" card, we denote its atomic card by $\mathsf{atom}(\frac{c}{?}) = \mathsf{atom}(\frac{?}{c}) = \mathsf{atom}(\frac{\mathsf{ss}(c)}{c}) = c$, and denote its visible symbol by $\mathsf{top}(\frac{c}{?}) = c$, $\mathsf{top}(\frac{?}{c}) = ?$, and $\mathsf{top}(\frac{\mathsf{ss}(c)}{c}) = \mathsf{ss}(c)$, respectively. We say that a d-tuple $\Gamma = (\alpha_1, \alpha_2, \ldots, \alpha_d)$ consisting of d cards from a standard deck $\mathcal{D}$ is a *sequence* if $[\mathsf{atom}(\alpha_1), \mathsf{atom}(\alpha_2), \ldots, \mathsf{atom}(\alpha_d)] = \mathcal{D}$.

We denote the set of all (possible) sequences from a standard deck $\mathcal{D}$ by

$$\mathsf{Seq}^{\mathcal{D}} := \{\Gamma \mid \Gamma \text{ is a sequence of } \mathcal{D}\}.$$

We extend the use of $\mathsf{top}(\cdot)$ to a sequence: given a sequence $\Gamma = (\alpha_1, \alpha_2, \ldots, \alpha_d)$, we write $\mathsf{top}(\Gamma) = (\mathsf{top}(\alpha_1), \mathsf{top}(\alpha_2), \ldots, \mathsf{top}(\alpha_d))$, and we call it the *visible sequence* of Γ. We also define the *visible sequence set* $\mathsf{Vis}^{\mathcal{D}}$ as

$$\mathsf{Vis}^{\mathcal{D}} := \{\mathsf{top}(\Gamma) \mid \Gamma \in \mathsf{Seq}^{\mathcal{D}}\}.$$

5.2 Protocols

We present a formal description of a protocol. As seen below, starting from an initial sequence, a protocol specifies an action to be applied to a current sequence step by step, depending on its internal state and the visible sequence.

A *protocol* (having a *finite state control* and a *table* on which a single sequence is put) is formally specified with a quadruple $\mathcal{P} = (\mathcal{D}, U, Q, A)$:

- $\mathcal{D}$ is a deck;
- $U \subseteq \mathsf{Seq}^{\mathcal{D}}$ is an *input set*;
- Q is a *state set* having an initial state $q_0 \in Q$ and a final state $q_{\mathsf{f}} \in Q$;
- $A : (Q \backslash \{q_{\mathsf{f}}\}) \times \mathsf{Vis}^{\mathcal{D}} \to Q \times \mathsf{Action}$ is an *action function*, where Action is the set of the following actions:
  - (turn, T) for $T \subseteq \{1, 2, \ldots, |\mathcal{D}|\}$;
  - (perm, π) for $\pi \in S_{|\mathcal{D}|}$, where S_i denotes the symmetric group of degree i;
  - $(\mathsf{shuf}, \Pi, \mathcal{F})$ for $\Pi \subseteq S_{|\mathcal{D}|}$ and a probability distribution $\mathcal{F}$ on Π. If $\mathcal{F}$ is uniform, we omit it and write this action as (shuf, Π);

---

[2] It should be noted that $\mathcal{D}$ can be any set of cards taken from 52 playing cards.

- (hopen, T) for $T \subseteq \{1, 2, \ldots, |\mathcal{D}|\}$;
- (hclose, T) for $T \subseteq \{1, 2, \ldots, |\mathcal{D}|\}$;

Given a current sequence $\Gamma = (\alpha_1, \alpha_2, \ldots, \alpha_{|\mathcal{D}|})$, each action in Action transforms the current sequence Γ into the next sequence Γ' as follows.

- (turn, T): $\Gamma' = (\beta_1, \beta_2, \ldots, \beta_{|\mathcal{D}|})$ such that

$$\beta_i = \begin{cases} \mathsf{swap}(\alpha_i) & \text{if } i \in T, \\ \alpha_i & \text{otherwise,} \end{cases}$$

for every i, $1 \le i \le |\mathcal{D}|$, where $\mathsf{swap}(\frac{c}{?}) = \frac{?}{c}$ and $\mathsf{swap}(\frac{?}{c}) = \frac{c}{?}$ for an atomic card c;
- (perm, π): $\Gamma' = (\alpha_{\pi^{-1}(1)}, \alpha_{\pi^{-1}(2)}, \ldots, \alpha_{\pi^{-1}(|\mathcal{D}|)})$;
- (shuf, $\Pi, \mathcal{F}$): Γ' resulting from applying action (perm, π) to Γ, where π is a permutation drawn from Π according to the probability distribution $\mathcal{F}$;
- (hopen, T): $\Gamma' = (\beta_1, \beta_2, \ldots, \beta_{|\mathcal{D}|})$ such that for every i, $1 \le i \le |\mathcal{D}|$,

$$\beta_i = \begin{cases} \frac{\mathsf{ss}(c_i)}{c_i} & \text{if } i \in T, \\ \alpha_i & \text{otherwise,} \end{cases}$$

where α_j for every $j \in T$ must be a face-down card $\alpha_j = \frac{?}{c_j}$.
- (hclose, T): $\Gamma' = (\beta_1, \beta_2, \ldots, \beta_{|\mathcal{D}|})$ such that for every i, $1 \le i \le |\mathcal{D}|$,

$$\beta_i = \begin{cases} \frac{?}{c_i} & \text{if } i \in T, \\ \alpha_i & \text{otherwise,} \end{cases}$$

where α_j for every $j \in T$ must be a "half-open" card $\alpha_j = \frac{\mathsf{ss}(c_j)}{c_j}$ (to which hopen has been applied).

6   Formal Description of Our Protocol

In this section, we show a formal description of our proposed AND protocol based on the computational model formalized in Sect. 5.

6.1   Pseudocode

The following is a pseudocode of our protocol, where we define $RC_{1,2,3,4} = \{(1\,2\,3\,4)^i \mid 1 \le i \le 4\}$. Its deck is $[(3, \clubsuit), (3, \heartsuit), (9, \clubsuit), (9, \heartsuit)]$.

input set:

$$\left\{ \left(\frac{?}{(3, \clubsuit)}, \frac{?}{(9, \heartsuit)}, \frac{?}{(9, \clubsuit)}, \frac{?}{(3, \heartsuit)} \right), \left(\frac{?}{(3, \clubsuit)}, \frac{?}{(9, \heartsuit)}, \frac{?}{(3, \heartsuit)}, \frac{?}{(9, \clubsuit)} \right), \right.$$

$$\left. \left(\frac{?}{(9, \heartsuit)}, \frac{?}{(3, \clubsuit)}, \frac{?}{(9, \clubsuit)}, \frac{?}{(3, \heartsuit)} \right), \left(\frac{?}{(9, \heartsuit)}, \frac{?}{(3, \clubsuit)}, \frac{?}{(3, \heartsuit)}, \frac{?}{(9, \clubsuit)} \right) \right\}$$

$(\mathsf{shuf}, \mathsf{RC}_{1,2,3,4})$

$(\mathsf{turn}, \{1\})$

if visible seq. $= ((3, \clubsuit), ?, ?, ?)$ or $((3, \heartsuit), ?, ?, ?)$ **then**

  $(\mathsf{hopen}, \{2\})$

  if visible seq. $= ((3, \clubsuit), \clubsuit, ?, ?)$ or $((3, \heartsuit), \heartsuit, ?, ?)$ **then** $a \wedge b = 1$

  else $a \wedge b = 0$

if visible seq. $= ((9, \clubsuit), ?, ?, ?)$ or $((9, \heartsuit), ?, ?, ?)$ **then**

  $(\mathsf{hopen}, \{4\})$

  if visible seq. $= ((9, \clubsuit), ?, ?, \clubsuit)$ or $((9, \heartsuit), ?, ?, \heartsuit)$ **then** $a \wedge b = 1$

  else $a \wedge b = 0$

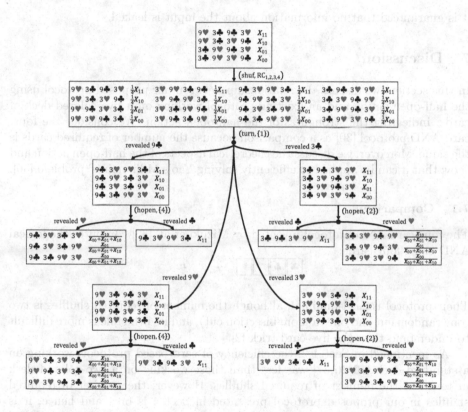

Fig. 7. The KWH-tree for the proposed AND protocol. Here, $(\mathsf{hopen}, \{i\})$ indicates that only the suit of the i-th card is to be disclosed.

6.2 Correctness and Security

We depict the *KWH-tree* [9,13,48] of our protocol in Fig. 7, by which we can confirm that the correctness and security of the protocol are satisfied.

The KWH-tree is a tree-like diagram that shows the transitions of possible sequences of cards along with their respective polynomials (or monomials) in a box, where actions to be applied to the sequence are appended to an edge. In the figure, the probability of $(a, b) = (x, y)$ is denoted by X_{xy}. A polynomial annotating a sequence in a box such as $1/4X_{00}$ represents the conditional probability that the current sequence is the one next to the polynomial, given what can be observed so far on the table. Because the sum of all polynomials in each box (except for the bottom-most boxes) is equal to

$$\sum_{x,y \in \{0,1\}} X_{xy},$$

it is guaranteed that no information about the input is leaked.

7   Discussion

In this section, we discuss the relationship between our proposed protocol using the half-open action and an existing protocol working on a two-colored deck of cards, indicating the strongness of the half-open action. We consider the four-card AND protocol [30] as a comparison because the number of required cards is the same. Moreover, we discuss the theoretical aspects of the half-open action and show that it can be applied to efficiently solving Yao's Millionaires' problem [50].

7.1   Comparison

The four-card AND protocol invented by Mizuki et al. [30] computes the logical AND using four cards:

$$\boxed{\clubsuit}\,\boxed{\clubsuit}\,\boxed{\heartsuit}\,\boxed{\heartsuit} \;\rightarrow\; a \wedge b.$$

Their protocol is card-minimal, although the number of required shuffles is two (one random cut and one random bisection cut), and its principle is more difficult to understand than the five-card trick [2].

As mentioned in Sect. 1.2, the efficiency of card-based protocols working on a standard deck [10,26,37] was less than those working on a two-colored deck in terms of the number of required shuffles. However, the number of required shuffles in our proposed protocol presented in Sect. 4 is one, and hence, it is more efficient than the existing AND protocol working on a two-colored deck [30] (except for the need of the half-open action), which is the first result of card-based cryptography history. This implies that the half-open action is useful, and we could obtain similar results for other protocols.

7.2   Theoretical Aspects

If we replace the turning-over action with the half-open action, we can regard a playing card with a card having only a suit because only a suit is always revealed on the front of the playing card. For example, face-down ♣ and A♣ can be regarded as identical as follows:

$$\boxed{?} \overset{\text{Turn}}{\to} \boxed{♣} = \boxed{?} \overset{\text{Half-open}}{\to} \boxed{♣}.$$

That is, we can regard a standard deck as a four-colored deck, and hence, any i-card protocol working on two-colored deck can be implemented using a standard deck for $i \le 26$. For example, to implement the five-card trick [2] introduced in Sect. 1.1, it suffices to use the half-open action thrice (instead of using the turning-over action). In summary, the half-open action solves the need for a customized deck to implement the existing protocols using red and black cards.

A more striking example is to solve Yao's Millionaire' problem [50]. The problem determines whether $a < b$ or not without revealing any information more than necessary for $a, b \in \{1, 2, \ldots, m\}$ and some natural number $m \ge 2$. Miyahara et al. [23] in 2020 proposed a card-based millionaire protocol working on a standard deck, although its efficiency is less than the one working on a two-colored deck. We show that it can be improved almost as efficiently as working on a two-colored deck in the next subsection.

7.3   Millionaire Protocol Using Half-Open

Our millionaire protocol requires only one more half-open action compared to the existing solution [23] working on a two-colored deck of cards. Our protocol proceeds as follows.

1. Alice holds m club cards and $m - 1$ heart cards. She places a sequence of $m + 1$ face-down cards in which the cards from the first to a-th are clubs and the remaining cards are hearts:[3]

Bob holds m spade cards and one diamond card. He places a sequence of $m + 1$ face-down cards below Alice's sequence, in which the b-th card is a diamond and the remaining cards are spades:

Observe that the suit of the card above the Bob's diamond card (i.e., the b-th card in Alice's sequence) determines only whether $a < b$ or not, i.e., the suit is a heart if $a < b$, and the suit is a club if $a \ge b$.

---

[3] It is sufficient for the suits to satisfy the above arrangement: the orders of the numbers (written on the cards) can be arbitrary.

2. Fix the two cards in each column to make $m + 1$ piles, and then apply a random cut to the piles (denoted by $< \cdots | \cdots | \cdots >$):

$$\left\langle \begin{array}{|c|c|} \hline ? \\ \hline ? \\ \hline \end{array} \begin{array}{|c|} \hline ? \\ \hline ? \\ \hline \end{array} \cdots \begin{array}{|c|} \hline ? \\ \hline ? \\ \hline \end{array} \right\rangle \rightarrow \begin{array}{cccc} \boxed{?}\boxed{?} \cdots \boxed{?} \\ \boxed{?}\boxed{?} \cdots \boxed{?} \end{array}.$$

Note that Alice's and Bob's sequences are randomly shifted, but their offsets are the same.

3. Reveal all the cards in Bob's sequence. Then, one diamond card should appear. Let $r \in \{1, \ldots, m+1\}$ denote the position of the revealed diamond card. Note that information about the value of b does not leak to Alice because r is a uniformly distributed random value.

4. Apply the half-open action to the r-th card in Alice's sequence. If a heart is revealed ♥, then we have $a < b$; otherwise ♣, we have $a \geq b$.

5. If $a \geq b$, we can further determine whether equality holds or not by applying the half-open action to the $(r + 1)$-st card in Alice's sequence (i.e., $(b + 1)$-st card). If a heart is revealed ♥, then we have $a = b$; otherwise ♣, we have $a > b$.

This is our millionaire protocol working on a standard deck of cards using the half-open action. The numbers of required cards and shuffles are $3m + 1$ and one, respectively. Remember that we use the half-open action in Step 4 of our protocol presented in Sect. 7.3. Note that if we just reveal the card in Step 4 instead of using the half-open action, information about the value of b leaks to Alice because she knows where she placed the revealed card in Step 1. We need a more complicated subprotocol if we do not use the half-open action.

The existing millionaire protocol working on a standard deck [23] employs such a complicated subprotocol. This protocol requires $4m$ cards and four shuffles. Therefore, our protocol is more efficient in terms of the numbers of cards and shuffles.

8   Conclusion

In this paper, we proposed a simple two-input AND protocol that requires only one shuffle, which is the trivial lower bound on the number of required shuffles[4]. The key is to consider a new action, called the half-open action, by which players can know only a suit of a face-down card. Surprisingly, this simple action contributes to the significant improvement on the numbers of required shuffles and cards, compared to the existing AND protocols.

Card-based cryptography has evolved steadily [27,28] as new concepts, techniques, and/or applications (such as the random bisection cut [33], permutation manipulation [1,6,7], zero-knowledge proofs for pencil puzzles [4,17,40–45], private operations [20,35,36,38], graph theoretical methods [22,25], information

---

[4] Single-shuffle protocols have attracted attention recently [16,47].

leakage analysis [29,48], new physical tools [18,24,34], and comparison with Turing complexity [3,12]) were found. We believe that the half-open action will open a new vista in this research field: This paper is a first step toward developing efficient protocols based on the half-open/close actions.

Acknowledgements. We thank the anonymous referees, whose comments have helped us improve the presentation of the paper. We thank Hiroto Koyama, who invented the half-open action, for his cooperation in preparing an earlier Japanese draft version of this paper. This work was supported in part by JSPS KAKENHI Grant Numbers JP21K11881 and JP18H05289.

References

1. Crépeau, C., Kilian, J.: Discreet solitary games. In: Stinson, D.R. (ed.) CRYPTO 1993. LNCS, vol. 773, pp. 319–330. Springer, Heidelberg (1994). https://doi.org/10.1007/3-540-48329-2_27
2. Boer, B.: More efficient match-making and satisfiability *The Five Card Trick*. In: Quisquater, J.-J., Vandewalle, J. (eds.) EUROCRYPT 1989. LNCS, vol. 434, pp. 208–217. Springer, Heidelberg (1990). https://doi.org/10.1007/3-540-46885-4_23
3. Dvořák, P., Koucký, M.: Barrington plays cards: the complexity of card-based protocols. In: Bläser, M., Monmege, B. (eds.) Theoretical Aspects of Computer Science. LIPIcs, vol. 187, pp. 26:1–26:17. Schloss Dagstuhl, Dagstuhl (2021). https://doi.org/10.4230/LIPIcs.STACS.2021.26
4. Gradwohl, R., Naor, M., Pinkas, B., Rothblum, G.N.: Cryptographic and physical zero-knowledge proof systems for solutions of Sudoku puzzles. Theor. Comput. Syst. **44**(2), 245–268 (2009). https://doi.org/10.1007/s00224-008-9119-9
5. Haga, R., Hayashi, Y., Miyahara, D., Mizuki, T.: Card-minimal protocols for three-input functions with standard playing cards. In: Batina, L., Daemen, J. (eds.) AFRICACRYPT 2022. LNCS, vol. 13503, pp. 448–468. LNCS, Springer, Cham (2022). https://doi.org/10.1007/978-3-031-17433-9_19
6. Ibaraki, T., Manabe, Y.: A more efficient card-based protocol for generating a random permutation without fixed points. In: Mathematics and Computers in Sciences and in Industry (MCSI), pp. 252–257 (2016). https://doi.org/10.1109/MCSI.2016.054
7. Ishikawa, R., Chida, E., Mizuki, T.: Efficient card-based protocols for generating a hidden random permutation without fixed points. In: Calude, C.S., Dinneen, M.J. (eds.) UCNC 2015. LNCS, vol. 9252, pp. 215–226. Springer, Cham (2015). https://doi.org/10.1007/978-3-319-21819-9_16
8. Kastner, J., et al.: The minimum number of cards in practical card-based protocols. In: Takagi, T., Peyrin, T. (eds.) ASIACRYPT 2017. LNCS, vol. 10626, pp. 126–155. Springer, Cham (2017). https://doi.org/10.1007/978-3-319-70700-6_5
9. Koch, A.: Cryptographic protocols from physical assumptions Ph.D. thesis, Karlsruhe Institute of Technology (2019). https://doi.org/10.5445/IR/1000097756
10. Koch, A., Schrempp, M., Kirsten, M.: Card-based cryptography meets formal verification. In: Galbraith, S.D., Moriai, S. (eds.) ASIACRYPT 2019. LNCS, vol. 11921, pp. 488–517. Springer, Cham (2019). https://doi.org/10.1007/978-3-030-34578-5_18
11. Koch, A., Schrempp, M., Kirsten, M.: Card-based cryptography meets formal verification. New Gener. Comput. **39**(1), 115–158 (2021). https://doi.org/10.1007/s00354-020-00120-0

12. Koch, A., Walzer, S.: Private function evaluation with cards. New Gener. Comput. **40**, 115–147 (2022). https://doi.org/10.1007/s00354-021-00149-9
13. Koch, A., Walzer, S., Härtel, K.: Card-based cryptographic protocols using a minimal number of cards. In: Iwata, T., Cheon, J.H. (eds.) ASIACRYPT 2015. LNCS, vol. 9452, pp. 783–807. Springer, Heidelberg (2015). https://doi.org/10.1007/978-3-662-48797-6_32
14. Koyama, H., Miyahara, D., Mizuki, T., Sone, H.: A secure three-input AND protocol with a standard deck of minimal cards. In: Santhanam, R., Musatov, D. (eds.) CSR 2021. LNCS, vol. 12730, pp. 242–256. Springer, Cham (2021). https://doi.org/10.1007/978-3-030-79416-3_14
15. Koyama, H., Toyoda, K., Miyahara, D., Mizuki, T.: New card-based copy protocols using only random cuts. In: ASIA Public-Key Cryptography Workshop, pp. 13–22. ACM, New York (2021). https://doi.org/10.1145/3457338.3458297
16. Kuzuma, T., Toyoda, K., Miyahara, D., Mizuki, T.: Card-based single-shuffle protocols for secure multiple-input AND and XOR computations. In: ASIA Public-Key Cryptography, pp. 1–8. ACM, New York (2022, to appear). https://doi.org/10.1145/3494105.3526236
17. Lafourcade, P., Miyahara, D., Mizuki, T., Robert, L., Sasaki, T., Sone, H.: How to construct physical zero-knowledge proofs for puzzles with a "single loop" condition. Theor. Comput. Sci. **888**, 41–55 (2021). https://doi.org/10.1016/j.tcs.2021.07.019
18. Lafourcade, P., Mizuki, T., Nagao, A., Shinagawa, K.: Light cryptography. In: Drevin, L., Theocharidou, M. (eds.) WISE 2019. IFIP AICT, vol. 557, pp. 89–101. Springer, Cham (2019). https://doi.org/10.1007/978-3-030-23451-5_7
19. Manabe, Y., Ono, H.: Card-based cryptographic protocols with a standard deck of cards using private operations. In: Cerone, A., Ölveczky, P.C. (eds.) ICTAC 2021. LNCS, vol. 12819, pp. 256–274. Springer, Cham (2021). https://doi.org/10.1007/978-3-030-85315-0_15
20. Manabe, Y., Ono, H.: Card-based cryptographic protocols with malicious players using private operations. New Gener. Comput. **40**, 67–93 (2022). https://doi.org/10.1007/s00354-021-00148-w
21. Marcedone, A., Wen, Z., Shi, E.: Secure dating with four or fewer cards. Cryptology ePrint Archive, Report 2015/1031 (2015)
22. Miyahara, D., Haneda, H., Mizuki, T.: Card-based zero-knowledge proof protocols for graph problems and their computational model. In: Huang, Q., Yu, Yu. (eds.) ProvSec 2021. LNCS, vol. 13059, pp. 136–152. Springer, Cham (2021). https://doi.org/10.1007/978-3-030-90402-9_8
23. Miyahara, D., Hayashi, Y., Mizuki, T., Sone, H.: Practical card-based implementations of Yao's millionaire protocol. Theor. Comput. Sci. **803**, 207–221 (2020). https://doi.org/10.1016/j.tcs.2019.11.005
24. Miyahara, D., Komano, Y., Mizuki, T., Sone, H.: Cooking cryptographers: Secure multiparty computation based on balls and bags. In: Computer Security Foundations Symposium, pp. 1–16. IEEE, New York (2021). https://doi.org/10.1109/CSF51468.2021.00034
25. Miyamoto, K., Shinagawa, K.: Graph automorphism shuffles from pile-scramble shuffles. New Gener. Comput. **40**, 199–223 (2022). https://doi.org/10.1007/s00354-022-00164-4
26. Mizuki, T.: Efficient and secure multiparty computations using a standard deck of playing cards. In: Foresti, S., Persiano, G. (eds.) CANS 2016. LNCS, vol. 10052, pp. 484–499. Springer, Cham (2016). https://doi.org/10.1007/978-3-319-48965-0_29
27. Mizuki, T.: Preface: special issue on card-based cryptography. New Gener. Comput. **39**(1), 1–2 (2021). https://doi.org/10.1007/s00354-021-00127-1

28. Mizuki, T.: Preface: special issue on card-based cryptography 2. New Gener. Comput. **40**, 47–48 (2022). https://doi.org/10.1007/s00354-022-00170-6

29. Mizuki, T., Komano, Y.: Information leakage due to operative errors in card-based protocols. Inf. Comput., 1–15 (2022). https://doi.org/10.1016/j.ic.2022.104910,in press

30. Mizuki, T., Kumamoto, M., Sone, H.: The five-card trick can be done with four cards. In: Wang, X., Sako, K. (eds.) ASIACRYPT 2012. LNCS, vol. 7658, pp. 598–606. Springer, Heidelberg (2012). https://doi.org/10.1007/978-3-642-34961-4_36

31. Mizuki, T., Shizuya, H.: A formalization of card-based cryptographic protocols via abstract machine. Int. J. Inf. Secur. **13**(1), 15–23 (2013). https://doi.org/10.1007/s10207-013-0219-4

32. Mizuki, T., Shizuya, H.: Computational model of card-based cryptographic protocols and its applications. IEICE Trans. Fundam. **E100.A**(1), 3–11 (2017). https://doi.org/10.1587/transfun.E100.A.3

33. Mizuki, T., Sone, H.: Six-card secure AND and four-card secure XOR. In: Deng, X., Hopcroft, J.E., Xue, J. (eds.) FAW 2009. LNCS, vol. 5598, pp. 358–369. Springer, Heidelberg (2009). https://doi.org/10.1007/978-3-642-02270-8_36

34. Murata, S., Miyahara, D., Mizuki, T., Sone, H.: Public-PEZ cryptography. In: Susilo, W., Deng, R.H., Guo, F., Li, Y., Intan, R. (eds.) ISC 2020. LNCS, vol. 12472, pp. 59–74. Springer, Cham (2020). https://doi.org/10.1007/978-3-030-62974-8_4

35. Nakai, T., Misawa, Y., Tokushige, Y., Iwamoto, M., Ohta, K.: Secure computation for threshold functions with physical cards: power of private permutations. New Gener. Comput. **40**, 95–113 (2022). https://doi.org/10.1007/s00354-022-00153-7

36. Nakai, T., Misawa, Y., Tokushige, Y., Iwamoto, M., Ohta, K.: How to solve Millionaires' problem with two kinds of cards. New Gener. Comput. **39**(1), 73–96 (2021). https://doi.org/10.1007/s00354-020-00118-8

37. Niemi, V., Renvall, A.: Solitaire zero-knowledge. Fundam. Inf. **38**(1,2), 181–188 (1999), https://doi.org/10.3233/FI-1999-381214

38. Ono, H., Manabe, Y.: Card-based cryptographic logical computations using private operations. New Gener. Comput. **39**(1), 19–40 (2020). https://doi.org/10.1007/s00354-020-00113-z

39. Pass, R., Shelat, A.: A course in cryptography (2010). http://www.cs.cornell.edu/~rafael/

40. Robert, L., Miyahara, D., Lafourcade, P., Mizuki, T.: Card-based ZKP for connectivity: applications to Nurikabe, Hitori, and Heyawake. New Gener. Comput. **40**, 149–171 (2022). https://doi.org/10.1007/s00354-022-00155-5

41. Robert, L., Miyahara, D., Lafourcade, P., Libralesso, L., Mizuki, T.: Physical zero-knowledge proof and NP-completeness proof of Suguru puzzle. Inf. Comput. **285**, 1–14 (2022). https://doi.org/10.1016/j.ic.2021.104858

42. Ruangwises, S.: Two standard decks of playing cards are sufficient for a ZKP for Sudoku. New Gener. Comput. **40**, 49–65 (2022). https://doi.org/10.1007/s00354-021-00146-y

43. Ruangwises, S., Itoh, T.: Physical zero-knowledge proof for Numberlink puzzle and k vertex-disjoint paths problem. New Gener. Comput. **39**(1), 3–17 (2020). https://doi.org/10.1007/s00354-020-00114-y

44. Ruangwises, S., Itoh, T.: Physical zero-knowledge proof for Ripple Effect. Theor. Comput. Sci. **895**, 115–123 (2021). https://doi.org/10.1016/j.tcs.2021.09.034

45. Sasaki, T., Miyahara, D., Mizuki, T., Sone, H.: Efficient card-based zero-knowledge proof for Sudoku. Theor. Comput. Sci. **839**, 135–142 (2020). https://doi.org/10.1016/j.tcs.2020.05.036

46. Shinagawa, K.: Card-based cryptography with dihedral symmetry. New Gener. Comput. **39**(1), 41–71 (2021). https://doi.org/10.1007/s00354-020-00117-9

47. Shinagawa, K., Nuida, K.: A single shuffle is enough for secure card-based computation of any Boolean circuit. Discret. Appl. Math. **289**, 248–261 (2021). https://doi.org/10.1016/j.dam.2020.10.013

48. Takashima, K., Miyahara, D., Mizuki, T., Sone, H.: Actively revealing card attack on card-based protocols. Nat. Comput., 1–13 (2021, in press). https://doi.org/10.1007/s11047-020-09838-8

49. Ueda, I., Miyahara, D., Nishimura, A., Hayashi, Y., Mizuki, T., Sone, H.: Secure implementations of a random bisection cut. Int. J. Inf. Secur. **19**(4), 445–452 (2019). https://doi.org/10.1007/s10207-019-00463-w

50. Yao, A.C.: Protocols for secure computations. In: Foundations of Computer Science, pp. 160–164. IEEE Computer Society, Washington, DC (1982). https://doi.org/10.1109/SFCS.1982.88

Streaming Submodular Maximization with the Chance Constraint

Shufang Gong, Bin Liu(✉) ⓘ, and Qizhi Fang

School of Mathematical Sciences, Ocean University of China, Qingdao 266100,
Shandong, China
binliu@ouc.edu.cn

Abstract. Submodular optimization plays a significant role in combinatorial problems due to its diminishing marginal return property. Many artificial intelligence and machine learning problems can be cast as submodular maximization problems with applications in object detection, data summarization, and video summarization. In this paper, we consider the problem of monotone submodular function maximization in the streaming setting with the chance constraint. Using mainly the idea of guessing the threshold, we propose streaming algorithms and prove good approximation guarantees and computational complexity. In our experiments, we demonstrate the efficiency of our algorithm on synthetic data for the influence maximization problem and indicate that even if the strong restriction of chance constraint is imposed, we can still get a good solution. To the best of our knowledge, this is the first paper to study the problem of monotone submodular function maximization with chance constraint in the streaming model.

Keywords: Submodular maximization · Streaming algorithm · Chance constraint

1 Introduction

The submodular function has the property of diminishing marginal gains and can be used to describe many problems in the real world. The objective functions of some well-known combinatorial optimization problems including Max-Cut [26], Max-Dicut [8], Influence Maximization [19], Max-Coverage [20] and Max-Bisection [2] are submodular. Formally, given a ground set V, a non-negative set function $f : 2^V \rightarrow \mathbb{R}_{\geq 0}$ is called *submodular* iff for any subsets $S \subseteq T \subseteq V$ and element $e \in V \backslash T$, we have $f(S \cup \{e\}) - f(S) \geq f(T \cup \{e\}) - f(T)$. The submodular function f is *monotone* (*nondecreasing*) iff for any subsets $S \subseteq T$ we have $f(S) \leq f(T)$. The submodular function f is *normalized* iff

---

This work was supported in part by the National Natural Science Foundation of China (11971447, 11871442), and the Fundamental Research Funds for the Central Universities.

M. Li and X. Sun (Eds.): IJTCS-FAW 2022, LNCS 13461, pp. 129–140, 2022.
https://doi.org/10.1007/978-3-031-20796-9_10

$f(\emptyset) = 0$. Denote by $f_A(B) = f(A \cup B) - f(A)$ the marginal contribution of subset B to subset A.

The submodular maximization problems with constraints are generally NP-hard, e.g., submodular maximization with the cardinality, knapsack, matriod or routing constraints, etc [22,31]. Previous work has commonly studied the case where the probability of satisfying the constraint is 1. But such a requirement usually causes the solution to over-emphasize the worst-case scenario. To alleviate this, we naturally relax it to the case where the solution breaks the constraint with a low probability. More specifically, S satisfying the chance constraint is defined as $Pr[W(S) > B] \leq \alpha$ holds, where each element $e \in V$ has a random weight $W(e)$, α is a given failure probability, B is a given budget and $W(S) = \sum_{e \in S} W(e)$ is a linear function. In this paper, we mainly study that the weights of all element obey a uniform distribution with the same dispersion. Optimization problems with chance constraint are extensively used in finance and engineering owing to the uncertainties in price, demand, supply, energy availability, safety requirement etc [3,5,27]. Recently, the submodular problem with chance constraint attracts much attention [7,17,24].

Modern datasets derived every second from industry and science such as data summarization, sensor data and social networks have shown exponential growth. The amount of data is huge while the computer storage space is limited, then the traditional greedy methods are no longer suitable to solve streaming setting. Therefore, this motivates the research on the streaming model. In the streaming model, data arrives one by one and the algorithm needs make irrevocable decisions about whether to store the current data before the next data arrives. Badanidiyuru et al. [1] proposed four metrics to measure the quality of streaming algorithms, which are the number of passes over the dataset, the memory complexity, the query complexity per element and the approximation ratio.

Consequently, in the context of industrial science and massive data, it is imperative to study the problem of submodular maximization in the streaming setting with chance constraint. Our problem can be reformulated as

$$\max_{S \subseteq V} \ f(S)$$
$$\text{s.t.} \quad Pr[W(S) > B] \leq \alpha, \tag{1}$$

where $f : 2^V \to \mathbb{R}_{\geq 0}$ is a normalized monotone submodular function, each element $e \in V$ has a random weight $W(e)$, α is a given failure probability, B is a given budget and $W(S) = \sum_{e \in S} W(e)$ is a linear function.

Our Contribution. In this paper, we investigate the problem of maximizing monotone submodular with chance constraint in the streaming model. Inspired by Doerr et al. [7], we mainly focus on two distributions of weights of elements. One is that the weight of each element is uniformly and independently distributed in $[a - \delta, a + \delta]$, and the other is that the weight of each element is uniformly and independently distributed in $[a(e) - \delta, a(e) + \delta]$, $a(e)$ is related to element e and dispersion δ is a fixed constant. Clearly the former is a special case of the latter. For the first distribution, based on ideas from the Sieve-Streaming algorithm proposed by Badanidiyuru et al. [1], we design a single-pass $(1/2 - \varepsilon)$ approximation streaming algorithm with memory complexity $O(B/\varepsilon)$ and query complexity per element $O(\log B/\varepsilon)$. For the second distribution, by extending ideas from the DynamicMRT algorithm introduced by Huang et al. [15], we design a single-pass $(1/3 - \varepsilon)$ approximation streaming algorithm with memory complexity $O(B/\varepsilon)$ and query complexity per element $O(\log B/\varepsilon)$. To the best of our knowledge, this is the *first* paper to investigate the problem of submodular function maximization with chance constraint in the streaming model.

The rest of this paper is organized as follows. In Sect. 3, we introduce two tail inequalities and two surrogate functions related to chance constraint. In Sect. 4, we propose a streaming algorithm called UIIDW for uniform independently identically distribution weights. In Sect. 5, we design a streaming algorithm called UWSD for uniform weights with the same dispersion. In Sect. 6, some concluding remarks and further research are given. Due to page limit, the numerical experiments are placed in Appendix.

2   Related Work

In the following, we summarize part of the relevant theoretical work on submodular optimization, which provides an innovative guarantee for the research in this paper.

Submodular Function with the Different Constraints. The submodular maximization problems with constraints are generally NP-hard. The constraints that have been extensively studied in previous work are the cardinality constraint, the knapsack constraint and the matriod constraint, etc. In the following, we summarize part of the classical theoretical work on submodular optimization with different constraints. For maximizing a monotone submodular function with cardinality constraint, Nemhauser et al. [22] devised a simple discrete greedy algorithm to achieve the $1 - 1/e$ approximation, and furthermore, this result is shown to be tight [23]. As far as the knapsack constraint is concerned, Sviridenko [25] enumerated all subsets S of size 3 and augmented them using the greedy algorithm, eventually obtained the $1 - 1/e$ approximation. For the matriod constraint, a simple discrete greedy algorithm gives the $1/2$ approximation [11] and later Calinescu et al. [4] improved the approximation ratio to $1 - 1/e$ using a continuous greedy algorithm. In addition, Filmus et al. [8] demonstrated that there no exist $1 - 1/e + \varepsilon$ polynomial-time approximation algorithms via a reduction from the max-k-cover problem for any $\varepsilon > 0$.

Chance Constraint. The chance constraint involves random components and requires that the probability of an event breaking the constraint cannot exceed a given threshold value. The uncertainty comes from the fact that the weights of elements in the ground set are random. Under the chance constraint, the study of the objective function extends from linear function [29,30] to submodular function [7,12,24]. As a rule, even if we fully know the probability distribution of uncertain data in advance, the chance constraint is hard to evaluate efficiently. To address this issue, Doerr et al. [7] used the tail inequalities to construct easily computable surrogate functions to replace chance constraint $Pr[W(S) > B] \leq \alpha$. Chance constraint holds as long as the surrogate function holds. Note that the chance constraint is not equivalent to the surrogate function. With this idea, for Problem (1) in general setting, Doerr et al. [7] utilized the greedy algorithm to achieve arbitrarily close to the $1 - 1/e$ approximation. Later, Neumann et al. [24] presented an evolutionary multi-objective algorithm that could achieve the same approximation ratio in less time and experimental results also showed that there is a significant improvement in the quality of the solution compared to the greedy algorithm.

Streaming Model. In the streaming setting, for maximizing a monotone submodular function with the cardinality constraint, Badanidiyuru et al. [1] designed the first single-pass $(1/2 - \varepsilon)$ approximation streaming algorithm with memory complexity $O(k \log k/\varepsilon)$ and query complexity per element $O(\log k/\varepsilon)$ based on threshold and parallel ideas. Subsequently, Kazemi et al. [18] raised the lower bound on the optimal value estimation, improving the memory complexity to $O(k/\varepsilon)$ while keeping other parameters the same. In hardness result, Norouzi-Ford et al. [10] proved that the above approximation ratio has reached the optimum when the elements arrive in any given order and the memory complexity does not exceed $O(k \log k)$. As far as the knapsack constraint is concerned, Huang et al. [15] proposed the first single-pass $(1/3 - \varepsilon)$ approximation streaming algorithm. Later, Huang and Kakimura. [14] designed a single-pass $(0.4 - \varepsilon)$ approximation streaming algorithm. For maximizing a submodular function with p-matriod constraint, Chekuri et al. [6] designed a single-pass $\Omega(1/p)$ approximation streaming algorithm with memory complexity $O(k \log k)$, where k is an upper bound on the cardinality of the desired set. Later, for this problem, using the random sampling technology, Feldman et al. [9] proposed a streaming algorithm with $1/4p$ approximation, memory complexity $O(k)$, and query complexity per element $O(km/p)$ without having to scan all elements completely, where m represents the number of matroids needed to define p-matriod constraint. Further, Xu et al. [16,28] generalized the results of submodular optimization to non-submodular optimization in the streaming setting.

3   Preliminaries

We first introduce Chernoff bound and Chebyshev's inequality two tail inequalities. Later Doerr et al. [7] utilized them to give two surrogate functions of the chance constraint.

Lemma 1. [13] (Multiplicative Chernoff Bound). *Independent random variables* $U_1, U_2, \cdots, U_m$ *satisfy* $U_i \in [0,1]$ *for any* $i = 1, 2, \cdots, m$. *We denote* $U = \sum_{i=1}^{n} U_i$. *When* $\varepsilon \in [0,1]$, *it holds*

$$Pr[U \geq (1+\varepsilon)E[U]] \leq e^{-\frac{\varepsilon^2 E[U]}{3}}.$$

Lemma 2. [21] (One-Sided Chebyshev's Inequality). *Denote by U a random variable with expectation $E[U]$ and variance $Var[U]$. For any $\lambda > 0$, it holds*

$$Pr[U \geq E[U] + \lambda] \leq \frac{Var[U]}{Var[U] + \lambda^2}.$$

Lemma 3. [7] (Surrogate function based on Multiplicative Chernoff Bound). *For any $e \in V$, weight $W(e)$ is independently and uniformly distributed in $[a(e) - \delta, a(e) + \delta]$. For any $S \subseteq V$, if*

$$B - E[W(S)] \geq \sqrt{6\delta^2 |S| \ln(\frac{1}{\alpha})},$$

then $Pr[W(S) > B] \leq \alpha$.

Lemma 4. [7] (Surrogate function based on One-Sided Chebyshev's Inequality). *For any $e \in V$, weight $W(e)$ is independently and uniformly distributed in $[a(e) - \delta, a(e) + \delta]$. For any $S \subseteq V$, if*

$$B - E[W(S)] \geq \sqrt{\frac{(1-\alpha)|S|\delta^2}{3\alpha}},$$

then $Pr[W(S) > B] \leq \alpha$.

4   Streaming Algorithm: Uniform Independently Identically Distribution Weights (UIIDW)

In this section, we discuss the case that the weights $W(e)$ of all elements are uniformly and independently distributed in $[a - \delta, a + \delta]$ ($0 \leq \delta \leq a$) and propose the UIIDW algorithm along with the relevant detailed analysis. According to Lemma 3 and Lemma 4, we naturally define k^* to verify whether subset S satisfies the chance constraint. For Chernoff-based and Chebyshev-based surrogate functions, let $k^* = \max\{k \in \mathbb{Z}_+ | k + \frac{\sqrt{6\delta^2 k \ln(\frac{1}{\alpha})}}{a} \leq \frac{B}{a}\}$ and $k^* = \max\{k \in \mathbb{Z}_+ | k + \sqrt{\frac{(1-\alpha)k\delta^2}{3\alpha}} \leq \frac{B}{a}\}$ respectively. That is, if $|S| \leq k^*$, then S satisfies the chance constraint.

The main line of Algorithm 1 is: Set a suitable threshold and calculate the marginal contribution of the arriving element to the current set. If the element brings a gain exceed the threshold, then it is added to the current solution, otherwise the element is discarded. Specifically, we choose the threshold $\frac{f(OPT)}{2k^*}$ to evaluate the quality of the arriving element. But since $f(OPT)$ is unknown in advance, it needs to be guessed. We find that $\max_{e \in V} f(e) \leq f(OPT) \leq k^* \max_{e \in V} f(e)$, but $\max_{e \in V} f(e)$ is still unknown. We can solve it in Line 6 and the value of $\max_{e \in V} f(e)$ can be obtained at some iteration. Due to the need for analysis, the upper bound is set to $2k^* \max_{e \in V} f(e)$. For the lower bound, to improve the memory complexity, we take a maximum between the current maximum singleton value and the feasible function value which narrowes the guessing range of $f(OPT)$. Therefore, there exists a $v \in I = \{(1 + \varepsilon)^i | i \in \mathbb{Z}_+\}$ which satisfies $(1 - \varepsilon)f(OPT) \leq v \leq f(OPT)$. Then, run the algorithm in parallel for different guesses v of $f(OPT)$.

Denote e_i as the i-th arrived element, m_i as the maximum singleton value among the first i elements. From the UIIDW algorithm, we can see that e_i is not taken into account by set S_v where $v > 2k^*m_i$. The next lemma tells us that even if e_i is taken into account, it will be discarded due to its low marginal contribution. Therefore, we can assume that each S_v is checked from the beginning of data streaming. Due to page limit, the details of proofs in this section are provided in Appendix.

Algorithm 1. UIIDW

Require: the ground set V, the monotone submodular function f, the parameter
  $\varepsilon \in (0, 1)$, α, a, δ
1: Chernoff-based:
  $k^* = \max\{k \in \mathbb{Z}_+ | k + \frac{\sqrt{6\delta^2 k \ln(\frac{1}{\alpha})}}{a} \leq \frac{B}{a}\}$
  (or Chebyshev-based:
  $k^* = \max\{k \in \mathbb{Z}_+ | k + \sqrt{\frac{(1-\alpha)k\delta^2}{3\alpha}} \leq \frac{B}{a}\}$)
2: $I \leftarrow \{(1 + \varepsilon)^i | i \in \mathbb{Z}_+\}$
3: For each $v \in I, S_v \leftarrow \phi$
4: $m_0 \leftarrow 0, LB \leftarrow 0$
5: **while** element e_i is arriving **do**
6:     $m_i \leftarrow \max\{m_{i-1}, f(\{e_i\})\}$
7:     $LB \leftarrow \max\{m_i, LB\}$
8:     $I_i \leftarrow \{(1 + \varepsilon)^j | \frac{LB}{1+\varepsilon} \leq (1 + \varepsilon)^j \leq 2k^*m_i\}$
9:     Delete all S_v such that $v \notin I_i$
10:     **for** $v \in I_i$ **do**
11:         **if** $f(S_v \cup \{e_i\}) - f(S_v) \geq \frac{v}{2k^*}$ and $|S_v| < k^*$ **then**
12:             $S_v \leftarrow S_v \cup \{e_i\}$ and $LB \leftarrow \max\{LB, f(S_v)\}$
13:         **end if**
14:     **end for**
15: **end while**
16: return $S \leftarrow \arg\max_{v \in I_n} f(S_v)$

Lemma 5. *For any* $i \in \{1, 2, \cdots, n\}$, *if* $v > 2k^*m_i$, *then* e_i *always satisfies* $f(\{e_i\}) < \frac{v}{2k^*}$.

By Lemma 5, we get the conclusion that $f(S_v \cup \{e_i\}) - f(S_v) = f_{S_v}(\{e_i\}) \leq f(\{e_i\}) < \frac{v}{2k^*}$ when $v > 2k^*m_i$. Denote S^* as the optimal solution and OPT as the optimal value of Problem (1). In the following, we argue that there is an upper bound on $|S^*|$.

Lemma 6. *When* $\alpha \leq \frac{1}{2}$, *we have* $|S^*| \leq \frac{B}{a}$.

Note that α is a small positive real number and is usually set to $\alpha \leq 0.1$. Hence, $|S^*| \leq \frac{B}{a}$ is holds.

Lemma 7. *In the UIIDW algorithm, the memory complexity is $O(\frac{k^*}{\varepsilon})$ and the query complexity per element is $O(\frac{\log k^*}{\varepsilon})$ where $\varepsilon \in (0,1)$.*

Observe easily that there exists $\widehat{v} \in I_n \cap [(1-\varepsilon)f(S^*), f(S^*)]$. Denote $S_{\widehat{v}}$ as the set corresponding to $\widehat{v}$ at the end of the UIIDW algorithm.

Theorem 1. *For any $e \in V$, $W(e)$ is independently and uniformly distributed in $[a - \delta, a + \delta]$ $(0 \leq \delta \leq a)$, $\varepsilon \in (0,1)$. When $B \to +\infty$, the UIIDW algorithm satisfies the following properties.*

- *It outputs a feasible set and achieves almost the $\frac{1}{2}$ approximation.*
- *It requires one pass, memory complexity $O(\frac{k^*}{\varepsilon})$ and query complexity per element $O(\frac{\log k^*}{\varepsilon})$ where $k^* = \max\{k \in \mathbb{Z}_+ | k + \frac{\sqrt{6\delta^2 k \ln(\frac{1}{\alpha})}}{a} \leq \frac{B}{a}\}$ (Chernoff-based) or $k^* = \max\{k \in \mathbb{Z}_+ | k + \sqrt{\frac{(1-\alpha)k\delta^2}{3\alpha}} \leq \frac{B}{a}\}$ (Chebyshev-based).*

According to the above results, we observe that Algorithm 1 needs less storage space and queries for small α and high δ.

5   Streaming Algorithm: Uniform Weights with the Same Dispersion

In this section, another case is discussed, where the weight $W(e)$ of each element is uniformly and independently distributed in $[a(e) - \delta, a(e) + \delta]$ $(0 \leq \delta \leq \min_{e \in V} a(e)$ and $1 \leq \min_{e \in V} a(e))$. We propose the UWSD algorithm along with the relevant detailed analysis. The UWSD algorithm first performs pre-processing steps to remove singletons that do not meet the chance constraint. The main ideas of the UWSD algorithm are similar to the UIIDW algorithm. Furthermore, to make the following analysis easier, we try to adopt a fixed value $\widehat{B}$ to verify whether subset S satisfies the chance constraint. For Chernoff-based and Chebyshev-based surrogate functions, let $\widehat{B} = B - \sqrt{6\delta^2 B \ln(\frac{1}{\alpha})}$ and $\widehat{B} = B - \sqrt{\frac{(1-\alpha)\delta^2 B}{3\alpha}}$ respectively. That is, if $E[W(S)] \leq \widehat{B}$, it holds that S satisfies the chance constraint.

The main line of Algorithm 2 is: Set a suitable threshold and calculate the marginal contribution of the arriving element to the current set. If the element brings a gain exceed the threshold, then it is added to the current solution, otherwise the element is discarded. Specially, we choose the threshold $\frac{2f(OPT)}{3\widehat{B}}$ to evaluate the quality of the arriving element. The guessing process of $f(OPT)$ is similar to Sect. 4.

Due to page limit, the details of proofs in this section are provided in Appendix.

Algorithm 2. UWSD

Require: the set V, the monotone submodular function f, the parameter $\varepsilon \in (0,1)$,
  B, α, $a(e)$, δ
1: Chernoff-based:
  $\widehat{B} = B - \sqrt{6\delta^2 B \ln(\frac{1}{\alpha})}$
  (or Chebyshev-based:
  $\widehat{B} = B - \sqrt{\frac{(1-\alpha)\delta^2 B}{3\alpha}}$)
2: $I \leftarrow \{(1+\varepsilon)^i | i \in \mathbb{Z}_+\}$
3: For each $v \in I, S_v \leftarrow \phi$
4: $m_0 \leftarrow 0, LB \leftarrow 0$
5: **while** element e_i is arriving **do**
6:   **if** $\frac{a(e_i)+\delta-B}{2\delta} \leq \alpha$ **then**
7:     $m_i \leftarrow \max\{m_{i-1}, f(\{e_i\})\}$
8:     $LB \leftarrow \max\{m_i, LB\}$
9:     $I_i \leftarrow \{(1+\varepsilon)^j | \frac{LB}{1+\varepsilon} \leq (1+\varepsilon)^j \leq \frac{3m_i\widehat{B}}{2}\}$
10:    Delete all S_v such that $v \notin I_i$
11:    **for** $v \in I_i$ **do**
12:      **if** $\frac{f(S_v \cup \{e_i\})-f(S_v)}{E[W(e_i)]} \geq \frac{2v}{3\widehat{B}}$ and $E[W(S \cup \{e_i\})] \leq \widehat{B}$ **then**
13:        $S_v \leftarrow S_v \cup \{e_i\}$ and $LB = \max\{LB, f(S_v)\}$
14:      **end if**
15:    **end for**
16: **end if**
17: **end while**
18: **return** $\dot{S} \leftarrow \arg\max\{\max_{v \in I_n} f(S_v), m_n\}$

Lemma 8. *For any* $i \in \{1,2,\cdots,n\}$, *if* $v > \frac{3m_i\widehat{B}}{2}$, *then* e_i *always satisfies*
$\frac{f(\{e_i\})}{E[W(e_i)]} < \frac{2v}{3\widehat{B}}$.

By Lemma 8, we get the conclusion that $\frac{f(S_v \cup \{e_i\})-f(S_v)}{E[W(e_i)]} = \frac{f_{S_v}(e_i)}{E[W(e_i)]} \leq$
$\frac{f(\{e_i\})}{E[W(e_i)]} < \frac{2v}{3\widehat{B}}$ when $v > \frac{3m_i\widehat{B}}{2}$. Next, we can assume that each S_v is checked
from the beginning of data streaming. The following lemma shows that there is
an upper bound on the expectation of the weight of the optimal solution S^*.

Lemma 9. *When* $\alpha \leq \frac{1}{2}$, *we have* $E[W(S^*)] \leq B$.

Note that α is a small positive real number and is usually set to $\alpha \leq 0.1$.
Hence, $E[W(S^*)] \leq B$ is holds.

Lemma 10. *Let* S *be the any current set during the execution of the UWSD
algorithm and* v *is a guess of OPT corresponding to this* S. *It holds* $f(S) \geq \frac{2vE[W(S)]}{3\widehat{B}}$.

The following result is implied by the proof of Lemma 10.

Corollary 1. *For any current set* S *in the UWSD algorithm, if an element
satisfies threshold condition, i.e.,* $\frac{f(S \cup \{e\})-f(S)}{E[W(e)]} \geq \frac{2v}{3\widehat{B}}$, *then we have the fact that*
$f(S \cup \{e\}) \geq \frac{2v \cdot E[W(S \cup \{e\})]}{3\widehat{B}}$.

It is clear that there exists a $\widehat{v} \in I_n \cap [(1 - \varepsilon)f(S^*), f(S^*)]$ and the definition of $S_{\widehat{v}}$ is same as before. Denote S' as the current set of $S_{\widehat{v}}$ during the execution of the UWSD algorithm. It implies that $S' \subseteq S_{\widehat{v}}$ and we reach the following conclusion.

Lemma 11. *For the current set S' of $S_{\widehat{v}}$ in the UWSD algorithm, if an element does not satisfy the threshold condition, i.e.,$\frac{f(S' \cup \{e\}) - f(S')}{E[W(e)]} < \frac{2\widehat{v}}{3\widehat{B}}$, then it holds that $f_{S_{\widehat{v}}}(e) < \frac{2\widehat{v} \cdot E[W(e)]}{3\widehat{B}}$.*

If $e \in S^*$ but $e \notin S_{\widehat{v}}$, this implies that e does not satisfy the threshold condition or its addition would destroy the budget $\widehat{B}$. The latter is defined below.

Definition 1. *If an element $e \in S^*/S_{\widehat{v}}$ satisfies the following two conditions*

1. *$f(S \cup \{e\}) - f(S) \geq \frac{2\widehat{v}}{3\widehat{B}}E[W(e)]$,*
2. *$E[W(S \cup \{e\})] > \widehat{B}$ and $E[W(S)] \leq \widehat{B}$,*

we call e a negative element, where S is the current set corresponding to $\widehat{v}$ just before e arrives.

Lemma 12. *When $B \rightarrow +\infty$ and there are no negative elements, the UWSD algorithm outputs a feasible solution $\dot{S}$ and achieves almost the $\frac{1}{3}$ approximation.*

Lemma 13. *In the UWSD algorithm, the memory complexity is $O(\frac{\widehat{B}}{\varepsilon})$ and the query complexity per element is $O(\frac{\log \widehat{B}}{\varepsilon})$ where $\varepsilon \in (0, 1)$.*

Theorem 2. *For any $e \in V$, $W(e)$ is independently and uniformly distributed in $[a(e) - \delta, a(e) + \delta]$ ($0 \leq \delta \leq \min_{e \in V} a(e)$ and $1 \leq \min_{e \in V} a(e)$). $\varepsilon \in (0, 1)$. When $B \rightarrow +\infty$, the UWSD algorithm satisfies the next properties*

- *It outputs a feasible set and achieves almost the $\frac{1}{3}$ approximation.*
- *It requires one pass, memory complexity $O(\frac{\widehat{B}}{\varepsilon})$ space and query complexity per element $O(\frac{\log \widehat{B}}{\varepsilon})$ where $\widehat{B} = B - \sqrt{6\delta^2 B \ln(\frac{1}{\alpha})}$ (Chernoff-based) or $\widehat{B} = B - \sqrt{\frac{(1-\alpha)\delta^2 B}{3\alpha}}$) (Chebyshev-based).*

According to the above results, we observe that Algorithm 2 needs less storage space and queries for small α and high δ.

6   Conclusion

In this paper, we study streaming algorithms for the problem of monotone submodular function maximization with the chance constraint and obtain the following two main results.

1. For uniform independently identically distribution weights, we design a single-pass streaming $(1/2-\varepsilon)$-approximation algorithm called UIIDW with memory complexity $O(B/\varepsilon)$ and query complexity per element $O(\log B/\varepsilon)$.
2. For uniform weights with the same dispersion, we propose a single-pass streaming $(1/3 - \varepsilon)$-approximation algorithm called UWSD with memory complexity $O(B/\varepsilon)$ and query complexity per element $O(\log B/\varepsilon)$ since k^*, $\widehat{B}$ and B are of the same order of magnitude.

References

1. Badanidiyuru, A., Mirzasoleiman, B., Karbasi, A.: Streaming submodular maximization: massive data summarization on the fly. In: 20th ACM SIGKDD international conference on Knowledge discovery and data mining, pp. 671–680, Assiciation for Computing Machinery, New York, NY, USA (2014)
2. Bazgan, C., Gourves, L., Monnot, J.: Approximation with a fixed number of solutions of some multi objective maximization problems. J. Discrete Algorithms **22**, 19–29 (2013)
3. Berning, A.-W., Girard, A., Kolmanovsky, I.: Rapid uncertainty propagation and chance-constrained path planning for small unmanned aerial vehicles. Adv. Control Appl. Eng. Ind. Syst. **2**(1), e23 (2020)
4. Calinescu, G., Chekuri, C., Pál, M., Vondrák, J.: Maximizing a Submodular Set Function Subject to a Matroid Constraint (Extended Abstract). In: Fischetti, M., Williamson, D.P. (eds.) IPCO 2007. LNCS, vol. 4513, pp. 182–196. Springer, Heidelberg (2007). https://doi.org/10.1007/978-3-540-72792-7_15
5. Chekuri, C., Gupta, S., Quanrud, K.: Streaming Algorithms for Submodular Function Maximization. In: Halldórsson, M.M., Iwama, K., Kobayashi, N., Speckmann, B. (eds.) ICALP 2015. LNCS, vol. 9134, pp. 318–330. Springer, Heidelberg (2015). https://doi.org/10.1007/978-3-662-47672-7_26
6. Chekuri, C., Gupta, S., Quanrud, K.: Streaming algorithms for submodular function maximization. In: 42nd International Colloquium on Automata, Languages, and Programming, pp.318-330. Springer, Heidelberg (2015)
7. Doerr, B., Doerr, C., Neumann, A.: Optimization of chance-constrained submodular functions. In: 34th AAAI Conference on Artificial Intelligence. **34**, pp. 1460–1467. Association for the Advancement of Artifical Intelligence, New York, NY, USA (2020)
8. Feige, U.: A threshold of ln n for approximating set cover. J. ACM **45**(4), 634–652 (1998)
9. Feldman, M., Karbasi, A., Kazemi, E.: Do less, get more: streaming submodular maximization with subsampling. Advances in Neural Information Processing Systems, 31 (2018)
10. Feldman, M., Norouzi-Fard, A., Svensson, O.: The one-way communication complexity of submodular maximization with applications to streaming and robustness. In: 52nd Annual ACM SIGACT Symposium on Theory of Computing, pp. 1363–1374, Association for Computing Machinery, Chicago, IL, USA (2020)
11. Fisher, M.-L., Nemhauser, G.-L., Wolsey, L.-A.: An analysis of approximations for maximizing submodular set functions-II. In: Balinski, M.L., Hoffman, A.J. (eds.) Polyhedral combinatorics. Springer, Heidelberg (1978). https://doi.org/10.1007/BFb0121195

12. Frick, D., Sessa, P.-G., Wood, T.-A.: Exploiting structure of chance constrained programs via submodularity. Automatica **105**, 89–95 (2019)
13. Hagerup, T., Rub, C.: A guided tour of Chernoff bounds. Inf. Process. Lett. **33**(6), 305–308 (1990)
14. Huang, C.-C., Kakimura, N.: Improved streaming algorithms for maximizing monotone submodular functions under a knapsack constraint. Algorithms **83**(3), 879–902 (2021)
15. Huang, C.-C., Kakimura, N., Yoshida, Y.: Streaming algorithms for maximizing monotone submodular functions under a knapsack constraint. Algorithms **82**(4), 1006–1032 (2020)
16. Jiang, Y., Wang, Y., Xu, D.: Streaming algorithm for maximizing a monotone non-submodular function under d-knapsack constraint. Optimization Letters **14**(5), 1235–1248 (2020)
17. Joung, S., Lee, K.: Robust optimization-based heuristic algorithm for the chance-constrained knapsack problem using submodularity. Optimization Letters **14**(1), 101–113 (2020)
18. Kazemi, E., Mitrovic, M., Zadimoghaddam, M.: Submodular streaming in all its glory: tight approximation, minimum memory and low adaptive complexity. In: 36th International Conference on Machine Learning, pp. 3311–3320. International Machine Learning Society, Long Beach, California, USA (2019)
19. Kempe, D., Kleinberg, J., Tardos, E.: Maximizing the spread of influence through a social network. In: 9th ACM SIGKDD international conference on Knowledge discovery and data mining, pp. 137–146. Association for Computing Machinery, Washington, DC, USA (2003)
20. Khuller, S., Moss, A., Naor, J.-S.: The budgeted maximum coverage problem. Inf. Process. Lett. **70**(1), 39–45 (1999)
21. Marshall, A.-W., Olkin, I.: A one-sided inequality of the Chebyshev type. The Ann. Math. Stat. **31**, 488–491 (1960)
22. Nemhauser, G.-L., Wolsey, L. A., Fisher M L.: An analysis of approximations for maximizing submodular set functions-I. Math. program. **14**(1), 265–294 (1978)
23. Nemhauser, G.-L., Wolsey, L.-A.: Best algorithms for approximating the maximum of a submodular set function. Math. Oper. Res. **3**(3), 177–188 (1978)
24. Neumann, A., Neumann, F.: Optimising monotone chance-constrained submodular functions using evolutionary multi-objective algorithms. In: Bäck, T., et al. (eds.) PPSN 2020. LNCS, vol. 12269, pp. 404–417. Springer, Cham (2020). https://doi.org/10.1007/978-3-030-58112-1_28
25. Sviridenko, M.: A note on maximizing a submodular set function subject to a knapsack constraint. Oper. Res. Lett. **32**(1), 41–43 (2004)
26. Trevisan, L., Sorkin, G.-B., Sudan, M.: Gadgets, approximation, and linear programming. J. Comput **29**(6), 2074–2097 (2000)
27. Wang, B., Dehghanian, P., Zhao, D.: Chance-constrained energy management system for power grids with high proliferation of renewables and electric vehicles IEEE Trans. Smart Grid **11**(3), 2324–2336 (2019)
28. Wang, Y., Xu, D., Wang, Y.: Non-submodular maximization on massive data streams. J. Global Optim. **76**(4), 729–743 (2020)
29. Xie, Y., Neumann, A., Neumann, F.: Specific single-and multi-objective evolutionary algorithms for the chance-constrained knapsack problem. In: 22th Genetic and Evolutionary Computation Conference, pp. 271–279. Association for Computing Machinery, Cancún Mexico (2020)

30. Xie, Y., Harper, O., Assimi, H.: Evolutionary algorithms for the chance-constrained knapsack problem. In: 21th Genetic and Evolutionary Computation Conference, pp. 338–346. Association for Computing Machinery, Prague, Czech Republic (2019)
31. Zhang, H., Vorobeychik, Y.: Submodular optimization with routing constraints. In: 30th AAAI conference on artificial intelligence, pp. 819–826. Association for the Advancement of Artifical Intelligence, Phoenix, Arizona, USA (2016)

Colorful Graph Coloring

Zhongyi Zhang[✉] and Jiong Guo

School of Computer Science and Technology, Shandong University, Jinan, China
202015106@mail.sdu.edu.cn, jguo@sdu.edu.cn

Abstract. Given a simple graph G and a positive integer d, the Colorful Graph Coloring problem (CGC) asks for the minimum number of colors needed to color the "coloring elements" of G, such that for every "colorful element" of G, the coloring elements in its "neighborhood" have at least d colors. Both coloring elements and colorful elements of G can be vertices and edges, which means that there are four variants of CGC.

With both coloring and colorful elements being vertices, we use Vertex-Coloring Vertex-Colorful (VCVC) to denote the variant of CGC, which asks for the minimum number of colors needed to color the vertices such that for every vertex v, the vertices in the closed neighborhood of v are colored by at least d colors. The Vertex-Coloring Edge-Colorful variant (VCEC) colors the vertices such that every edge is incident to vertices of at least d colors. Clearly, this variant is meaningful only for $d \leq 2$ with simple graphs as input, and with $d = 2$ is equivalent to the classical Graph Coloring problem. The Edge-Coloring Vertex-Colorful variant (ECVC) demands for every vertex v that the edges incident to v are colored by at least d colors. Finally, the Edge-Coloring Edge-Colorful variant (ECEC) colors the edges such that the "closed neighborhood" of every edge contains d distinctly colored edges. The closed neighborhood of an edge e contains e and all edges sharing endpoints with e.

Motivated by the extensive research on Graph Coloring and the applications of Colorful Graph Coloring in resource allocation, we initialize the complexity study of VCVC, ECVC, and ECEC and achieve the following results. VCVC is polynomial-time solvable for $d \leq 2$ and becomes NP-hard for $d \geq 3$. We also present a parameterized algorithm for VCVC with treewidth as parameter. ECVC is NP-hard only in the case that d is set equal to the minimum degree of vertices $\delta(G)$. If $d \neq \delta(G)$, then ECVC is polynomial-time solvable. Based on this, we show that in the case of $d = \delta(G)$, we can compute a coloring with $d + 1$ colors in polynomial time, providing an absolute approximation with an additive term one. Moreover, we prove that ECEC is NP-hard with $d \geq 4$, and solvable in polynomial time with $d \leq 3$. Finally, we present integer linear programming formulations for the problems.

Keywords: Graph coloring · NP-hard · Absolute approximation · Parameterized algorithms

© The Author(s), under exclusive license to Springer Nature Switzerland AG 2022
M. Li and X. Sun (Eds.): IJTCS-FAW 2022, LNCS 13461, pp. 141–161, 2022.
https://doi.org/10.1007/978-3-031-20796-9_11

1 Introduction

The Graph Coloring problem, which given an undirected graph $G = (V, E)$ and an integer $k > 0$, asks for an assignment of the vertices in V to k colors, such that adjacent vertices have different colors, is one of the most prominent problems in Computer Science and Discrete Mathematics. If a graph has such a coloring, then G is k-colorable. The minimum value of k such that G is k-colorable is called the chromatic number of G. Most standard textbooks on graph theory have chapters on Graph Coloring [1,8]. Concerning its computational complexity, it is well-known that Graph Coloring is NP-hard for $k \geq 3$ [2] and inapproximable within a ratio of n^ε unless $P = NP$ [4].

A lot of variants of Graph Coloring have been introduced and studied, such as Edge Coloring, List Coloring, Total Coloring, etc. Edge Coloring seeks for the minimum number of colors needed to color the edges such that no two edges sharing an endpoint have the same color. Edge Coloring is NP-hard [2] but there is an absolute approximation algorithm with an additive term one [7].

Here, we study a new variant of Graph Coloring, called Colorful Graph Coloring (CGC), which given a simple graph G and a positive integer d, asks for the minimum number of colors needed to color the "coloring elements" of G, such that for every "colorful element" of G, the coloring elements in its "neighborhood" have at least d colors. Both coloring elements and colorful elements of G can be vertices and edges, which means that there are four variants of this problem.

With both coloring and colorful elements being vertices, we use Vertex-Coloring Vertex-Colorful (VCVC) to denote the variant of Colorful Graph Coloring, which asks for the minimum number of colors needed to color the vertices such that for every vertex v, the vertices in the closed neighborhood of v are colored by at least d colors. The Vertex-Coloring Edge-Colorful variant (VCEC) colors the vertices such that every edge is incident to vertices of at least d colors. Clearly, this variant is meaningful only for $d \leq 2$ with simple graphs as input, and with $d = 2$ is equivalent to the classical Graph Coloring problem. The Edge-Coloring Vertex-Colorful variant (ECVC) demands for every vertex v that the edges incident to v are colored by at least d colors. Finally, the Edge-Coloring Edge-Colorful variant (ECEC) colors the edges such that the "closed neighborhood" of every edge contains d distinctly colored edges. The closed neighborhood of an edge e contains e and all edges sharing endpoints with e.

CGC can have applications in the area of resource allocation [3,6]. Considering a city plan task, where given a set of locations and a set of facilities, we want to build a facility in every location, such that every location has as diverse facilities as possible in its nearby locations. By using vertices to denote locations and adding an edge between two vertices representing that the corresponding two locations are nearby, we can formulate the city plan task as VCVC.

Motivated by the extensive research on Graph Coloring and the applications of Colorful Graph Coloring in resource allocation, we initialize the complexity study of VCVC, ECVC and ECEC and achieve the following results: VCVC is polynomial-time solvable for $d \leq 2$ and becomes NP-hard for every $d \geq 3$. We also present a parameterized algorithm for VCVC with treewidth as parameter.

ECVC is NP-hard only in the case that d is set equal to the minimum degree of vertices $\delta(G)$. If $d \neq \delta(G)$, then ECVC is polynomial-time solvable. Based on this, we show that in the case of $d = \delta(G)$, we can compute a coloring with $d+1$ colors in polynomial time, providing an absolute approximation with an additive term one. Moreover, we prove that ECEC is NP-hard with $d \geq 4$, and solvable in polynomial time with $d \leq 3$. Finally, we present integer linear programming formulations for the problems.

Preliminary. We consider only undirected, simple graphs $G = (V, E)$ with $n = |V|$ and $m = |E|$. Let $N(v)$ and $N[v]$ be the open and closed neighborhoods of a vertex $v \in V$, respectively. Similarly, we use $N(e)$ to denote the "open" neighborhood of an edge $e \in E$, which contains the edges sharing endpoints with e, and $N[e]$ to denote the "closed" neighborhood of e, that is, $N[e] = N(e) \cup \{e\}$. Moreover, the degree of $v \in V$ (or $e \in E$) in G is set equal to $deg_G(v) = |N(v)|$ (or $deg_G(e) = |N(e)|$), and $\Delta(G) = \max_{v \in V} deg_G(v)$, $\delta(G) = \min_{v \in V} deg_G(v)$.

A coloring of the vertices in G is a function $c : V \to C$, where C is the set of colors. Here, $c(v)$ for $v \in V$ is the color assigned to v. The coloring of the edges and $c(e)$ for $e \in E$ can be defined analogously. The elements to be colored are called the coloring elements. The variants Vertex-Coloring Vertex-Colorful (VCVC) and Vertex-Coloring Edge-Colorful (VCEC) have vertices as coloring elements, while Edge-Coloring Vertex-Colorful (ECVC) and Edge-Coloring Edge-Colorful (ECEC) have edges as coloring elements. The "colorful elements" of VCVC and ECVC are the vertices and VCEC and ECEC have edges as colorful elements. We define the "coloring neighborhood" of the colorful elements for the four variants as follows. In VCVC, the coloring neighborhood of a vertex $v \in V$ is the closed neighborhood of v. In VCEC, the coloring neighborhood of an edge $e \in E$ contains exactly the two endpoints of e. In ECVC, the coloring neighborhood of a vertex v is the set of edges incident to v. In ECEC, the coloring neighborhood of an edge e is the closed neighborhood of e. Given a coloring c of the coloring elements, we use $D_G(y)$ for a colorful element y to denote the set of colors assigned by c to the coloring elements in the coloring neighborhood of y, and $d_G(y) = |D_G(y)|$. All four variants have an integer $d \geq 1$ as input, called the "diversity bound". Given a coloring c, a colorful element y is called d-colorful, if $d_G(y) \geq d$. A coloring c is d-colorful, if all colorful elements are d-colorful. The problem studied in this paper can be defined as follows.

Colorful Graph Coloring (CGC)
INPUT: $G = (V, E)$ and two integers $k \geq d \geq 1$.
OUTPUT: Is there a d-colorful coloring c for G with at most k colors?

Due to lack of space, some proofs are deferred to Appendix.

2   Vertex-Coloring Vertex-Colorful

We first show that Vertex-Coloring Vertex-Colorful (VCVC) is NP-complete if and only if $d \geq 3$ and then present a parameterized algorithm with respect to the treewidth. Recall that VCVC asks for a coloring of the vertices with at most k colors, such that all vertices are d-colorful.

Theorem 1. *VCVC with $d \leq 2$ is solvable in $O(m+n)$ time, and NP-complete for every $d \geq 3$.*

In the following, we present a parameterized algorithm for VCVC with treewidth tw as parameter.

Definition 1. *A tree decomposition of a graph $G = (V, E)$ consists of a tree T and a mapping function X. For each node a in T, $X(a)$ corresponds to a subset of V. For each edge $e = \{u, v\} \in E$, there exists a node a in T with $\{u, v\} \subseteq X(a)$. For each vertex $v \in V$, the nodes in $\{a \in T | v \in X(a)\}$ form a subtree of T. In a tree decomposition, $\max_{a \in T} |X(a)| - 1$ is called its width. The treewidth tw of a graph G is equal to the minimum width of all tree decompositions of G.*

As in many tree decomposition-based algorithms, we use the so-called nice tree decomposition [5].

Definition 2. *A nice tree decomposition is a special kind of tree decomposition, which has the following additional properties:*

1. *T is rooted at node r.*
2. *$X(r) = \emptyset$ and for every leaf l, $X(l) = \emptyset$.*
3. *Every node b in T has at most two children.*
4. *If b has two children b_1 and b_2, then $X(b) = X(b_1) = X(b_2)$, and b is called a union node.*
5. *If b has only one child b_1 then one of the following two conditions is true:*
   (a) *$X(b_1) \subset X(b)$ and $|X(b)| = |X(b_1)| + 1$ (b is called an introduce node).*
   (b) *$X(b) \subset X(b_1)$ and $|X(b_1)| = |X(b)| + 1$ (b is called a forget node).*

Theorem 2. *VCVC can be solved in $O(n \cdot tw^{4tw^2})$ time.*

Proof. If graph G has a tree decomposition of width tw, then it also has a nice tree decomposition of the same width. Given a tree decomposition, we can construct a nice tree decomposition with the same width in polynomial time.

We describe the dynamic programming algorithm for VCVC, based on a nice tree decomposition (T, X). For a node $a \in T$, we define $Y[a] = \bigcup_{b \in T(a)} X(b)$, where $T(a)$ denotes the subtree of T rooted at a, and $Y(a) = Y[a] \setminus X(a)$.

Given a set C of k colors, we define the states of a node $a \in T$ with $X(a) = \{b_1, b_2, \cdots, b_w\}$. A state is a vector of $2w + 1$ components, $S = (a, s_1, s_2, \cdots, s_{2w})$, where $s_i \in C$ for $1 \leq i \leq w$, and $s_i \subseteq C$ with $w < i \leq 2w$. Note that there are at most $k^w \cdot 2^{kw}$ different states for node a. We aim to compute the function f_a for every node a, mapping all states S of a to $\{0, 1\}$, where $f_a(S) = 1$ means that there exists a coloring of the vertices in $Y[a]$, such that all vertices in $Y(a)$ are d-colorful and for each $b_i \in X(a)$, b_i is colored with s_i and its neighbors in $Y[a]$ are colored with the colors in s_{w+i}. Then, G has a d-colorful coloring, if $f_r(S) = 1$ for at least one state S of the root r; otherwise, no such coloring exists. The function f_a is computed bottom-up.

Leaf node. Since $X(a) = \emptyset$ for each leaf node $a \in T$, there is only one state $S = (a)$ and we set $f_a(S) = 1$.

Introduce node. Consider an introduce node a with $X(a) = \{b_1, b_2, \cdots, b_{w+1}\}$, where its only child is denoted as a' with $X(a') = \{b_1, b_2, \cdots, b_w\}$. Let $Z = \{b_i \in X(a') | b_i$ is adjacent to $b_{w+1}\}$. We set $f_a(S) = 1$ where $S = (a, s_1, \cdots, s_{w+1}, s_{w+2}, \cdots, s_{2(w+1)})$ if there exists a state $S' = (a', s'_1, \cdots, s'_w, s'_{w+1}, \cdots, s'_{2w})$ of a' such that:

1. $f_{a'}(S') = 1$;
2. For $1 \le i \le w$, $s_i = s'_i$;
3. For each $b_i \in Z$, $s_{i+(w+1)} = s'_{i+w} \cup \{s_{w+1}\}$;
4. For each $b_i \notin Z$, $s_{i+(w+1)} = s'_{i+w}$;
5. $s_{2(w+1)} = (\bigcup_{b_i \in Z}\{s_i\}) \cup \{s_{w+1}\}$.

Forget node. Consider a forget node a with $X(a) = \{b_1, b_2, \cdots, b_w\}$, where its only child is denoted as a' with $X(a') = \{b_1, b_2, \cdots, b_{w+1}\}$. We set $f_a(S) = 1$ where $S = (a, s_1, \cdots, s_w, s_{w+1}, \cdots, s_{2w})$ if there exists a state $S' = (a', s'_1, \cdots, s'_{w+1}, s'_{w+2}, \cdots, s'_{2(w+1)})$ of a' such that:

1. $f_{a'}(S') = 1$;
2. For $1 \le i \le w$, $s_i = s'_i$;
3. For $w < i \le 2w$, $s_i = s'_{i+1}$;
4. $|s'_{2(w+1)}| \ge d$.

Union node. Consider a union node a with $X(a) = \{b_1, b_2, \cdots, b_w\}$, where its children are denoted as a' and a'' with $X(a') = X(a'') = X(a)$. We set $f_a(S) = 1$ where $S = (a, s_1, \cdots, s_w, s_{w+1}, \cdots, s_{2w})$ if there exist a state $S' = (a', s'_1, \cdots, s'_w, s'_{w+1}, \cdots, s'_{2w})$ of a' and a state $S'' = (a'', s''_1, \cdots, s''_w, s''_{w+1}, \cdots, s''_{2w})$ of a'' such that:

1. $f_{a'}(S') = f_{a''}(S'') = 1$;
2. For $1 \le i \le w$, $s_i = s'_i = s''_i$;
3. For $w < i \le 2w$, $s_i = s'_i \cup s''_i$.

To prove the running time, note that in the above algorithm, we use s_i with $w < i \le 2w$ in a state S to record the colors in the closed neighborhood of a vertex $b_i \in X(a)$. Thus, if in the bottom-up process, we have $|s_i| \ge d$ at a node a, meaning that b_i becomes d-colorful, then s_i can be replaced by a "colorful symbol" and ignored by further computation. Therefore, for a node a with $X(a) = \{b_1, b_2, \cdots, b_w\}$, we have at most $k^w \cdot (\sum_{i=0}^{d} \binom{k}{i})^w$ different states.

It is clear that a VCVC instance $(G = (V, E), k, d)$ with $d > \delta(G) + 1$ is a no-instance. We prove that for any instance with $d \le \delta(G) + 1$, $(d - 1) \cdot tw + 1$ colors suffice. Since s_i with $w < i \le 2w$ and $|s_i| \ge d$ is ignored, there are at most $(d - 1) \cdot w$ different colors in $\bigcup_{i=w+1}^{2w}\{s_i\}$ and at most w different colors in $\bigcup_{i=1}^{w}\{s_i\}$ in a state S. If $k \ge d \cdot tw + 1$, then for each introduce node, we can color the added vertex with a color other than all the colors in $\bigcup_{i=1}^{2w}\{s_i\}$. Then for each node a, $X(a) = \{b_1, b_2, \cdots, b_w\}$ will be colored with distinct colors whether adjacent or not. For each $v \in V$, v will be colored with a different color from all the previous neighbors, and the neighbors added later will be colored with a new color until $d(v) \ge d$, which means v is d-colorful. Since $|N[v]| = deg_G(v) + 1 \ge \delta(G) + 1 \ge d$, the coloring is d-colorful. Thus, the instances with $k \geq d \cdot tw + 1$ are yes-instances and we need only to consider the instances with $k \leq d \cdot tw$.

The running time of the algorithm is $O(n \cdot (k^w \cdot \sum_{i=0}^{d} \binom{k}{i}^w)^2)$ time. By $\sum_{i=0}^{d} \binom{k}{i} \leq k^d$, $w \leq tw + 1$, $d \leq \delta(G) + 1 \leq tw + 1$ and $k \leq d \cdot tw$, the algorithm runs in $O(n \cdot tw^{4tw^2})$ time.   □

3   Edge-Coloring Vertex-Colorful

The Edge-Coloring Vertex-Colorful (ECVC) variant asks for a coloring of the edges with k colors such that for every vertex v, the edges incident to v are colored by at least d colors. Clearly, $d \leq \delta(G)$, with $\delta(G)$ denoting the minimum degree. We show that ECVC is NP-hard only in the case $d = \delta(G)$ and otherwise polynomial-time solvable. Further, for the NP-hard case, we achieve an absolute approximation algorithm with additive term one. In both the polynomial-time algorithm and the approximation algorithm, we apply the absolute approximation algorithm for Edge Coloring by V. Vizing [7]. Given a graph $G = (V, E)$, Edge Coloring asks for the minimum number of colors needed to color the edges, such that no two adjacent edges have the same color. Clearly, at least $\Delta(G)$ many colors are needed, with $\Delta(G)$ being the maximum degree of vertices. The algorithm from [7] provides in cubic time a coloring with at most $\Delta(G) + 1$ colors.

Theorem 3. *ECVC with $d < \delta(G)$ is solvable in cubic time.*

Proof. Given an ECVC-instance $(G = (V, E), k, d)$ with $d < \delta(G)$ and $k \geq d$, we apply the following algorithm:

Step 1: Set $G_1 = G$, $i = 1$;
   While $\exists v : deg_{G_i}(v) > d + 1$ do
   Let e_i be an arbitrary edge $\{u, v\}$ incident to v;
   $G_{i+1} = G_i \setminus \{e_i\}; i = i + 1$;
   END While.
Step 2: Use the approximation algorithm by V. Vizing [7] to color the resulting graph G' from Step 1, such that no two adjacent edges have the same color. Since $\Delta(G') = d + 1$, the color set contains at most $d + 2$ colors $c_1, c_2, \cdots, c_{d+2}$.
Step 3: Suppose that there are l edges removed from Step 1. Add these edges back in the reversed order of their removals. That is, we start with G_{l+1}, add e_l to G_{l+1} to get G_l, and finally arrive at G. During this, we color each added edge and compute for each vertex $v \in V$ its unused color set $\varphi_{G_i}(v)$, that is, the set of colors not used by the edges incident to v in G_i. For $1 \leq i \leq l$, we assume that the edge $e_i = \{u, v\}$ was removed in Step 1 due to $deg_{G_i}(v) > d + 1$ and distinguish the following cases:
Case 1: $deg_{G_{i+1}}(u) \geq d + 1$: color e_i with an arbitrary color in $\{c_1, c_2, \cdots, c_d\}$;
Case 2: $\varphi_{G_{i+1}}(u) = \{c_{d+1}, c_{d+2}\}$: color e_i with an arbitrary color in $\{c_1, c_2, \cdots, c_d\}$;
Case 3: $deg_{G_{i+1}}(u) < d + 1$ and $\varphi_{G_{i+1}}(u) \neq \{c_{d+1}, c_{d+2}\}$: color e_i with an arbitrary color in $\varphi_{G_{i+1}}(u) \cap \{c_1, c_2, \cdots, c_d\}$.

Step 4: Recolor the edges colored with c_{d+1} or c_{d+2} with colors in $\{c_1, c_2, \cdots, c_d\}$:
Note that in a Edge Coloring, the edges of each two colors formed some disjoint cycles and paths [7]. Consider the subgraph formed by the edges with c_{d+1} and c_{d+2}, which consists of disjoint cycles $(v_0, s_0, v_1, \cdots, s_j, v_0)$ or paths $(v_0, s_0, v_1, \cdots, s_j, v_{j+1})$, where v_i are the vertices and s_i are the edges. For every $0 \leq i \leq j$, if $c_r \in \varphi_G(v_i)$ with $1 \leq r \leq d$, then recolor s_i with c_r, otherwise, recolor s_i with an arbitrary color in $\{c_1, c_2, \cdots, c_d\}$.

After Step 3 of the above algorithm, all edges are present and colored and Step 4 reduced the number of colors. Note that $\Delta(G') = d + 1$ and the approximation algorithm in Step 2 returns a coloring with at most $d + 2$ colors, $c_1, c_2, \cdots, c_{d+2}$. Observe that the edges added in Step 3 are colored with colors in $\{c_1, c_2, \cdots, c_d\}$. Thus, Step 4 affects only edges in G'. Next, we prove the resulting coloring is d-colorful. To this end, we prove first that the coloring after Step 3 is d-colorful. Hereby, we prove by an induction that for each $1 \leq i \leq l+1$, the following properties always hold for all vertices $v \in V$:

1. If $deg_{G_i}(v) < d + 1$, then $|D_{G_i}(v)| = deg_{G_i}(v)$.
2. If $deg_{G_i}(v) \geq d + 1$, then $|D_{G_i}(v)| \geq d + 1$ or $\varphi_{G_i}(v) = \{c_{d+1}, c_{d+2}\}$.

For $i = l + 1$, there is no vertex v with $deg_{G_{l+1}}(v) > d + 1$. If $deg_{G_{l+1}}(v) < d + 1$, then $|D_{G_{l+1}}(v)| = deg_{G_{l+1}}(v)$ is guaranteed by the correctness of the approximation algorithm for Edge Coloring [7]. By the same reason we also know $|D_{G_{l+1}}(v)| = d + 1$ for all vertices v with $deg_{G_{l+1}}(v) = d + 1$. Suppose the properties hold for G_{i+1}. We arrive at G_i by adding $e_i = \{u, v\}$ to G_{i+1}. We only have to examine the properties for u and v. Again, we assume e_i was removed in Step 1 due to $deg_{G_i}(v) > d + 1$. Then, $deg_{G_{i+1}}(v) \geq d + 1$ and by the induction assumption, $|D_{G_{i+1}}(v)| \geq d + 1$ or $\varphi_{G_{i+1}}(v) = \{c_{d+1}, c_{d+2}\}$. Since Step 3 only adds edges with colors in $\{c_1, c_2, \cdots, c_d\}$, it holds that $|D_{G_i}(v)| \geq d + 1$ or $\varphi_{G_i}(v) = \{c_{d+1}, c_{d+2}\}$. Concerning u, we assume $deg_{G_i}(u) \leq d + 1$, since, otherwise, it is handled in the same way as v. Then, $deg_{G_{i+1}}(u) < d + 1$, since $e_i = \{u, v\}$. By the induction assumption, $|D_{G_{i+1}}(u)| = deg_{G_{i+1}}(u) < d + 1$. Case 2 or 3 of Step 3 applies to e_i. If Case 2 applies, then $D_{G_{i+1}}(u) = \{c_1, c_2, \cdots, c_d\}$, $deg_{G_{i+1}}(u) = d$, and $deg_{G_i}(u) = d+1$. Clearly, $\varphi_{G_i}(u) = \{c_{d+1}, c_{d+2}\}$ and the properties hold. If Case 3 applies, $|D_{G_i}(u)| = |D_{G_{i+1}}(u)| + 1 = deg_{G_{i+1}}(u) + 1 = deg_{G_i}(u)$, meaning that the properties also hold.

Consider the properties for G_1, that is, the original graph G. We have $\delta(G) > d$. The property (2) holds for all vertices v: $|D_G(v)| \geq d + 1$ or $\varphi_G(v) = \{c_{d+1}, c_{d+2}\}$, meaning that all vertices are d-colorful.

We proceed then proving that Step 4 does not violate the d-colorful requirement. Recall that Step 4 only changes the colors of the edges, which are colored with c_{d+1} or c_{d+2} and thus, are in G'. These edges form disjoint cycles or paths. Let $(v_0, s_0, v_1, \cdots, s_j, v_0)$ or $(v_0, s_0, v_1, \cdots, s_j, v_{j+1})$ be such a cycle or path, where the edges are colored alternatively by c_{d+1} and c_{d+2}. After Step 3, all vertices on the cycle or path are d-colorful. By property (2), $|D_G(v_i)| \geq d+1$ and $|\varphi_G(v_i)| \leq 1$ for all $0 \leq i \leq j$. If $|\varphi_G(v_i)| = 0$, then v_i remains d-colorful after Step 4.

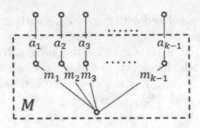

Fig. 1. A binding gadget.

If $\varphi_G(v_i) = \{c_r\}$ with $1 \leq r \leq d$, then Step 4 recolors the edge e_i with c_r and thus, v_i remains d-colorful. Note that $\varphi_G(v_i) \cap \{c_{d+1}, c_{d+2}\} = \emptyset$ for all vertices on the cycle or the internal vertices on the path. The endpoints of the paths must have $\varphi_G(v_i) = \{c_{d+1}\}$ or $\varphi_G(v_i) = \{c_{d+2}\}$ and have already all colors in $\{c_1, c_2, \cdots, c_d\}$ in their neighborhood. They remain d-colorful after Step 4.

Concerning the running time, since Step 2 is doable in cubic time, and Steps 1, 3, and 4 can be finished in $O(m + n)$ time, this algorithm runs in cubic time. □

Theorem 4. *ECVC with $d = \delta(G)$ is NP-complete.*

Next, we prove that in the case of $d = \delta(G)$, $d+1$ colors suffice for a d-colorful coloring, resulting in an absolute approximation with an additive term one.

Theorem 5. *ECVC with $d = \delta(G)$ can be approximated in cubic time with additive term one.*

4 Edge-Coloring Edge-Colorful

We prove that Edge-Coloring Edge-Colorful (ECEC) is NP-hard for every $d \geq 4$. If $d \leq 2$, then ECEC can be solved by similar approaches as in the proof of Theorem 1 for VCVC. We then prove that it is solvable in polynomial time with $d = 3$. Recall that ECEC colors edges and the colorful elements are edges.

Theorem 6. *ECEC is NP-complete for every $d \geq 4$.*

Proof. ECEC is clearly in NP. The hardness is proven by a reduction from Graph Coloring. Given a Graph Coloring instance $(G = (V, E), k)$ with $k \geq 4$, we construct an ECEC-instance $(G' = (V', E'), k, d = k)$. Assume the color set $C = \{c_1, c_2, \cdots, c_k\}$. For each vertex $v \in V$, we construct a vertex gadget, which consists of $deg_G(v)$ many "binding" gadgets and $deg_G(v)$ many "connecting" gadgets. The gadgets are connected alternatively in a cycle. Each binding gadget looks like an "extended" $(k-1)$-star, which consists of one center, $k-1$ degree-2 vertices, and $k-1$ leaves; see Fig. 1 for an illustration. The edges incident to the center, denoted as $m_1, m_2, \cdots, m_{k-1}$, are called shoulder edges, and the edges incident to the leaves, denoted as $a_1, a_2, \cdots, a_{k-1}$, are called arm

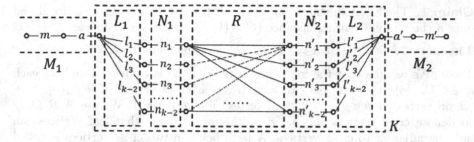

Fig. 2. A connecting gadget with two binding gadgets M_1 and M_2.

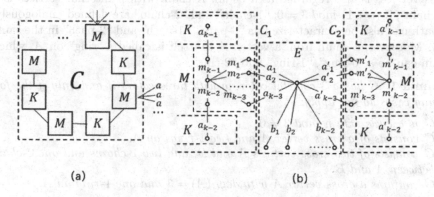

(a)   (b)

Fig. 3. (a) The cycle of binding and connecting gadgets for one vertex $v \in V$. M: binding gadgets and K: connecting gadgets. (b) Edge gadget for $e = \{u, v\}$.

edges. A connecting gadget connects two binding gadgets and has three parts, as shown in Fig. 2. In the middle is a complete bipartite graph R with $2(k - 2)$ vertices, $k - 2$ on each side. On the both sides of R are two size-$(k - 2)$ matchings. Then, a binding gadget is connected to a connecting gadget by adding $k - 2$ edges between one leaf of the binding gadget and the $k - 2$ vertices of one matching of the connecting gadget; see Fig. 2 for an illustration. One connecting gadget can connect two binding gadgets. The connecting and binding gadgets corresponding to one vertex $v \in V$ are organized in a cycle by adding a connecting gadget between two binding gadgets as shown in Fig. 3 (a). Note that each binding gadget has $k - 2$ "free" leaves, that is, leaves not connected to connecting gadgets. They are used to "bind" edge gadgets.

For each edge $e = \{u, v\} \in E$, we create a $(k - 2)$-star with one center and $k - 2$ leaves. And we "merge" the center with the free leaves of one binding gadget created for u. By merging two vertices, we delete these two vertices and add one new vertex, which is adjacent to the neighbors of the deleted vertices. Also, the free leaves of one binding gadget of v are also merged with this center, as shown in Fig. 3 (b). Note that each binding gadget of each vertex is "merged" with exactly one edge gadget. The construction needs clearly polynomial time.

Claim 1. The ECEC-instance $(G' = (V', E'), k, d = k)$ is a yes-instance if and only if the Graph Coloring instance $(G = (V, E), k)$ is a yes-instance. □

Theorem 7. *ECEC with $d = 3$ is solvable in polynomial time.*

Proof. We assume that the input graph $G = (V, E)$ is connected and for each $e \in E$, $deg_G(e) \geq 2$. We call a vertex v a "cross vertex" if $deg_G(v) \geq 3$, "chain vertex" if $deg_G(v) = 2$, "leaf vertex" if $deg_G(v) = 1$. We use $A, B, C \ldots$ to denote cross vertices and $a, b, c \ldots$ the others. Note that leaf vertices can only be adjacent to cross vertices. A path between two cross vertices is called a "chain", if all internal vertices are chain vertices. A single edge between two cross vertices is also regarded as a chain. A chain with n internal vertices is a "0-chain", if $(n + 1) \bmod 3 = 0$; "1-chain" and "2-chain" are defined analogously. A path can also be characterized as "0-path", "1-path" and "2-path" in the same way. Moreover, a chain from a vertex A to itself is called a "ring" on A, which again can be a "0-ring", "1-ring" and "2-ring".

Lemma 1. *G has a 3-colorful coloring with three colors if and only if the following cases do not apply:*

1. *G is a cycle with $n \bmod 3 \neq 0$.*
2. *G consists of a cross vertex A and two 1-rings on A.*
3. *G consists of two cross vertices A and B, and two 1-chains and one 0-chain between A and B.*
4. *G contains a cross vertex A with $deg_G(A) = 3$ and one 1-ring on A.*

The "only if"-direction is easy to prove. In case (1), if $n = 5$, then we need at least five colors to achieve 3-colorfulness; if $n \neq 5$, then at least four colors are needs. Cases (2), (3) and (4) require also at least four colors. All four cases can be examined in polynomial time.

In the following, we show that, in the absence of the above cases, we can find in polynomial time a 3-colorful coloring for G with three colors. We apply firstly a preprocessing to eliminate rings. For each 0-ring or 2-ring on a cross vertex A, we remove its vertices with the only exception with A, and add two leaf vertices as neighbors of A. For each 1-ring on A, we replace it by one leaf neighbor of A. The correctness of this preprocessing is obvious. Note that since cases (2) and (4) do not apply, we still have $deg_G(e) \geq 2$ for every $e \in E$.

Next, we use the following two steps to construct the 3-colorful coloring. First, we find in polynomial time a special subgraph G' of G. Then, we extend the coloring of G' to entire graph. A 3-colorful coloring is a beautiful coloring of G, if for each $v \in V$ with $deg_G(v) \geq 2$, the edges incident to v have at least two colors. The proofs of the next claims give the details of the two steps.

Claim 1. Given a connected subgraph G' of G, whose leaf vertices are also leaf vertices of G, and a beautiful coloring c of G' with three colors, we can construct in polynomial time a beautiful coloring of G with three colors.

Claim 2. We can always find a non-empty subgraph G' of G, which has a beautiful coloring with three colors, and whose leaf vertices are also leaf vertices in G. □

5   Integer Linear Programming

In the following, we present an integer programming formulation for VCVC. The formulations for ECVC and ECEC can be constructed in a similar way: create a coloring variable for every coloring element and construct the following constraints for the neighborhood of every colorful element.

Given a VCVC-instance $(G = (V, E), k, d)$, we first create for each vertex v a variable $c(v) \in \{1, 2, \ldots, k\}$ to store the color of v. Then, we consider the neighborhood of v, $N[v] = \{v, u_1, \ldots, u_l\}$ and create two variables $\alpha(x, y) \in \{0, 1\}$ and $\beta(x, y) \in \{0, 1\}$ for each pair of x, y in $N[v]$. Moreover, we create $l + 1$ variables $\gamma(x) \in \{0, 1\}$ for $x \in N[v]$. Next, we add the following constraints for each vertex v.

First, for each pair of x, y in $N[v]$, we have two constraints as follows:

$$\begin{cases} c(x) - c(y) + (k+1)\alpha(x,y) + (k+1)(1 - \beta(x,y)) - 1 \geq 0 \\ c(x) - c(y) + (k+1)(1 - \alpha(x,y)) + (k+1)(1 - \beta(x,y)) - 1 \geq 0 \end{cases}$$

These two constraints guarantee that $\beta(x, y) = 1$ implies $c(x) \neq c(y)$. The next $2l + 3$ constraints require that there are d vertices in $N[v]$, which are colored distinctly. With $\gamma(x) = 1$ for $x \in N[v]$, we mean that x is one of these d vertices. Thus, we have the follows constraint:

$$\sum_{x \in N[v]} \gamma(x) = d$$

The following $2l + 2$ constraints guarantee that those d vertices x with $\gamma(x) = 1$ are colored distinctly:

$$\begin{cases} \sum_{x \in N[v] \setminus \{v\}} \beta(x, v) + (l+1)(1 - \gamma(v)) \geq d - 1 \\ \sum_{x \in N[v] \setminus \{u_1\}} \beta(x, u_1) + (l+1)(1 - \gamma(u_1)) \geq d - 1 \\ \ldots \\ \sum_{x \in N[v] \setminus \{u_l\}} \beta(x, u_l) + (l+1)(1 - \gamma(u_l)) \geq d - 1 \\ \sum_{x \in N[v] \setminus \{v\}} \beta(x, v) - (l+1)\gamma(v) \leq 0 \\ \sum_{x \in N[v] \setminus \{u_1\}} \beta(x, u_1) - (l+1)\gamma(u_1) \leq 0 \\ \ldots \\ \sum_{x \in N[v] \setminus \{u_l\}} \beta(x, u_l) - (l+1)\gamma(u_l) \leq 0 \end{cases}$$

For a vertex $x \in N[v]$, which is not one of the d distinctly colored vertices, i.e., $\gamma(x) = 0$, the second half of these $2l + 2$ constraints makes sure that $\beta(x, y) = 0$ for each $y \in N[v]$ and $x \neq y$. If x is one of the distinctly colored vertices, i.e., $\gamma(x) = 1$, then the first half of these constraints guarantees the existence of other $d - 1$ vertices $y_1, y_2, \ldots, y_{d-1} \in N[v]$ satisfying $\beta(x, y_i) = 1$ for $1 \leq i \leq d - 1$, meaning that x is colored differently from $y_1, y_2, \ldots, y_{d-1}$. Adding together, we can conclude a satisfying assignment to $c(v)$, $\beta(x, y)$, $\gamma(x)$, $\alpha(x, y)$ corresponds to a coloring $c(v)$ of the vertices such that $N[v]$ for each vertex v contains d distinctly colored vertices.

6   Conclusion

We studied the classical complexity of three variants of the Graph Coloring problem. The next natural step is to consider the approximability of the problems. Moreover, the complexity of the problems on special graph classes is an interesting research direction for future work. Finally, as for the classical Graph and Edge Coloring, the lower and upper bounds on the color numbers for the CGC-variants might be another research topic.

A   Proof of Sect. 2

Proof (Proof of Theorem 1). For $d = 1$, VCVC needs only one color and is trivial. In the case of $d = 2$, we show that two colors suffice. First, we color the vertices with two colors arbitrarily. Then, if all vertices have two colors c_1 and c_2 in their closed neighborhoods, then we are done; otherwise, let v be a vertex whose neighbors are all colored with the same color as v, say c_1. We change the color of v to c_2. Clearly, v satisfies the 2-colorful requirement. Moreover, it is easy to observe that changing the color of v can only affect its neighbors, which have the color c_1. Thus, if a vertex u is 2-colorful before changing the color of v, then u remains 2-colorful afterwards. This means that each color-change increases the number of 2-colorful vertices by at least one and the color-change can be applied at most n times. The first step, coloring the vertices arbitrarily, is clearly doable in $O(n)$ time. And we can find all 2-colorful vertices in $O(m+n)$ time. Altogether, the algorithm needs $O(m + n)$ time.

VCVC is clearly in NP. Its NP-hardness for $d \geq 3$ is achieved by a reduction from Graph Coloring. Given a Graph Coloring instance $(G = (V, E), k)$ with $k \geq 3$, we construct a VCVC-instance $(G' = (V', E'), k, d = k)$ as follows. For each $v \in V$, we create a complete graph K_v with k vertices $v_0, v_1, \cdots, v_{k-1}$ in G'. For each edge $e = \{u, v\} \in E$, we add $k + 1$ vertices $e_0, e_1, \cdots, e_k$ to G', where $e_1, e_2, \cdots, e_k$ form a complete graph K_e and e_0 is connected to $e_1, e_2, \cdots, e_{k-3}$ by $k - 3$ edges. Furthermore, we add two edges $\{e_0, v_0\}$ and $\{e_0, u_0\}$ to G'. Then, $|V'| = k \cdot |V| + (k + 1) \cdot |E|$ and $|E'| = (k - 1) \cdot |E| + \frac{k(k-1)}{2} \cdot (|V| + |E|)$. The construction of G' is clearly in polynomial time.

Next, we prove the equivalence between the instances. If there is a proper coloring of G with k colors, then for each $v \in V$, we color the corresponding vertex $v_0 \in V'$ with the same color as v. The other $k - 1$ vertices in K_v are colored distinctly with the remaining $k-1$ colors. By doing so, the vertices in K_v are all k-colorful. Next, for each edge $e = \{u, v\}$, we color the corresponding vertex e_0 by a color, which is different from the colors of u and v. Then, the vertices $e_1, e_2, \cdots, e_{k-3}$ in the corresponding K_e are colored with the $k-3$ colors, which are different from the colors of u, v, and e_0, and e_{k-2}, e_{k-1}, e_k are colored with the colors of u, v, e_0, respectively. The vertices in K_e are colored distinctly and thus are k-colorful. The vertex e_0 has a color other than all its neighbors $u_0, v_0, e_1, e_2, \cdots, e_{k-3}$ and is also k-colorful. This means that (G', k, k) is a yes-instance of VCVC.

If (G', k, k) has a k-colorful coloring, then we color the vertices v in G with the same colors of their corresponding vertices v_0 in G'. For each edge $e = \{u, v\} \in E$, we can conclude from the fact that the corresponding vertex e_0 is k-colorful, all neighbors of e_0 are colored distinctly, meaning that u_0 and v_0 are colored with different colors. Thus, the coloring of G is proper, and uses k colors.   □

B   Proof of Sect. 3

Proof (Proof of Theorem 4). ECVC is clearly in NP. Its NP-hardness for $d = \delta(G)$ is achieved by a reduction from Edge Coloring. Given an Edge Coloring instance $(G = (V, E), k)$ with $k = \Delta(G)$, we construct an ECVC-instance $(G' = (V', E'), k, d = k)$ with $\delta(G') = k$ as follows. First, we construct a complete bipartite graph with $2k$ vertices $\{a_1, a_2, \cdots, a_k, b_1, b_2, \cdots, b_k\}$, where $\{a_1, a_2, \cdots, a_k\}$ forms one side and $\{b_1, b_2, \cdots, b_k\}$ the other side. Then, for each $v \in V$, we add a vertex v' to V', and for each $e = \{u, v\} \in E$, we add an edge $e' = \{u', v'\}$ to E'. Finally, for each $v \in V$, we add $r_v = k - deg_G(v)$ edges between v' and $\{a_1, a_2, \cdots, a_{r_v}\}$. Since $k = \Delta(G)$, $r_v \geq 0$. The construction of G' is clearly in polynomial time.

Next, we prove the equivalence between the instances. If there is a proper coloring of G with k colors $C = \{c_1, c_2, \ldots, c_k\}$, then for each $e \in E$, we color the corresponding edge $e' \in E'$ with the same color as e. For each $v \in V$, it is clearly that v has $deg_G(v)$ many different colors assigned to the edges incident to v in G, say $c_1, c_2, \cdots, c_{k-r_v}$. Then we color the r_v edges between v' and $\{a_1, a_2, \cdots, a_{r_v}\}$ distinctly with the colors in $C \setminus \{c_1, c_2, \cdots, c_{k-r_v}\}$. The vertices $v' \in V'$, which correspond to the vertices $v \in V$, are k-colorful. Finally, we color the edges between a_i and b_j with c_s with $s = [(i + j - 1) \bmod k] + 1$, so that $a_1, a_2, \cdots, a_k, b_1, b_2, \cdots, b_k$ are also k-colorful. This means that (G', k, k) is a yes-instance of ECVC.

If (G', k, k) has a k-colorful coloring, then we color the edges e in E with the same color as their corresponding edges e' in E'. For each vertex $v \in V$, we can conclude from the fact that the corresponding vertex v' is k-colorful, all the edges incident to v' are colored distinctly, meaning that all edges incident to v are colored with different colors. Thus, the coloring of G is proper and uses k colors.   □

Proof (Proof of Theorem 5). The following algorithm returns in polynomial time a d-colorful coloring with $d+1$ colors for a given graph $G = (V, E)$ with $\delta(G) = d$ and is based on the same idea as the algorithm in the proof of Theorem 3.

Step 1: Set $G_1 = G$, $i = 1$;
   While $\exists v : deg_{G_i}(v) > d$ do
   Let e_i be an arbitrary edge $\{u, v\}$ incident to v;
   $G_{i+1} = G_i \setminus \{e_i\}$;
   $i = i + 1$;
   END While.
Step 2: Use the approximation algorithm by V. Vizing [7] to color the resulting graph G' from Step 1, such that no two adjacent edges have the same color. Since $\Delta(G') = d$, the color set contains at most $d + 1$ colors.
Step 3: Suppose that there are l edges removed from Step 1. Add these edges back in the reversed order of their removals. That is, we start with G_{l+1}, add e_l to G_{l+1} to get G_l, and finally arrive at G. During this, we color each added edge and compute for each vertex $v \in V$ its unused color set $\varphi_{G_i}(v)$, that is, the set of colors not used by the edges incident to v in G_i. For $1 \leq i \leq l$, we assume that the edge $e_i = \{u, v\}$ was removed in Step 1 due to $deg_{G_i}(v) > d$ and distinguish the following cases:
Case 1: $deg_{G_{i+1}}(u) \geq d$: color e_i with an arbitrary color;
Case 2: $deg_{G_{i+1}}(u) < d$: color e_i with an arbitrary color in $\varphi_{G_{i+1}}(u)$.

Next, we prove the resulting coloring is d-colorful. As in the proof of Theorem 3, we prove that for each $1 \leq i \leq l+1$, the following properties always hold for all vertices $v \in V$:

1. If $deg_{G_i}(v) < d$, then $|D_{G_i}(v)| = deg_{G_i}(v)$.
2. If $deg_{G_i}(v) \geq d$, then $|D_{G_i}(v)| \geq d$.

For $i = l + 1$, there is no vertex v with $deg_{G_{l+1}}(v) > d$. If $deg_{G_{l+1}}(v) < d$, then $|D_{G_{l+1}}(v)| = deg_{G_{l+1}}(v)$ is guaranteed by the correctness of the approximation algorithm for Edge Coloring [7]. By the same reason we also know $|D_{G_{l+1}}(v)| = d$ for all vertices v with $deg_{G_{l+1}}(v) = d$. Suppose the properties hold for G_{i+1}. We arrive at G_i by adding $e_i = \{u, v\}$ to G_{i+1}. We only have to examine the properties for u and v. Again, we assume e_i was removed in Step 1 due to $deg_{G_i}(v) > d$. Then, $deg_{G_{i+1}}(v) \geq d$ and by the induction assumption, $|D_{G_{i+1}}(v)| \geq d$, it holds that $|D_{G_i}(v)| \geq d$. Concerning u, we assume $deg_{G_i}(u) \leq d$, since, otherwise, it is handled in the same way as v. Then, $deg_{G_{i+1}}(u) < d$, since $e_i = \{u, v\}$. By the induction assumption, $|D_{G_{i+1}}(u)| = deg_{G_{i+1}}(u) < d$. Case 2 of Step 3 applies to e_i. $|D_{G_i}(u)| = |D_{G_{i+1}}(u)| + 1 = deg_{G_{i+1}}(u) + 1 = deg_{G_i}(u)$, and the properties also hold.

Consider the properties for G_1, that is, the original graph G. We have $\delta(G) = d$. The property (2) holds for all vertices v, $|D_G(v)| \geq d$, meaning that all vertices are d-colorful.

As in the proof of Theorem 3, the algorithm runs clearly in cubic time and we have the following conclusion.   □

C   Proof of Sect. 4

Proof (Proof of Claim 1 of Theorem 6). Before showing the equivalence between the instances, we prove a property of the ECEC-instance.

First, we claim that if there is a k-colorful coloring for $G' = (V', E')$, then all arm edges of all binding gadgets corresponding to one vertex $v \in V$ are colored with the same color. Observe that each shoulder edge e in Fig. 1 has exactly $k-1$ adjacent edges, among them $k-2$ shoulder edges and one arm edge in the same binding gadget. Thus e and these adjacent edges should be colored distinctly. This means the arm edge has a color different from the $k-1$ colors of the shoulder edges. This is true for all shoulder edges, meaning that the arm edges of one binding gadgets have the same color. See Fig. 2 for a connecting gadget with two binding gadgets, where there are one arm edge a and one shoulder edge m from one binding gadget and one arm edge a' and one shoulder edge m' from the other. Suppose $c(a) = c_1$, $c(a') = c_2$, $c(m) = c_3$, $c(m') = c_4$. Note a is only adjacent to m and the $k-2$ edges $l_1, l_2, \cdots, l_{k-2}$ are between the binding gadget and the connecting gadget. Thus, to make a k-colorful, $l_1, l_2, \cdots, l_{k-2}$ need to be colored with $k-2$ distinct colors, that is, $C \setminus \{c_1, c_3\}$. Then, to make l_i with $1 \le i \le k-2$ k-colorful, the matching edge n_i has to be colored with the same color as m, that is, c_3. The same argument applies to a', m', l'_i and n'_i, that is, $c(n'_i) = c_4$.

Further, the $k-2$ edges in the complete bipartite graph R, which are adjacent to one matching edge n_i, are colored with $k-2$ distinct colors, where one has to be colored with c_1 and none is colored with c_3. The same can be concluded for the $k-2$ R-edges adjacent to n'_i, with one edge colored with c_2 and none with c_4. If $c_3 \ne c_4$, then all edges in R have colors from $C \setminus \{c_3, c_4\}$. By $k \ge 4$, there are at least two matching edges on both sides of R, meaning at least one edge l_i is colored with a color in $C \setminus \{c_1, c_3, c_4\}$. Then, for the corresponding matching edge n_i, the adjacent $k-2$ R-edges have only $k-3$ colors available, a contradiction to the conclusion that these edges are colored distinctly. Therefore, $c_3 = c_4$. Similarly, if $c_1 \ne c_2$, then there is an edge l_i colored with c_2. This means that the R-edges adjacent to n_i cannot by colored with c_2. As shown above, every edge n'_i has an adjacent edge in R which is colored with c_2. Then, there are $k-2$ R-edges colored with c_2. Since no two R-edges with the same color can be adjacent, these $k-2$ edges form a matching, meaning that n_i has an adjacent R-edge colored with c_2, a contradiction. Thus, $c_1 = c_2$. We can conclude that all arm edges of the vertex gadgets of one vertex $v \in V$ have the same color.

Suppose there is a k-coloring C of G. We color the arm edges of vertex $v \in V$ with $c(v)$. The shoulder edges in one binding gadget of v are colored distinctly with the colors in $C \setminus \{c(v)\}$. Hereby, we color the shoulder edges, whose adjacent arm edges are connected to the same connecting gadget, with the same color, saying c'. This is possible due to $k \ge 4$. Then, we color the edges n_i and n'_i for $1 \le i \le k-2$ with c' and l_i and l'_i distinctly with colors in $C \setminus \{c(v), c'\}$. Suppose that $C \setminus \{c(v), c'\} = \{c_{a_1}, c_{a_2}, \cdots, c_{a_{k-2}}\}$ and l_i and l'_i are colored with c_{a_i}. We color the edge in R, which is between n_i and n'_j, with $c(v)$ if $i = j$

and with c_{a_r} with $r = [(2i - j - 1) \bmod (k - 2)] + 1$ if $i \neq j$. Finally, the edges in the edge gadget for an edge $e = \{u, v\}$ are colored distinctly with the colors in $C \setminus \{c(u), c(v)\}$. The resulting coloring clearly is k-colorful.

Concerning the reversed direction, we already know that all arm edges of a vertex $v \in V$ are colored with the same color. We then color v with this color. The colorful edges b_i in the edge gadget for an edge $e = \{u, v\}$ guarantee that the arm edges of u and the arm edges of v have different colors, resulting in $c(u) \neq c(v)$ and a proper k-coloring.   □

Proof (Proof of Claim 1 of Theorem 7). We extend G' and c iteratively, each step adding at least one edge. The correctness and running time follow from the following description of a single step.

If $G' \neq G$, then there must be a vertex A, which is a cross vertex in G and whose incident edges are partially colored by c, since G is connected and every leaf vertex in G' is a leaf vertex in G. By the same argument and the definition of beautiful colorings, there are at least two edges incident to A which are assigned different colors by c. Let e and f be the two edges, and a be one edge incident to A which is not colored by c. We extend G' by adding at least a to G'. Hereby, we distinguish the following cases.

Case 1. There is an uncolored path P between A and a vertex $B \in G'$, which passes through a. An uncolored path means that all its internal vertices are not in G'. Note that it might be $A = B$. Since G' is connected and all leaf vertices in G' are leaf vertices in G, B must be a cross vertex in G and at least two edges incident to B are colored by c with two different colors; let g and h be these two edges and b denote the edge of P incident to B. See Fig. 4 for an illustration. If $\{c(e), c(f)\} = \{c(g), c(h)\}$, say $c(e) = c(g) = c_1$ and $c(f) = c(h) = c_2$, then we color path P with $(c_1, c_3, c_2, \ldots, c_3, c_2)$ if P is a 0-path; with $(c_3, c_1, c_2, \ldots, c_2, c_3)$ if P is 1-path; with $(c_3, c_2, c_1, \ldots, c_3, c_2)$ if P is 2-path. If $\{c(e), c(f)\} \neq \{c(g), c(h)\}$, say $c(e) = c_1$, $c(f) = c(g) = c_2$ and $c(h) = c_3$, then we color path P with $(c_3, c_2, c_1, \ldots, c_2, c_1)$ if P is a 0-path; with $(c_3, c_2, c_1, \ldots, c_1, c_3)$ if P is a 1-path; with $(c_3, c_1, c_2, \ldots, c_3, c_1)$ if P is a 2-path.

Fig. 4. There is an uncolored path between A and B.

The determination of path P is clearly doable in polynomial time and the coloring of P is trivial. It remains to show that the resulting subgraph G'' and the new coloring c' satisfy the precondition of the claim. Since the vertices added to G' are degree-2 vertices in G'', all leaf vertices in G'' are leaf vertices in G' and thus leaf vertices in G. The 3-colorfulness and conditions for beautiful colorings are clearly fulfilled by c'. The proofs of running time and correctness are trivial in all cases and thus are omitted in the following.

Case 2. There is no uncolored path P between A and vertices in G' passing through a. We distinguish further cases.

Case 2.1. The other endpoint of a is a leaf vertex in G. We color a with a color different from $c(e)$ and $c(f)$.

Case 2.2. The other endpoint x of a is not a leaf vertex. Note that the removal of a disconnects x and G'. Then, we remove G', a, and all leaf vertices from G and start a depth-first search from x. Consider the rightmost path P from x to a leaf in the resulting depth-first search tree. Let $B_1, \ldots, B_l$ be the cross vertices in this path according to the order of their occurrences from top to bottom. Let b be the edge between B_l and its father in this path. We color a with the color different from $c(e)$ and $c(f)$ and color P till B_l alternatively with $c(e)$, $c(f)$ and $c(a)$. Further subcases are considered.

Case 2.2.1. B_l has at least two leaf vertices as neighbors. Let g and h be the two corresponding edges. Then, add a, the subpath of P between x and B_l and B_l's two leaf neighbors to G' and color g and h with colors different from $c(b)$.

Case 2.2.2. B_l has only one leaf vertex as neighbor. Let g be the corresponding edge. B_l must have another incident edge h, which is part of a chain between B_l and a cross vertex B_i on P. Let p and q be the two edges on P, which are incident to B_i, and r be the edge of the chain, which is incident to B_i, as shown in Fig. 5.

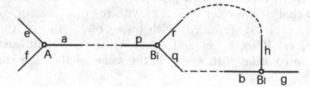

Fig. 5. g incident to a leaf vertex, h part of a chain to a cross vertex.

We add the subpath of P between x and B_l, the chain between B_i and B_l and the edges a and g to G'. If $c(b) \in \{c(p), c(q)\}$, say $c(b) = c(p) = c_1$ and $c(q) = c_2$, then we have the following alternatives: if the chain between B_i and B_l is a 0-chain, then color g with c_3 and the chain with $(c_2, c_1, c_3, \ldots, c_1, c_3)$; if 1-chain, then color g with c_2 and the chain with $(c_3, c_1, c_2, \ldots, c_2, c_3)$; if 2-chain, then color g with c_3 and the chain with $(c_2, c_3, c_1, \ldots, c_2, c_3)$. If $c(b) \notin \{c(p), c(q)\}$, say $c(b) = c_1$, $c(p) = c_2$ and $c(q) = c_3$, then we have the following possibilities: if the chain between B_i and B_l is a 0-chain, then color g with c_2 and the chain with $(c_3, c_2, c_1, \ldots, c_2, c_1)$; if 1-chain, then color g with c_3 and the chain with $(c_2, c_3, c_1, \ldots, c_1, c_2)$; if 2-chain, then color g with c_3 and the chain with $(c_2, c_1, c_3, \ldots, c_2, c_1)$.

Case 2.2.3. B_l has no leaf vertex as neighbor. Then, let g and h be two incident edges of B_l different from b. Clearly, both g and h are part of chains between B_l and some cross vertices on P. Let g be on a chain C_1 between B_l and B_j and h on a chain C_2 between B_l and B_i. See Fig. 6 for an illustration. Observe that there is a subgraph G_0 consisting of two cross vertices B_j and B_l, and three disjoint paths between them, $(z, \ldots, g)$, $(y, \ldots, b)$ and $(x, \ldots, q, r, \ldots, h)$. The path from A to B_i connects G' and G_0. We now add G_0 and this path to G'. We distinguish further cases. The coloring strategy of most cases is to firstly construct a beautiful coloring to G_0, and the coloring of the path between A and B_i is then trivial, since the combination of two beautiful colorings for two subgraphs obviously gives a beautiful coloring of the union of the two subgraphs. For the brevity of presentation, we give only one case as an example. If there are two 2-paths between B_l and B_j and the remaining one is a 0-path, then we can color the 2-path with $(c_1, c_2, c_3, \ldots, c_1, c_2)$ and $(c_2, c_3, c_1, \ldots, c_2, c_3)$, respectively, and the 0-path with $(c_3, c_2, c_1, \ldots, c_2, c_1)$. This is clearly a beautiful coloring. The only tricky case is that there are two 1-paths between B_l and B_j and the remaining one is a 0-path. Here, we need to first color the path from A to B_i, and then color the paths between B_j and B_l accordingly. The concrete coloring depends on which of the three paths is the 0-path. Suppose $c(p) = c_1$. In the case that $(x, \ldots, q, r, \ldots, h)$ is one 1-path, suppose $(z, \ldots, g)$ is the other 1-path. We color the 2-path $(q, \ldots, x, z, \ldots, g, h, \ldots, r)$ with $(c_2, c_3, c_1, \ldots, c_2, c_3)$; then the coloring of the path $(y, \ldots, b)$ is trivial. Finally, if $(x, \ldots, q, r, \ldots, h)$ is the 0-path, we color $(q, \ldots, x)$ with $(c_2, c_3, c_1, \ldots, c_3, c_1)$, $(r, \ldots, h)$ with $(c_3, c_2, c_1, \ldots, c_2, c_1)$, $(z, \ldots, g)$ with $(c_2, c_3, c_1, \ldots, c_1, c_2)$, $(y, \ldots, b)$ with $(c_3, c_1, c_2, \ldots, c_2, c_3)$ if both $(q, \ldots, x)$ and $(r, \ldots, h)$ are 0-chains. We color $(q, \ldots, x)$ with $(c_2, c_3, c_1, \ldots, c_1, c_2)$, $(r, \ldots, h)$ with $(c_3, c_2, c_1, \ldots, c_3, c_2)$, $(z, \ldots, g)$ with $(c_1, c_2, c_3, \ldots, c_3, c_1)$, $(y, \ldots, b)$ with $(c_3, c_1, c_2, \ldots, c_2, c_3)$ if $(q, \ldots, x)$ and $(r, \ldots, h)$ are a 1-chain and a 2-chain, respectively. The coloring with $i = j$ is the same as the case that $(q, \ldots, z)$ is a 0-chain.

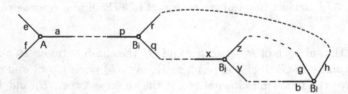

Fig. 6. Both g and h are part of chains to cross vertices.

In summary, we have corresponding colorings for all above cases, which are doable in polynomial time. The resulting new subgraphs satisfy the precondition

of the claim, and the resulting colorings are beautiful. Thus, the claim is correct.
□

Proof (Proof of Claim 2 of Theorem 7). Firstly, if G has no cross vertex, that is G is a circle. Since we have ruled out case (1), so G can only be C_{3k}, which obviously has a beautiful coloring.

Assume that G has only one cross vertex A. As we have eliminated all rings, so G can only be a star consisting of A and at least three leaf vertices. It is simple to color the edges with three different colors.

Fig. 7. There are two chains between A and B.

Fig. 8. There are three chains and one leaf vertex.

Assume that G has two cross vertices A and B. If there is only one chain between A and B, then each of A and B has at least two adjacent leaf vertices. We color the chain firstly, and then color these leaf edges with two colors other than the color of the end edge of the chain. If there are two chains between A and B, let $(e, \ldots, g)$ and $(f, \ldots, h)$ be these two chains, and l and r be two edges connecting leaf vertices, as shown in Fig. 7. If there are two 0-chains or a 1-chain and a 2-chain, then we have a C_{3k}, easy to color. If there are two 1-chains or a 0-chain and a 2-chain, we color l with c_1, $(e, \ldots, g, h, \ldots, f)$ with $(c_2, c_3, c_1, \ldots, c_2, c_3)$. If there are two 2-chains, we color l with c_1, r with c_2, $(e, \ldots, g)$ with $(c_2, c_3, c_1, \ldots, c_2, c_3)$, and $(f, \ldots, h)$ with $(c_3, c_1, c_2, \ldots, c_3, c_1)$. If there is a 0-chain and a 1-chain, we color l with c_1, r with c_2, $(e, \ldots, g)$ with $(c_2, c_3, c_1, \ldots, c_3, c_1)$, and $(f, \ldots, h)$ with $(c_3, c_1, c_2, \ldots, c_2, c_3)$. If there are at least three chains between A and B, then as shown in the proof of Claim 1, only the case that two 1-chains and a 0-chain cannot be colored with a beautiful coloring. Since we have ruled out Case (3), at least one leaf vertex or another chain should exist. If there is an edge l connecting A and a leaf vertex, then let $(e, \ldots, g)$ and $(f, \ldots, h)$ be the two 1-chains, $(p, \ldots, q)$ be the 0-chain, as shown in Fig. 8. We color l with c_1, $(e, \ldots, g)$ with $(c_2, c_3, c_1, \ldots, c_1, c_2)$, $(f, \ldots, h)$ with $(c_3, c_2, c_1, \ldots, c_1, c_3)$, and $(p, \ldots, q)$ with $(c_3, c_2, c_1, \ldots, c_2, c_1)$. If there is another chain between A and B, then this new chain forms a C_{3k} or a three-1-chain-structure with the three existing chains. Both C_{3k} and three-1-chain-structure have beautiful colorings.

In the following, we discuss the case that G has at least three cross vertices. Let A and B be the two cross vertices with the maximum number of vertex-disjoint paths between them in G. If there is only one path between A and B, then G is a tree. Hereby, we can w.l.o.g. assume A and B have the maximum distance among all pairs of cross vertices. Then, each of A and B has two leaf vertices as neighbors. We can use the same way as in the previous paragraph to color the path and the four leaf edges, giving the subgraph G'.

Suppose that there are two paths between A and B, as shown in Fig. 7. If each of A and B has a leaf vertex as neighbor, we can use the same coloring as in the case that G has only two cross vertices; otherwise, suppose B does not have a leaf vertex as neighbor. Let r be another edge adjacent to B. We apply the procedure used in the proof of Case 2.2 in Claim 1, removing r, starting a depth-first search from the other endpoint x of r, finding a subgraph F and coloring it. If $c(r) \in \{c(g), c(h)\}$, say $c(r) = c(g) = c_1$ and $c(h) = c_2$, we swap all the c_1 colors with c_3 colors in F, such that it is a beautiful coloring.

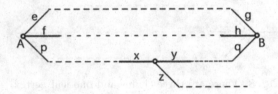

Fig. 9. z is a bridge.

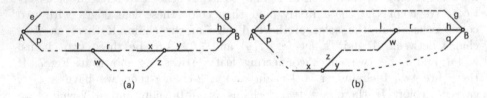

(a)   (b)

Fig. 10. z is not a bridge.

Finally, if there are at least three paths between A and B, we only need to discuss the case that there are two 1-paths and a 0-path. Let $(e, \ldots, g)$, $(f, \ldots, h)$ and $(p, \ldots, q)$ be the three paths. There must be another edge incident to this subgraph. Let z be this edge, and x and y be the two edges on the paths incident to z. If z is a bridge of G, as shown in Fig. 9, then we can apply the same procedure as in the previous paragraph to the bridge z. We can color the three paths with the same coloring as in the proof of Case 2.2.3 in Claim 1. If z is not a bridge, then there exists a path between two (A, B)-paths or between two vertices of the same (A, B)-path. Let w be the other end edge of this path, and l and r be the two edges on the path between A and B adjacent to w. There are two cases, namely, l and r on the same (A, B)-path as x and y as shown in Fig. 10 (a) and on another one as shown in Fig. 10 (b). We can firstly rule out the case that three paths exist between two cross vertices and do not form a (1-path,1-path,0-path)-structure. For the case in Fig. 10 (b), $(f, \ldots, l)$, $(r, \ldots, h)$, $(p, \ldots, x)$ and $(y, \ldots, q)$ can only be 0-paths or 1-paths. But no matter whether $(w, \ldots, z)$ is 0-path or 1-path, we have a C_{3k}. For the case in Fig. 10 (a), $(w, \ldots, z)$, $(r, \ldots, x)$, $(e, \ldots, g)$ and $(f, \ldots, h)$ should be 1-paths. We color $(r, \ldots, x)$ and $(e, \ldots, g)$ with $(c_2, c_3, c_1, \ldots, c_1, c_2)$, $(w, \ldots, z)$ and $(f, \ldots, h)$ with $(c_3, c_1, c_2, \ldots, c_2, c_3)$, $(p, \ldots, l)$ and $(y, \ldots, q)$ with $(c_1, c_2, c_3, \ldots, c_3, c_1)$, if both $(p, \ldots, l)$ and $(y, \ldots, q)$ are 1-paths. We color $(r, \ldots, x)$ and $(e, \ldots, g)$ with $(c_2, c_3, c_1, \ldots, c_1, c_2)$, $(w, \ldots, z)$ with $(c_3, c_1, c_2, \ldots, c_2, c_3)$, $(f, \ldots, h)$ with $(c_1, c_2, c_3, \ldots, c_3, c_1)$, $(p, \ldots, l)$ with $(c_3, c_2, c_1, \ldots, c_2, c_1)$, $(y, \ldots, q)$ with $(c_1, c_3, c_2, \ldots, c_1, c_3)$, if $(p, \ldots, l)$ and $(y, \ldots, q)$ are a 0-path and a 2-path, respectively.

In the case enumeration above, if any two cross vertices coincide, the coloring is the same as for the case that there is a 0-chain between them. Thus, we can always find such a subgraph. □

References

1. Diestel, R.: Graph Theory, 5th edn. Springer, Heidelberg (2016). https://doi.org/10.1007/978-3-662-53622-3
2. Garey, M.R., Johnson, D.S.: Computers and Intractability: A Guide to NP-Completeness. W. H. Freeman, San Francisco (1979)
3. Katoh, N., Ibaraki, T.: Resource Allocation Problems. The MIT Press, Cambridge (1988)
4. Lund, C., Yannakakis, M.: On the hardness of approximating minimization problems. In: STOC 1993, pp. 286–293 (1993)
5. Niedermeier, R.: Invitation to Fixed-Parameter Algorithms. Oxford Press, Oxford (2006)
6. Norton, M.S., Kelly, L.K.: Resource Allocation: Managing Money and People. Routledge, London (1997)
7. Vizing, V.G.: On an estimate of the chromatic class of a p-graph. Diskret. Analiz **3**, 23–30 (1964). (in Russian)
8. West, D.B.: Introduction to Graph Theory. Prentiss Hall, Englewood Cliffs (2000)

On the Transversal Number of Rank k Hypergraphs

Zhongzheng Tang[1] and Zhuo Diao[2(✉)]

[1] School of Science, Beijing University of Posts and Telecommunications,
Beijing 100876, China
tangzhongzheng@amss.ac.cn
[2] School of Statistics and Mathematics, Central University of Finance
and Economics, Beijing 100081, China
diaozhuo@amss.ac.cn

Abstract. For $k \geq 2$, let H be a hypergraph with rank k on n vertices and m edges. The transversal number $\tau(H)$ is the minimum number of vertices that intersect every edge. In this paper, the following conjecture is proposed: Is $\tau(H) \leq \frac{(k-1)m+1}{k}$? We prove the inequality in some special hypergraphs: (i) the inequality holds for $k = 2$ and $k = 3$. (ii) the inequality holds for the hypergraphs with the König Property. (iii) the inequality holds for the hypergraphs with maximum degree 2 and the extremal hypergraphs with equality holds are characterized.

Keywords: Transversal · Rank k · Maximum degree 2 · Extremal hypergraphs

1 Introduction

A hypergraph is a generalization of a graph in which an edge can join any number of vertices. A simple hypergraph is a hypergraph without multiple edges. Let $H = (V, E)$ be a simple hypergraph with vertex set V and edge set E. As for a graph, the order of H, denoted by n, is the number of vertices. The number of edges will be denoted by m. The rank is $r(H) = max_{e \in E}|e|$.

For each vertex $v \in V$, the degree $d(v)$ is the number of edges in E that contains v. We say v is an isolated vertex of H if $d(v) = 0$. Hypergraph H is k-regular if each vertex's degree is k $(d(v) = k, \forall v \in V)$. The maximum degree of H is $\Delta(H) = max_{v \in V} d(v)$. Hypergraph H is k-uniform if each edge contains exactly k vertices $(|e| = k, \forall e \in E)$. Hypergraph H is called linear if any two distinct edges have at most one common vertex. $(|e_1 \cap e_2| \leq 1, \forall e_1, e_2 \in E)$.

Let $k \geq 2$ be an integer. A cycle of length k, denoted as k-cycle, is a vertex-edge sequence $C = v_1 e_1 v_2 e_2 \cdots v_k e_k v_1$ with: $(1)\{e_1, e_2, \ldots, e_k\}$ are distinct edges

Supported by National Natural Science Foundation of China under Grant No.11901605, No.12101069, the disciplinary funding of Central University of Finance and Economics, the Emerging Interdisciplinary Project of CUFE, the Fundamental Research Funds for the Central Universities and Innovation Foundation of BUPT for Youth (500422309).

M. Li and X. Sun (Eds.): IJTCS-FAW 2022, LNCS 13461, pp. 162–175, 2022.
https://doi.org/10.1007/978-3-031-20796-9_12

of H. (2)$\{v_1, v_2, \ldots, v_k\}$ are distinct vertices of H. (3)$\{v_i, v_{i+1}\} \subseteq e_i$ for each $i \in [k]$, here $v_{k+1} = v_1$. We consider the cycle C as a sub-hypergraph of H with vertex set $\{v_i, i \in [k]\}$ and edge set $\{e_j, j \in [k]\}$. For any vertex set $S \subseteq V$, we write $H \setminus S$ for the sub-hypergraph of H obtained from H by deleting all vertices in S and all edges incident with some vertices in S. For any edge set $A \subseteq E$, we write $H \setminus A$ for the sub-hypergraph of H obtained from H by deleting all edges in A and keeping vertices. If S is a singleton set $\{s\}$, we write $H \setminus s$ instead of $H \setminus \{s\}$.

Given a hypergraph $H(V, E)$, a set of vertices $S \subseteq V$ is a vertex transversal if every edge has at least a vertex in S which means that $H \setminus S$ has no edges. The vertex transversal number is the minimum cardinality of a vertex transversal, denoted by $\tau(H)$. A set of edges $A \subseteq E$ is an edge cover if every vertex is adjacent to at least an edge in A. The edge covering number is the minimum cardinality of an edge cover, denoted by $\tau'(H)$. A set of edges $A \subseteq E$ is a matching if every two distinct edges have no common vertex. The matching number is the maximum cardinality of a matching, denoted by $\nu(H)$. In this paper, we consider the vertex transversal set in simple hypergraphs with rank k.

1.1 Known Results

Hypergraphs are systems of sets which are conceived as natural extensions of graphs. A subset S of vertices in a hypergraph H is a transversal (also called vertex cover or hitting set in many papers) if S has a nonempty intersection with every edge of H. The transversal number $\tau(H)$ of H is the minimum size of a transversal in H. Transversals in hypergraphs are well studied in the literature (see [5, 10, 11, 13–15, 17, 18]).

Chvátal and McDiarmid [5] established the following upper bound on the transversal number of a uniform hypergraph in terms of its order and size.

Theorem 1. *[5] For $k \geq 2$, if H is a k-uniform hypergraph on n vertices with m edges, then $\tau(H) \leq \frac{n + \lfloor \frac{k}{2} \rfloor m}{\lfloor \frac{3k}{2} \rfloor}$.*

Henning and Yeo [8] proposed the following question:

Conjecture 1. [8] For $k \geq 2$, let H be a k-uniform hypergraph on n vertices with m edges. If H is linear, then $(k + 1)\tau(H) \leq n + m$ holds for all $k \geq 2$?

The Chvátal and McDiarmid theorem implies that $(k + 1)\tau(H) \leq n + m$ holds for $k \in \{2, 3\}$ even without the linearity constraint imposed on H. Henning and Yeo [8] remarked that if H is not linear, then conjecture 1 is not always true, showing an example by taking $k = 4$ and letting $\overline{F_7}$ be the complement of the Fano plane F_7. Henning and Yeo [8] proved the following theorem which verified conjecture 1 for linear hypergraphs with maximum degree two:

Theorem 2. *[8] For $k \geq 2$, let H be a k-uniform linear hypergraph satisfying $\Delta(H) \leq 2$. Then, $(k+1)\tau(H) \leq n+m$ with equality if and only if each component of H consists of a single edge or is the dual of a complete graph of order $k + 1$ and k is even.*

Henning and Yeo [12] proposed the following conjecture in another paper:

Conjecture 2. [12] $\tau(H) \leq \frac{n}{k} + \frac{m}{6}$ holds for all uniform hypergraphs with maximum degree at most 3.

Henning and Yeo [12] showed that $\tau(H) \leq \frac{n}{k} + \frac{m}{6}$ holds when $k = 2$ and characterized the hypergraphs for which equality holds. Chvátal and McDiarmid [5] showed that $\tau(H) \leq \frac{n}{k} + \frac{m}{6}$ holds when $k = 3$. Henning and Yeo characterized the extremal hypergraphs. Henning and Yeo [12] showed that $\tau(H) \leq \frac{n}{k} + \frac{m}{6}$ holds when $\Delta(H) \leq 2$ and characterized the hypergraphs for which it holds with equality in that case.

1.2   Our Results

In this paper, for $k \geq 2$, we propose a conjecture as follows:

Conjecture 3. For every connected rank k hypergraph $H(V, E)$ with m edges, $\tau(H) \leq \frac{(k-1)m+1}{k}$.

By a simple operation of adding vertices, it is easy to show if the conjecture holds in k-uniform hypergraphs, then it holds in hypergraphs with rank k. To prove the conjecture, it only needs to consider k-uniform hypergraphs.

For k-uniform hypergraphs, Conjecture 3 and Conjecture 1 are related. If Conjecture 1 holds, combined with the relationship between vertex number and edge number of connected rank k hypergraphs: $n \leq (k-1)m + 1$, we have

$$(k+1)\tau(H) \leq n + m, n \leq (k-1)m + 1 \Rightarrow \tau(H) \leq \frac{n+m}{k+1} \leq \frac{km+1}{k+1},$$

which is a weaker result of Conjecture 3. The main content of the article is organized as follows:

- In Sect. 2, we transform the conjecture on hypergraphs with rank k to the conjecture on k-uniform hypergraphs. For the consequent sections, we prove Conjecture 3 holds in some special hypergraphs.
- In Sect. 3, we prove Conjecture 3 holds for $k = 2$.
- In Sect. 4, we prove Conjecture 3 holds for $k = 3$.
- In Sect. 5, we prove Conjecture 3 holds for the hypergraphs satisfying the König Property with $\tau(H) = \nu(H)$.
- In Sect. 6, we prove Conjecture 3 holds for the hypergraphs with maximum degree 2 and characterize the extremal hypergraphs with equality holds.

2   The Conjecture

Conjecture 3 is our central problem and restated as follows:

For every connected rank k hypergraph $H(V, E)$ with m edges, $\tau(H) \leq \frac{(k-1)m+1}{k}$.

The next lemma tells us to prove Conjecture 3, it just needs to focus on uniform hypergraphs. The basic method in the proof of Lemma 1 is frequently used later.

Lemma 1. *If the conjecture holds in connected k-uniform hypergraphs, then it holds in connected hypergraphs with rank k.*

Proof. Let H be a connected hypergraph with rank k. If H is not k-uniform, we can construct a connected k-uniform hypergraph H' by adding new vertices to each edge. As shown in Fig. 1, for each edge e in H, if $|e| < \cdot k$, add $k - |e|$ new vertices to form an edge e' in H'. We derive that edge number does not change and $\tau(H) = \tau(H')$, which completes the proof.

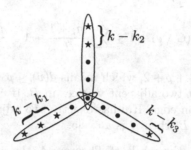

Fig. 1. Adding new vertices to form k-uniform hypergraphs

The following sections will consider Conjecture 3 in some special cases.

3 The Rank 2 Hypergraphs

In this section, we prove Conjecture 3 holds for the rank 2 hypergraphs. Such hypergraphs are actually general graphs.

Theorem 3. *For any connected graph $G(V, E)$ with m edges, $\tau(G) \leq \frac{m+1}{2}$.*

Proof. Suppose the theorem fails. Let us take out a counterexample $G = (V, E)$ with minimum number of edges, thus $\tau(G) > \frac{m+1}{2}$. For any vertex $v \in V$, let $C_1, C_2, \ldots, C_p$ be p components of $G \setminus v$ and C_i contains m_i edges for $i \in [p]$. Thus, $d(v) \geq p$. We have

$$\frac{m+1}{2} < \tau(G) \leq \tau(G \setminus v) + 1 \leq \sum_{i=1}^{p} \frac{m_i + 1}{2} + 1 = \frac{m - d(v) + p}{2} + 1 \leq \frac{m+2}{2}.$$

Then, we derive that $\tau(G) = \frac{m+2}{2}$, $p = d(v)$ and $\tau(C_i) = \frac{m_i+1}{2}$ for $i \in [p]$. Due to the arbitrariness of vertex v and the fact that $p = d(v)$, G contains no cycles. Thus, G is tree with $m + 1$ vertices and satisfies the König property [7] $\tau(G) = \nu(G)$. We have $\tau(G) = \nu(G) \leq \frac{m+1}{2}$, which is a contradiction.

Remark 1. For any connected graph, Theorem 3 implies a polynomial-time algorithm for computing a vertex transversal with cardinality no more than $\frac{m+1}{2}$.

Next, we establish a necessary condition of the extremal graphs G with m edges satisfying $\tau(G) = \frac{m+1}{2}$.

Theorem 4. *If a connected graph $G(V,E)$ with m edges satisfies that $\tau(G) = \frac{m+1}{2}$, then every block of G is an edge or a cycle, where a block is a maximal biconnected subgraph.*

Proof. For any block B of G, take arbitrarily a vertex v in B. Suppose that $C_1, C_2, \ldots, C_p$ are all components of $G \setminus v$ and C_i contains m_i edges. According to Theorem 3, we have

$$\frac{m+1}{2} = \tau(G) \le \tau(G \setminus v) + 1 \le \sum_{i=1}^{p} \frac{m_i + 1}{2} + 1 = \frac{m - d(v) + p + 2}{2}.$$

Thus, $m + 1 \le m - d(v) + p + 2$, which means $d(v) \le p + 1$. Since $d(v) \ge p$, we know that v has at most two adjacent vertices in B. If v has only one adjacent vertex in B, then B is an edge. If v has exactly two adjacent vertices in B, by the arbitrariness of v in B, then B is a cycle.

Remark 2. According to the result of Theorem 4, the extremal graphs belong to the partial 2-tree graph classes [3]. For the graphs with bounded tree width, the optimal vertex transversal can be computed in linear time in [2]. Then we derive a linear time algorithm to decide whether a m-edge graph G possesses the property $\tau(G) = \frac{m+1}{2}$.

4   The Rank 3 Hypergraphs

In this section, we prove Conjecture 3 holds for the rank 3 hypergraphs. This is an immediate corollary of the results by Chen [4]. Furthermore, Diao [6] characterize the extremal 3-uniform hypergraphs with equality holds. This demonstrates a polynomial-time algorithm to decide whether a rank 3 hypergraph is extremal.

Theorem 5. *[4] For every 3-uniform connected hypergraph $H(V,E)$ with m edges, $\tau(H) \le \frac{2m+1}{3}$.*

Theorem 6. *[6] For every 3-uniform connected hypergraph $H(V,E)$ with m edges, $\tau(H) = \frac{2m+1}{3}$ if and only if $H(V,E)$ is a hypertree with perfect matching.*

Combined with Lemma 1, the next two corollaries are derived immediately.

Corollary 1. *For every connected rank 3 hypergraph $H(V,E)$ with m edges, $\tau(H) \le \frac{2m+1}{3}$.*

Corollary 2. *For every connected rank 3 hypergraph $H(V,E)$ with m edges, $\tau(H) = \frac{2m+1}{3}$, then $H(V,E)$ is a hypertree with perfect matching.*

Remark 3. For a hypertree, there is a polynomial-time algorithm to compute the vertex transversal number. Thus for every connected hypergraph $H(V,E)$ with rank 3, it is decidable whether $\tau(H) = \frac{2m+1}{3}$ holds in polynomial time.

5 The Hypergraphs with König Property

In this section, we prove Conjecture 3 holds for the hypergraphs with the König Property. A hypergraph H has the König Property [1] if the transversal number is equal to the matching number: $\tau(H) = \nu(H)$.

Lemma 2. *For every connected rank k hypergraph $H(V, E)$ with n vertices and m edges, $n \leq (k-1)m + 1$.*

Proof. We prove this lemma by induction on m. When $m = 0$, $H(V, E)$ is an isolate vertex, $n \leq (k-1)m+1$ holds on. Assume this lemma holds on for $m \leq k$. When $m = k + 1$, take arbitrarily one edge e and consider the subgraph $H \setminus e$. Obviously, $H \setminus e$ has at most k components. Assume $H \setminus e$ has p components $H_i(V_i, E_i)$ with $n_i = |V_i|$ and $m_i = |E_i|$ for each $i \in [p]$. Then by induction, $n_i \leq (k-1)m_i + 1$ holds on. So we have

$$n = n_1 + \cdots + n_p \leq (k-1)m_1 + \cdots + (k-1)m_p + p = (k-1)(m-1) + p \leq (k-1)m + 1,$$

which completes the proof.

Lemma 3. *Let $H(V, E)$ be a k-uniform connected hypergraph with m edges. If H has the König Property, then $\tau(H) \leq \frac{(k-1)m+1}{k}$.*

Proof. According to the König Property and Lemma 2, we have the following inequalities:

$$\tau(H) = \nu(H) \leq \frac{n}{k} \leq \frac{(k-1)m+1}{k}.$$

According to Lemma 1 and Lemma 3, the next theorem is derived directly.

Theorem 7. *Let $H(V, E)$ be a connected rank k hypergraph with m edges. If H has the König Property, then $\tau(H) \leq \frac{(k-1)m+1}{k}$.*

Proof. Let H be a connected hypergraph with rank k. H has the König Property with $\tau(H) = \nu(H)$. If H is not k-uniform, we can construct a connected k-uniform hypergraph H' by adding new vertices to each edge. As shown in Fig. 1, for each edge e in H, if $|e| < k$, add $k - |e|$ new vertices to form an edge e' in H'. During this process, the edge number does not change and $\tau(H) = \tau(H'), \nu(H) = \nu(H')$. Thus, the new hypergraph H' maintains the König property with $\tau(H') = \nu(H')$. According to Lemma 3, we have

$$\tau(H') \leq \frac{(k-1)m+1}{k}, \tau(H) = \tau(H') \Rightarrow \tau(H) \leq \frac{(k-1)m+1}{k}.$$

6 The Hypergraphs with Maximum Degree 2

In this section, we prove Conjecture 3 holds for the hypergraphs with maximum degree 2 and characterize the extremal hypergraphs with equality holds. These results are proved by the dual hypergraphs. For a hypergraph $H(V,E)$, the dual hypergraph [1] $H^*(V^*,E^*)$ is a hypergraph whose vertices V^* correspond to the edges E of H and edges E^* correspond to the vertices V of H. Denote that $n=|V|$, $m=|E|$, $n^*=|V^*|$ and $m^*=|E^*|$. We have the following relationships of parameters between a hypergraph and its dual hypergraph: (i) $n^*=m$; (ii) $m^*=n$; (iii) $\tau(H)=\tau'(H^*)$.

6.1   The Bound of Hypergraphs with Maximum Degree 2

In this subsection, we prove Conjecture 3 holds for the hypergraphs with maximum degree 2. For a hypergraph $H(V,E)$ with maximum degree 2, its dual hypergraph is a multi-graph $G^*(V^*,E^*)$. The transversal number $\tau(H)$ corresponds to the edge covering number $\tau'(G^*)$. The edge covering number is related to the matching number by the theorem of Gallai [9]. The content of proof is organized as follows:

- An lower bound of matching number is proven by Lemmas 4, 5 and 6.
- An upper bound of edge covering number is proven by Lemma 7.
- Conjecture 3 for the hypergraphs with maximum degree 2 is proven by Theorem 9.

Theorem 8. *[9] [Gallai's Theorem]For every connected graph $G(V,E)$ with n vertices and the minimum degree $\delta(G)>0$, $\nu(G)+\tau'(G)=n$.*

Lemma 4. *For every tree $T(V,E)$ with m edges and $\Delta(T)\le k$, $\nu(T)\ge\frac{m}{k}$.*

Proof. We prove this lemma by contradiction. Let us take out the counterexample $T(V,E)$ with minimum edges. Thus $d(v)\le k$ for each $v\in V$ and $\nu(T)<\frac{m}{k}$. Obviously $T(V,E)$ has at least three vertices. The longest path in T is p, which connects one leaf v_1 to another leaf v_2, as shown in Fig. 2. v is the only adjacent vertex of v_1. The degree of v is $d(v)$ and $T\setminus v$ has $d(v)$ components, denoted as $\{T_i, 1\le i\le d(v)\}$.

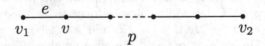

Fig. 2. The longest path p between leaves v_1 and v_2

Claim 1: $\nu(T\setminus v)\ge\frac{m-k}{k}$.
T is the counterexample with minimum edges, thus $\nu(T_i)\ge\frac{m_i}{k}$. Combined with $d(v)\le k$, we have the following inequality:

$$\nu(T\setminus v)=\sum_{1\le i\le d(v)}\nu(T_i)\ge\sum_{1\le i\le d(v)}\frac{m_i}{k}=\frac{m-d(v)}{k}\ge\frac{m-k}{k}.$$

Claim 2: $\nu(T) \geq \nu(T \setminus v) + 1$.

v_1 is a leaf in T. For every matching M in $T \setminus v$, $M \cup e(v_1, v)$ is a matching in T. According to these claims, we have the following inequality:

$$\nu(T) \geq \nu(T \setminus v) + 1 \geq \frac{m-k}{k} + 1 = \frac{m}{k},$$

which is a contradiction with $\nu(T) < \frac{m}{k}$.

Lemma 5. *For every tree $T(V, E)$ with n vertices and $\Delta(T) \leq k$, $\nu(T) \geq \frac{n-1}{k}$.*

Proof. Every tree $T(V, E)$ has $m = n - 1$. According to Lemma 4, $\nu(T) \geq \frac{n-1}{k}$ holds.

Lemma 6. *For every connected graph $G(V, E)$ with n vertices and $\Delta(G) \leq k$, $\nu(G) \geq \frac{n-1}{k}$.*

Proof. Take out a spanning tree T of G. According to Lemma 5, $\nu(G) \geq \nu(T) \geq \frac{n-1}{k}$.

Lemma 7. *For every connected graph $G(V, E)$ with n vertices and $\Delta(G) \leq k$, $\tau'(G) \leq \frac{(k-1)n+1}{k}$.*

Proof. According to Theorem 8 and Lemma 6, $\tau'(G) = n - \nu(G) \leq n - \frac{n-1}{k} = \frac{(k-1)n+1}{k}$ holds.

Theorem 9. *Let $H(V, E)$ be a connected hypergraph with m edges and rank k. If maximum degree of H is no more than 2, then $\tau(H) \leq \frac{(k-1)m+1}{k}$.*

Proof. Consider the dual hypergraph $H^*(V^*, E^*)$ of H whose vertices correspond to the edges of H and edges correspond to the vertices of H. A vertex transversal in H correspond to an edge cover in H^*. Thus we have

$$n^* = m, m^* = n, \tau(H) = \tau'(H^*), \tau(H) \leq \frac{(k-1)m+1}{k} \Leftrightarrow \tau'(H^*) \leq \frac{(k-1)n^*+1}{k}.$$

The rank of H is k, which means every edge of H contains at most k vertices. Thus every vertex's degree is at most k in H^*. The maximum degree of H is no more than 2, thus every edge of H^* contains at most 2 vertices. This means the hypergraph H^* is a multigraph, denoted by G^*. Delete the loops and multiedges to a simple graph. The deleting operations do not change the edge covering number. According to Lemma 7, $\tau'(G^*) \leq \frac{(k-1)n^*+1}{k}$, which means $\tau(H) \leq \frac{(k-1)m+1}{k}$.

6.2   The Extremal Hypergraphs with Maximum Degree 2

In this subsection, we characterize the extremal hypergraphs with maximum degree 2, meaning the equality $\tau(H) = \frac{(k-1)m+1}{k}$ holds. As shown before, the maximum degree restricts its dual hypergraph is a multi-graph. The transversal number of a hypergraph corresponds to the edge covering number of its dual multi-graph. For a graph, the edge covering number is related to the matching number by the theorem of Gallai [9]. The content of proof is organized as follows:

- An family of graphs called *k-star tree* is introduced by Definitions 1, 2, 3 and Lemma 8.
- The extremal graphs with matching number are characterized by Lemmas 9, 10 and 11.
- The extremal graphs with edge covering number are characterized by Lemma 12.
- The extremal maximum degree 2 hypergraphs with transversal number are characterized by Theorem 10.

Recall that k-star is a $(k+1)$-vertex tree with k leaves and the central vertex of a k-star is the k-degree vertex. Then we introduce the definition of k-star tree.

Definition 1. *For $k \geq 3$, a tree $T(V, E)$ is called a k-star tree if it satisfies*

- *Each vertex's degree is no more than k.*
- *The edges of T can be decomposed into several k-stars.*

Definition 2. *For a k-star tree $T(V, E)$, the central vertices of k-stars are called central vertices and other vertices are called noncentral vertices. The vertices connecting different k-stars are called adjacent vertices. Noncentral vertices are formed by adjacent vertices and leaves.*

Definition 3. *For a k-star tree $T(V, E)$, the structure tree describes the structure of T as formed by its k-stars. Let A denote the set of adjacent vertices of T, and B the set of its k-stars. Then, we have a natural tree $T'(A \cup B, E')$ on vertex set $A \cup B$ formed by the edges $e'(a, b) \in E'$ with $a \in A$, $b \in B$ and $a \in b$, which means the adjacent vertex a belongs to the k-star b in T. An example is shown in Fig. 3.*

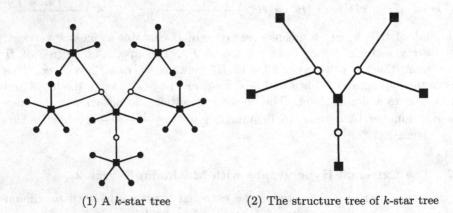

(1) A k-star tree   (2) The structure tree of k-star tree

Fig. 3. A k-star tree, where the squares, the hollow dots and the solid dots are central vertices, adjacent vertices and leaves, respectively.

Lemma 8. *For a k-star tree $T(V,E)$, there is a unique k-star decomposition of T.*

Proof. Let p be the number of k-stars in T. This proof can be finished by induction on p.

- When $p = 0$, T is an isolated vertex and there is a unique k-star decomposition.
- When $p = 1$, T is a k-star and there is a unique k-star decomposition.
- Assume the lemma holds for $p \leq t$. When $p = t+1$, let us consider the structure tree T' of the k-star tree T. v' is a leaf in T' and $S_{v'}$ is the corresponding k-star in T. v is the central vertex of $S_{v'}$, as shown in Fig. 4. $T \setminus v$ is also a k-star tree and the number of k-stars in $T \setminus v$ is exactly t. By induction, there is a unique k-star decomposition D in $T \setminus v$. Thus $D \cup S_{v'}$ is the unique k-star decomposition in T. The lemma holds for $p = t + 1$. Therefore, the lemma holds for every k-star tree.

Fig. 4. The central vertex v in k-star tree T

Lemma 9. *For every tree $T(V,E)$ with $m \geq 2$ edges and $\Delta(T) \leq k$, $\nu(T) = \frac{m}{k}$ if and only if T is a k-star tree.*

Proof. Necessity: $T(V,E)$ is a tree with $\Delta(T) \leq k$ and $\nu(T) = \frac{m}{k}$ holds. It needs to show T is a k-star tree. Since $m \geq 2$, $T(V,E)$ has at least three vertices. Take out a leaf u in T arbitrarily and v is the only adjacent vertex of u. The degree of v is $d(v)$ and $T \setminus v$ has $d(v)$ components, denoted as $\{T_i, 1 \leq i \leq d(v)\}$.
Claim 1: $\nu(T \setminus v) \geq \frac{m-k}{k}$.
According to Lemma 4, $\nu(T_i) \geq \frac{m_i}{k}$. Combined with $d(v) \leq k$, we have the following inequality:

$$\nu(T \setminus v) = \sum_{1 \leq i \leq d(v)} \nu(T_i) \geq \sum_{1 \leq i \leq d(v)} \frac{m_i}{k} = \frac{m - d(v)}{k} \geq \frac{m - k}{k}.$$

Claim 2: $\nu(T) \geq \nu(T \setminus v) + 1$.
u is a leaf in T. For every matching M in $T \setminus v$, $M \cup e(u,v)$ is a matching in T. According to these claims, we have the following inequality:

$$\nu(T) \geq \nu(T \setminus v) + 1 \geq \frac{m - k}{k} + 1 = \frac{m}{k}.$$

Combined with $\nu(T) = \frac{m}{k}$, we have $\nu(T \setminus v) = \frac{m-k}{k}$. This means the degree of v is exactly k and in $T \setminus v$, each component $\{T_i, 1 \leq i \leq d(v)\}$ satisfies $\nu(T_i) = \frac{m_i}{k}$ holds.

Take out $\{T_i, 1 \leq i \leq d(v)\}$ as T and repeat the above analysis process. Finally, there are some isolated vertices. Denote the deleted vertices as $\{v_j, 1 \leq j \leq t\}$. A k-star S_j is deleted when v_j is deleted. Thus the edges of T can be decomposed into several k-stars. According to Definition 1, T is a k-star tree.

Sufficiency: T is a k-star tree. It needs to show $\nu(T) = \frac{m}{k}$. Let p be the number of k-stars in T. We do induction on p.

- When $p = 0$, T is an isolated vertex and $\nu(T) = \frac{m}{k}$ holds.
- When $p = 1$, T is a k-star and $\nu(T) = \frac{m}{k}$ holds.
- Assume the sufficiency holds for $p \leq t$. When $p = t + 1$, let us consider the structure tree T' of the k-star tree T. v' is a leaf in T' and $S_{v'}$ is the corresponding k-star in T. v is the central vertex of $S_{v'}$, as shown in Fig. 4. $T \setminus v$ is also a k-star tree and the number of k-stars in $T \setminus v$ is exactly t. By induction, $\nu(T \setminus v) = \frac{m-k}{k}$ holds. Thus we have

$$\nu(T) = \nu(T \setminus v) + 1 = \frac{m-k}{k} + 1 = \frac{m}{k}.$$

The sufficiency holds for $p = t + 1$. Therefore, the sufficiency holds for every k-star tree.

Remark 4. The above proof also demonstrates a polynomial-time algorithm to decide whether a tree T is a k-star tree. In addition, If T is a k-star tree, the algorithm gives the unique k-star decomposition.

Lemma 10. *For every tree $T(V, E)$ with n vertices and $\Delta(T) \leq k$, $\nu(T) = \frac{n-1}{k}$ if and only if T is a k-star tree.*

Proof. Every tree $T(V, E)$ has $m = n - 1$ edges. According to Lemma 9, $\nu(T) = \frac{n-1}{k}$ if and only if T is a k-star tree.

Lemma 11. *For every connected graph $G(V, E)$ with n vertices and $\Delta(G) \leq k$, $\nu(G) = \frac{n-1}{k}$ if and only if G is a k-star tree.*

Proof. Sufficiency: G is a k-star tree. According to Lemma 10, $\nu(G) = \frac{n-1}{k}$ holds.

Necessity: $G(V, E)$ is a connected graph with $\Delta(G) \leq k$, $\nu(G) = \frac{n-1}{k}$. It needs to show G is a k-star tree. Take out arbitrarily a spanning tree T of G. According to Lemma 5, we have

$$\nu(G) \geq \nu(T) \geq \frac{n-1}{k}, \nu(G) = \frac{n-1}{k} \Rightarrow \nu(T) = \frac{n-1}{k}.$$

According to Lemma 10, T is a k-star tree. By arbitrariness, the next claim holds.

Claim 1: Each spanning tree in G is a k-star tree.

T is a k-star tree. The vertices are divided into *central* vertices and *noncentral* vertices. *Central* vertices are only connected with *noncentral* vertices and vice versa. Thus the next claim holds.

Claim 2: For each path p in T, *central* vertices and *noncentral* vertices are adjacent in p.

We will show G is T. This is proved by contradiction. Suppose there is an edge $e(u, v) \in G \setminus T$. T is a k-star tree. According to Definition 2, the vertices are divided into *central* vertices and *noncentral* vertices. *Noncentral* vertices are divided into *adjacent* vertices and *leaves*. Two cases are discussed as follows:

- u or v is a central vertex. Without loss of generality, suppose u is a central vertex. The degree of u is k in T. In $T \cup e(u, v)$, the degree of u is $k + 1$, which is a contradiction with any degree no more than k in G. This case is impossible.

- u and v are noncentral vertices. The unique $u - v$ path in T is denoted as p. w is the adjacent vertex of v in p. By **Claim 2**, w is a central vertex. Take out the longest path with start vertex w from $T \setminus e(v, w)$, denoted as $\tilde{p}(w, v_1)$. Thus, v_1 is a leaf in T. Consider the spanning tree $\tilde{T} = T \cup e(u, v) \setminus e(v, w)$, as shown in Fig. 5. By **Claim 1**, $\tilde{T}$ is also a k-star tree. $\tilde{p}(w, v_1)$ is a path in both T and $\tilde{T}$. By **Claim 2**, w is also a central vertex in $\tilde{T}$. This is a contradiction with $d(w) = k - 1 < k$ in $\tilde{T}$. This case is impossible.

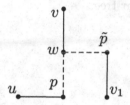

Fig. 5. The case when u and v are noncentral vertices

Above all, there is a contradiction for each cases. Thus our hypothesis does not hold and there is no edge $e(u, v) \in G \setminus T$. Thus G is T, which is a k-star tree.

Lemma 12. *For every connected graph $G(V, E)$ with n vertices and $\Delta(G) \leq k$, $\tau'(G) = \frac{(k-1)n+1}{k}$ holds if and only if G is a k-star tree.*

Proof. According to Theorem 8 and Lemma 11, $\tau'(G) = n - \nu(G) = n - \frac{n-1}{k} = \frac{(k-1)n+1}{k}$ holds if and only if G is a k-star tree.

Theorem 10. *Let $H(V, E)$ be a connected rank k hypergraph with m edges. If maximum degree of H is no more than 2, then $\tau(H) = \frac{(k-1)m+1}{k}$ if and only if the simple dual graph G^* is a k-star tree.*

Proof. Consider the dual hypergraph $H^*(V^*, E^*)$ of H whose vertices correspond to the edges of H and edges correspond to the vertices of H. A vertex transversal in H correspond to an edge cover in H^*. Thus we have

$$n^* = m, m^* = n, \tau(H) = \tau'(H^*), \tau(H) = \frac{(k-1)m+1}{k} \Leftrightarrow \tau'(H^*) = \frac{(k-1)n^*+1}{k}.$$

The rank of H is k, which means every edge of H contains at most k vertices. Thus every vertex's degree is at most k in H^*. The maximum degree of H is no more than 2, thus every edge of H^* contains at most 2 vertices. This means the hypergraph H^* is a multigraph, denoted by G^*. Delete the loops and multiedges to a simple graph. The deleting operations do not change the edge covering number. According to Lemma 12, $\tau'(G^*) = \frac{(k-1)n^*+1}{k}$ if and only if G^* is a k-star tree. This means $\tau(H) = \frac{(k-1)m+1}{k}$ if and only if the simple dual graph G^* is a k-star tree.

Remark 5. For a simple graph, there is a polynomial-time algorithm to compute the edge covering number [16]. Thus for every connected hypergraph $H(V, E)$ with maximum degree 2, it is decidable whether $\tau(H) = \frac{(k-1)m+1}{k}$ holds in polynomial time.

Remark 6. For a k-star tree as a dual graph, the primal hypergraph is a k-flower. The extremal hypergraphs are exactly k-flowers connected by common edges as shown in Fig. 6. Some 1-degree vertices are added in the edges which correspond to the deleted loops in a k-star tree.

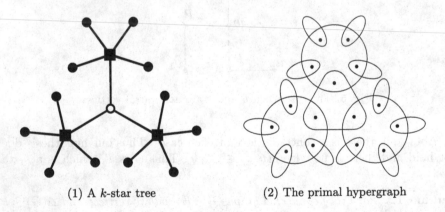

(1) A k-star tree   (2) The primal hypergraph

Fig. 6. A k-star tree and its primal hypergraph

References

1. Berge, C.: Hypergraphs. North-Holland, Paris (1989)
2. Bodlaender, H.L.: Dynamic programming on graphs with bounded treewidth. In: Lepistö, T., Salomaa, A. (eds.) ICALP 1988. LNCS, vol. 317, pp. 105–118. Springer, Heidelberg (1988). https://doi.org/10.1007/3-540-19488-6_110

3. Bodlaender, H.L.: A partial k-arboretum of graphs with bounded treewidth. Theor. Comput. Sci. **209**(1), 1–45 (1998)
4. Chen, X., Diao, Z., Hu, X., Tang, Z.: Covering triangles in edge-weighted graphs. Theory Comput. Syst. **62**(6), 1525–1552 (2018)
5. Chvátal, V., Mcdiarmid, C.: Small transversals in hypergraphs. Combinatorica **12**(1), 19–26 (1992)
6. Diao, Z.: On the vertex cover number of 3-uniform hypergraphs. J. Oper. Res. Soc. China **9**, 427–440 (2021)
7. Diestel, R.: Graph Theory, 4th Edition, Graduate texts in mathematics, vol. 173. Springer, New York (2012)
8. Dorfling, M., Henning, M.A.: Linear hypergraphs with large transversal number and maximum degree two. Eur. J. Comb. **36**, 231–236 (2014)
9. Gallai, T.: Uber extreme punkt-und kantenmengen, annales universitatis scientiarum budapestinensis de rolando eotvos nominatae. Sect. Math. **2**, 133–138 (1959)
10. Henning, M.A., Löwenstein, C.: Hypergraphs with large transversal number and with edge sizes at least four. Discrete Appl. Math. **10**(3), 1133–1140 (2012)
11. Henning, M.A., Yeo, A.: Total domination in 2-connected graphs and in graphs with no induced 6-cycles. J. Graph Theory **60**(1), 55–79 (2010)
12. Henning, M.A., Yeo, A.: Hypergraphs with large transversal number. Discrete Math. **313**, 959–966 (2013)
13. Henning, M.A., Yeo, A.: Lower bounds on the size of maximum independent sets and matchings in hypergraphs of rank three. J. Graph Theory **72**, 220–245 (2013)
14. Henning, M.A., Yeo, A.: Transversals and matchings in 3-uniform hypergraphs. Euro. J. Combinatorics **34**, 217–228 (2013)
15. Lai, F.C., Chang, G.J.: An upper bound for the transversal numbers of 4-uniform hypergraphs. J. Combinatorial Theory Ser. B **50**(1), 129–133 (1990)
16. Schrijver, A.: Combinatorial Optimization: Polyhedra and Efficiency, vol. 24. Springer, Heidelberg (2003)
17. Thomassé, S., Yeo, A.: Total domination of graphs and small transversals of hypergraphs. Combinatorica **27**(4), 473–487 (2007)
18. Tuza, Z.: Covering all cliques of a graph. Discret. Math. **86**(1–3), 117–126 (1990)

Exact Algorithms and Hardness Results for Geometric Red-Blue Hitting Set Problem

Raghunath Reddy Madireddy[1], Subhas C. Nandy[2], and Supantha Pandit[3(✉)]

[1] Birla Institute of Technology and Science Pilani, Hyderabad Campus,
Hyderabad, Telangana, India
raghunath@hyderabad.bits-pilani.ac.in
[2] Indian Statistical Institute, Kolkata, India
nandysc@isical.ac.in
[3] Dhirubhai Ambani Institute of Information and Communication Technology,
Gandhinagar, Gujarat, India
pantha.pandit@gmail.com

Abstract. We study geometric variations of the Red-Blue Hitting Set problem. Given two sets of objects R and B, colored red and blue, respectively, and a set of points P in the plane, the goal is to find a subset $P' \subseteq P$ of points that hits all blue objects in B while hitting the minimum number of red objects in R. We study this problem for various geometric objects. We present a polynomial-time algorithm for the problem with intervals on the real line. On the other hand, we show that the problem is NP-hard for axis-parallel unit segments. Next, we study the problem with axis-parallel rectangles. We give a polynomial-time algorithm for the problem when the rectangles are anchored on a horizontal line. The problem is shown to be NP-hard when all the rectangles intersect a horizontal line. Finally, we prove that the problem is APX-hard when the objects are axis-parallel rectangles containing the origin of the plane, axis-parallel rectangles where every two rectangles intersect exactly either zero or four times, axis-parallel line segments, axis-parallel strips, and downward shadows of segments. To achieve these APX-hardness results, we first introduce a variation of the Red-Blue Hitting Set problem in a set system, called the Special-Red-Blue Hitting Set problem. We prove that the Special-Red-Blue Hitting Set problem is APX-hard and we provide an encoding of each class of objects mentioned above as the Special-Red-Blue Hitting Set problem.

Keywords: Special Red-Blue Hitting Set · Anchored rectangles · Intervals · Polynomial-time algorithm · NP-hard · APX-hard

1 Introduction

The Hitting Set problem is a fundamental and well-studied problem in computer science and combinatorial optimization. Here, a set P of elements and a collection

© The Author(s), under exclusive license to Springer Nature Switzerland AG 2022
M. Li and X. Sun (Eds.): IJTCS-FAW 2022, LNCS 13461, pp. 176–191, 2022.
https://doi.org/10.1007/978-3-031-20796-9_13

S of subsets of P are given. The goal is to find a minimum size subset $P' \subseteq P$ such that each set in S is hit by some element of P'. In the geometric setting P represents a set of points and S represents a set of geometric objects in the plane. A well-studied variation of the Hitting Set problem is the Red-Blue Hitting Set problem. In this case, we are given a set P of elements and two collections, B and R, of subsets of P; the goal is to pick a subset of elements $P' \subseteq P$ which hits all the sets in B while hitting the minimum number of sets in R. In this paper, we consider a geometric variant of the problem which is given below.

> **Red-Blue Hitting Set ($RBHS$) Problem.** We are given a set of points P and two sets of objects R (red objects) and B (blue objects) in the plane. The goal is to pick a subset $P' \subseteq P$ that hits all the objects in B while hitting the minimum number of objects in R.

In this paper, we study the hardness and approximability of the $RBHS$ problem for the geometric objects such as intervals on the real line, line segments, and axis-parallel rectangles.

1.1 Previous Work

In the Set Cover problem, a set U and a collection S of subsets of 2^U are given, and the goal is to find a minimum size sub-collection $S' \subseteq S$ that covers all elements in U. Both Set Cover and Hitting Set problems are dual to each other in the classical setting[1]. The Hitting Set problem is NP-hard [11] and cannot be approximated better than $O(\log n)$ factor unless P= NP [9,14] (note that $n = |U|$). In the geometric setting the Set Cover problem is NP-hard even for simple geometric objects such as unit disk [10], unit squares [10], to name a few. Further, PTASes exist for unit disks [13,19] and axis-parallel unit squares [5,8]. Since the Set Cover and Hitting Set problems are dual of each other, the above results also hold for the Hitting Set problem for unit disks and unit squares. In [4], Chan and Grant proved that the Set Cover problem is APX-hard for several classes of objects such as axis-parallel rectangles containing a common point, axis-parallel strips, axis-parallel rectangles such that each pair intersects in either zero or four times, downward shadows of line segments, etc. They also proved that the Hitting Set problem is APX-hard for axis-parallel strips and rectangles intersecting zero or four times. Further, Madireddy and Mudgal [15] proved that the Hitting Set problem is APX-hard for rectangles that contain the origin of the plane and downward shadows of segments. The geometric Hitting Set and Set Cover problems are also NP-hard for half-strips anchored on two parallel lines [17,18].

Researchers have studied variations of the Set Cover problem due to its numerous applications and one such variations is the Red-Blue Set Cover problem ($RBSC$). It was first introduced by Carr et al. [3] for the set systems. Here, we are given two sets of elements R and B and a collection of subsets $S \subseteq 2^{R \cup B}$; the

---

[1] https://www8.cs.umu.se/kurser/TDBA77/VT06/algorithms/BOOK/BOOK5/NODE201.HTM.

goal is to pick a sub-collection $S' \subseteq S$ that covers all elements in B while covering the minimum number of elements in R. In [3], it is proved that, unless P= NP, the $RBSC$ problem cannot be approximated within $2^{\log^{1-\delta} n}$-factor in polynomial time, where $\delta = \frac{1}{\log^c \log n}$ with $c \leq \frac{1}{2}$ (note that $n = |U|$). In a geometric setting, Chan and Hu [5] first considered the $RBSC$ problem and proved that the problem is NP-hard for axis-parallel unit squares and provided a PTAS for the same. Further, it is known that the $RBSC$ problem is APX-hard when the objects are axis-parallel rectangles [21] of arbitrary size. In [16], APX-hardness results are given for the $RBSC$ problem for several classes of objects mentioned in [4]. Further, in [16], NP-hardness proofs are given for the $RBSC$ problem with axis-parallel rectangles intersecting a horizontal line and rectangles anchored on two parallel lines.

Dom et al. [7] studied the Red-Blue Hitting Set problem (in a set system) with *consecutive ones property*[2]. Based on whether red or blue sets or both the sets satisfy the consecutive ones property, the authors gave NP-hardness results or polynomial-time algorithms. Chang et al. [6] considered the same problem and gave an $O((m+n)k + k^2)$-time algorithm, (which improves the time bound $O(mnk^2)$ of [7]) for the $RBHS$ problem when the union of both red and blue sets satisfy the consecutive ones property where $m = |B|$, $n = |R|$, and $k = |P|$. We note that the polynomial-time algorithm in [6] also works for $RBHS$ problem with intervals on a real line such that no interval in $R \cup B$ completely covers another interval in $R \cup B$.

1.2   Our Contributions

We first provide an $O(m \log m + n \log n + k(m + n))$-time exact algorithm for $RBHS$ problem, where $m = |B|$, $n = |R|$, and $k = |P|$, when the objects in $R \cup B$ are intervals on the real line (Sect. 3.1). Further, for the case of line segments, we show that the problem is NP-hard even when the line segments are axis-parallel unit segments (Sect. 3.2).

Next, we consider the case when the objects in $R \cup B$ are axis-parallel rectangles. We give a polynomial-time algorithm for $RBHS$ problem when all the axis-parallel rectangles anchored along a horizontal line (see Sect. 4.1). On the other hand, we show that the $RBHS$ problem is NP-hard even when all the rectangles intersecting a horizontal line (see Sect. 4.2).

Finally, we show that $RBHS$ problem is APX-hard when the objects in $R \cup B$ are: (i) rectangles containing the origin, (ii) downward shadows of segments, (iii) axis-parallel strips, (iv) axis-parallel rectangles when every pair of rectangles intersect either zero or four times, and (v) axis-parallel line segments. In the process of proving the APX-hardness results for $RBHS$ problem, we define a restricted variant of the $RBHS$ problem in a set system, namely *Special Red Blue Hitting Set* (*SPECIAL-RBHS*) problem and show that the problem is APX-hard. By encoding the instance of *SPECIAL-RBHS* problem into instances of

---

[2] Let X be a set of elements and $\mathcal{C}$ be a collection of subsets of X. The collection $\mathcal{C}$ has the consecutive ones property if there exist a linear order of elements in X and a 0-1 matrix A, where $A_{ij} = 1$ if and only if i-th set in $\mathcal{C}$ contains the j-th element in X, such that all 1's in any row are consecutive.

RBHS problem, we show the *RBHS* problem is APX-hard for the above mentioned classes of objects.

Our work leaves two open questions (*i*) the complexity of the *RBHS* problem for axis-parallel lines and (*ii*) existence of a PTAS or APX-hardness for the *RBHS* problem with axis-parallel unit segments.

2 Preliminaries

In a formula in Conjunctive Normal Form (CNF), a clause is said to be *negative* if all its literals are negatively present in that clause. Otherwise, if all its literals are positively present, then the clause is called a *positive* clause. In *Monotone 3-SAT (M3SAT)* problem, a 3-CNF formula ϕ is given, where each clause is either positive or negative and contains exactly three literals, the objective is to decide whether ϕ is satisfiable. This problem is known to be NP-complete [12]. Now consider a planar embedding of the *M3SAT* problem, called the *Planar Monotone Rectilinear 3-SAT (PMR3SAT)* problem [2]. In this problem, for each variable or clause, a segment is considered. The variable segments are on a horizontal line L ordered from left to right. The positive (resp. negative) clause segments are placed below (resp. above) the line L and they are in different levels in the y-direction. Each clause connects to the three literals it contains by vertical connections. Finally, the embedding is drawn in such a way that it becomes planar. See an instance ϕ of the *PMR3SAT* problem in Fig. 1. de Berg and Khosravi [2] proved that this problem is NP-complete. For variable x_i, order the connections of positive clauses, that connect x_i, left to right. Let C_ℓ be a positive clause that connects to x_i through the ξ-th connection according to this order, then we say that C_ℓ is the ξ-th clause for x_i. For example, for the variable x_1, C_3 is the 1-st and C_1 is the 2-nd clause. A similar description is given for negative clauses by looking at the *PMR3SAT* embedding (Fig. 1) rotated 180°.

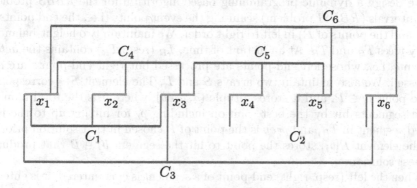

Fig. 1. Planar monotone rectilinear 3-SAT

3   The *RBHS* problem with lines and segments

In this section, we first present a polynomial-time algorithm for the *RBHS* problem where the elements in $R \cup B$ are intervals on the real line $\mathbb{R}$. Next, we show that the problem is NP-hard when the objects are axis-parallel unit segments. In the following, we assume that $m = |B|$, $n = |R|$, and $k = |P|$.

3.1   Intervals on a Real Line $\mathbb{R}$

Let R and B be the sets of n and m intervals of color *red* and *blue*, respectively, and P be the set of k points on the real line. We assume that (*i*) for every blue interval, there is at least one point in P that can hit it (otherwise the solution for this problem does not exist), (*ii*) the points in P are sorted, and (*iii*) the end points of intervals in R and B are sorted; each end-point is tagged with its color, whether it is left or right, and the coordinate of its other end-point.

Definition 1. *If a segment* $b = [\alpha, \beta] \in B$ *completely covers (spans over) another segment* $b' = [\alpha', \beta'] \in B$, *i.e.,* $\alpha \leq \alpha'$ *and* $\beta \geq \beta'$ *then* b *is said to be* dominated *by* b'.

We can get the dominated segments in B as follows. Scan the end-points of B in left to right order. While processing the left end-point of a segment, it is entered in a balanced binary tree T_B. When the right end-point of a segment t is processed, if its left end-point is the left-most element in T_B, it is a non-dominated segment, and is put in a set B'; otherwise, it is ignored. In both the cases t is deleted from T_B. Finally, B' is the set of not-dominated segments, and will be referred to as B. The segments in B are named as $\{b_1, b_2, \ldots, b_m\}$ in increasing order of their right end-points. The same naming holds if named with respect to their left end-point as we have eliminated the dominated set of points from B.

We design a dynamic programming based algorithm for the *RBHS* problem with intervals (*RBHSI* problem) scanning the event points (i.e., the end points of $B \cup R$ and the points of P) in left to right order. We maintain two height-balanced binary trees T_B and T_R. At an instant of time, T_B (resp. T_R) contains the *active* segments, i.e., whose left end-points are processed but right end-points are not processed. We also maintain two arrays S and Γ. The element $S[i]$ corresponds to the point $p_i \in P$, and it stores a tuple $(\sigma, prev)$, where σ is the total number of red segments hit by the best solution including p_i for hitting up to the last named segment in T_B, and $prev$ is the point of P chosen in that solution prior to p_i. The element $\Gamma[b_j]$ stores the point to hit the segment $b_j \in B$ that produces the best solution.

When the left (resp. right) end-point of a segment is encountered, it is entered in (resp. deleted from) the AVL Tree T_B or T_R depending on whether its color is *blue* or *red* respectively. When a point $p_i \in P$ is encountered, if p_i hits no blue segment (i.e., $T_B = \emptyset$), then p_i is ignored. Otherwise, we compute the cost of the best solution including p_i as follows:

Let $b_k, b_\ell \in B$ be the smallest and highest numbered segments in T_B, respectively. Let $\Gamma[b_{k-1}] = p_j$, and $S[j] = (\sigma'; p_\theta)$. We compute $\sigma = (\sigma' + C)$, where C denotes the number of red segments in T_R that are not hit by p_j but hit by p_i, and can be determined in $O(|T_R|)$ time. Next, we scan the array Γ backward, and update $\Gamma[\theta]$ with i for all $\theta = b_\ell, \ldots, b_k$ if the existing σ is such that $S[\Gamma[\theta]] \geq \sigma$.

The correctness follows from using the best solution of hitting the right-most segment that is not hit by the current point under processing, and the fact that after processing each point in P we are updating the best solution of hitting each blue segment in T_B by checking whether the current point under processing has the minimum σ value to hit that segment. The array Γ is always updated as B contains non-dominated segments. The time complexity of processing the endpoints of $R \cup B$ is $O(m \log m + n \log n)$. Processing each point $p_i \in P$ needs scanning the tree T_R to count C, and the tree T_B to update their associated tuple. Finally, following the *prev* pointers, we can identify the points in the optimum solution. Thus, we have the following result:

Theorem 1. *The proposed algorithm correctly solves the RBHSI problem in $O(m \log m + n \log n + k(m + n))$ time using $O(m + n + k)$ space.*

3.2   Axis-Parallel Line Segments

We prove that the *RBHS* problem is NP-hard even when each member of $R \cup B$ is an axis-parallel unit segment (*RBHS-ULS* problem), by giving a polynomial-time reduction from the *M3SAT* problem (see Sect. 2 for the definition). During the reduction, from an instance ϕ of the *M3SAT* problem, with n variables $x_1, x_2, \ldots, x_n$ and m clauses $C_1, C_2, \ldots, C_m$, an instance H_ϕ of the *RBHS-ULS* problem is generated. To make the reduction clearly visible, in the reduction, we take horizontal unit segments and vertical infinite lines. Next we compress the configuration vertically to make the vertical lines as unit segments.

Variable Gadget: For the variable x_i, the gadget (Fig. 2) consists of $4m + 4$ blue segments $\{b_1^i, b_2^i, \ldots, b_{4m+4}^i\}$, $4m + 4$ red segments $\{r_1^i, r_2^i, \ldots, r_{4m+4}^i\}$, and $4m + 4$ points $\{p_1^i, p_2^i, \ldots, p_{4m+4}^i\}$. The $2m + 1$ blue segments $\{b_2^i, b_3^i, \ldots, b_{2m+2}^i\}$, $2m + 2$ red segments $\{r_1^i, r_2^i, \ldots, r_{2m+2}^i\}$, and $2m + 2$ points $\{p_1^i, p_2^i, \ldots, p_{2m+2}^i\}$ are on a horizontal line. The $2m + 1$ blue segments $\{b_{2m+4}^i, b_{2m+5}^i, \ldots, b_{4m+4}^i\}$ and the $2m + 2$ red segments $\{r_{2m+3}^i, r_{2m+4}^i, \ldots, r_{4m+4}^i\}$, and $2m + 2$ points $\{p_{2m+3}^i, p_{2m+4}^i, \ldots, p_{4m+4}^i\}$ are on another horizontal line. There are two blue vertical lines b_1^i and b_{2m+3}^i. We place the blue and red segments and points in such a way that the point p_j^i hits exactly two blue segments b_j^i and b_{j+1}^i and exactly one red segment r_j^i, for $1 \leq j \leq 4m + 4$ (assuming $b_{4m+5}^i = b_1^i$) (Fig. 2). Observe that, there are exactly two sets of points; $P_1^i = \{p_1^i, p_3^i, \ldots, p_{4m+3}^i\}$ and $P_2^i = \{p_2^i, p_4^i, \ldots, p_{4m+4}^i\}$; such that each of them hits all the blue segments and half (i.e., $2m + 2$) of the red segments. We assume that, the set P_1^i interprets x_i to be false and P_2^i interprets x_i to be true.

The overall structure of the construction is shown in Fig. 2. The variable gadgets are placed vertically one after another in an identical way (i.e., all the j-th point; except 1-st, $(2m+2)$-th, $(2m+3)$-th, and $(4m+4)$-th; from n variables are on a vertical line, and left end points of j-th segment, except 1-st and $(m+3)$-th), are also on a vertical line. There are two regions G_1 and G_2 at the extreme left and extreme right of the construction. In the region G_1, m blue vertical lines $\{b_1^1, b_1^2, \ldots, b_1^m\}$ are arranged in a order from left to right. Similarly, in G_2, m blue vertical lines $\{b_{2m+3}^1, b_{2m+3}^2, \ldots, b_{2m+3}^m\}$ are arranged in an order from left to right. To the right of the point $p_{2\ell}^i$ (resp. $p_{2\ell+1}^i$) there is a dedicated region g_+^ℓ (resp. g_-^ℓ) where the gadget of C_ℓ is placed, $1 \leq \ell \leq m$.

Clause Gadgets and Its Placement: Let C_ℓ be a negative clause that contains variables x_i, x_j, and x_k. For C_ℓ, the gadget consists of a single vertical line b^ℓ. The line b^ℓ is placed inside the region g_-^ℓ. The three points $p_{2\ell+1}^i$, $p_{2\ell+1}^j$, and $p_{2\ell+1}^k$ are shifted horizontally and placed inside g_-^ℓ (see Fig. 2).

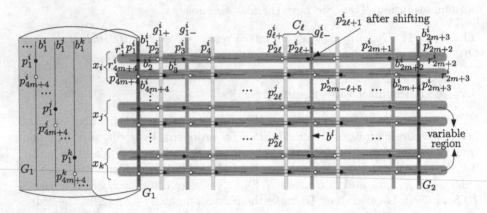

Fig. 2. Schematic construction of an instance of the *RBHS-ULS* problem from an instance of the *M3SAT* problem. Here, the strip G_1 is magnified at the left side of the figure. A similar structure as G_1 is there inside the strip G_2.

This completes the construction and the construction can be made in polynomial (in n and m) time. Observe that, in the construction the set R is a set of horizontal segments and the set B is a set of horizontal segments and vertical lines. By compressing the plane vertically such that vertical lines compressed to a vertical segment of unit length, all objects in $R \cup B$ become axis-parallel unit segments. We conclude the following theorem.

Theorem 2. *The RBHS-ULS problem is* NP-*hard.*

Proof. We prove that ϕ is satisfiable if and only if $2n(m+1)$ points hit all blue objects and $2n(m+1)$ red segments in H_ϕ.

Assume that ϕ is satisfiable and let there be a satisfying assignment. If x_i is true, the set P_2^i is selected, otherwise, the set P_1^i is selected. So a total $2n(m+1)$

points are selected across n variables that hits all blue segments and $2mn + 2n$ red segments in all variable gadgets. Now consider a negative clause C_ℓ, the three points $p_{2\ell+1}^i$, $p_{2\ell+1}^j$, and $p_{2\ell+1}^k$ are shifted vertically and placed on the blue vertical line b^ℓ. Since C_ℓ is satisfiable at least one of x_i, x_i and x_k is false and the corresponding point hits b^ℓ.

On the other hand assume that $2n(m + 1)$ points hit all blue segments and $2n(m+1)$ red segments in H_ϕ. At least $2m+2$ points are required to hit all blue segments in the gadget of x_i, $1 \leq i \leq n$. They also hit $2m+2$ red segments. Since the solution contains $2n(m+1)$ points and the variable gadgets are disjoint, there are exactly two sets of points, P_1^i and P_2^i such that each hits minimum number of red segments for x^i. Therefore, we set variable x_i to be true if P_2^i is selected, otherwise, we set x_i to be false. Now we claim that this assignment satisfies all the clauses. Let C_ℓ be a negative clause that contains variables x_i, x_j, and x_k. Since the segment b^ℓ is hit by the three points $p_{2\ell+1}^i$, $p_{2\ell+1}^j$, and $p_{2\ell+1}^k$, at least one of them must be in the solution. If $p_{2\ell+1}^\kappa$ hits b^ℓ, then the set P_1^κ that contains $p_{2\ell+1}^\kappa$ is selected, that makes x_i to be false. This implies that C_ℓ is satisfiable. □

4 The *RBHS* Problem with Axis-Parallel Rectangles

In this section we consider the objects in $R \cup B$ as axis-parallel rectangles. We first give a polynomial-time exact algorithm for the *RBHS* problem when the axis-parallel rectangles are anchored at the same side of a horizontal line (*RBHSAR* problem). Next, we prove that the *RBHS* problem is NP-hard even when axis-parallel rectangles stabbing a horizontal line (*RBHS-RSHL* problem).

4.1 Rectangles Anchored on a Horizontal Line

In this section, we give a dynamic programming algorithm for the *RBH-SAR* problem. Let $B = \{b_1, b_2, \ldots, b_m\}$ be a set of m blue rectangles, $R = \{r_1, r_2, \ldots, r_n\}$ be a set of n red rectangles in the plane that are anchored along the x-axis and lie above the x-axis. Also, let $P = \{p_1, p_2, \ldots, p_k\}$ be a set of k points in the plane lie above the x-axis. Further, assume that the point set $P = \{p_1, p_2, \ldots, p_k\}$ is sorted according to their x-coordinates. Add two points, p_0 at the extreme left and p_{k+1} at the extreme right, in the set P. Let P_{ij} be the set of points $\{p_i, p_{i+1}, \ldots, p_j\}$ and B_{ij} (resp. R_{ij}) be the set of blue (resp. red) rectangles whose left boundaries are to the right of p_i and right boundaries are to the left of p_j. Define a sub-problem $H(i, j)$ that includes the point-set P_{ij}, blue rectangle set B_{ij}, and red rectangle set R_{ij}. Denoting by b_h, the minimum height (top boundary has the minimum y-coordinate) rectangle in B_{ij}, we have the following result:

Lemma 1. *The optimum value $\pi(i, j)$ for the sub-problem $H(i, j)$ satisfy*

$$\pi(i, j) = \min_{\ell:p_\ell \in P \cap b_h} \{\chi + \pi(i, \ell) + \pi(\ell, j)\},$$

where χ is the number of red rectangles (in R) hit by p_ℓ.

Proof. Let $p_d \in P$ be a point that lies inside $b_h \in B_{ij}$. As the given rectangles in $R \cup B$ are anchored and lie above the x-axis, if we choose p_d in a solution S, all the rectangles in $R \cup B$ that are hit by p_d are removed, and the rectangles in $R_{ij} \cup B_{ij}$ that remain have their horizontal span is fully within either of the vertical strips defined by the point-pair $[p_i, p_\ell]$ and $[p_\ell, p_j]$.   □

Our aim is to recursively compute $\pi(0, k + 1)$. We consider a DP table of dimension $(k + 1) \times (k + 1)$. We consider a recursion stack, and the algorithm starts with pushing $\pi(0, k + 1)$ in the stack. While processing a top element $\pi(i, j)$ of the stack, if the value(s) $\pi(i, \ell)$ and $\pi(\ell, j)$ are previously computed, those values will be available in the DP table. We add that/those value(s) with χ, and in the recursion along this/these path(s) is/are terminated. The unsolved sub-problem(s) (if any) are pushed in the stack for further processing. Finally, when the recursion stack is empty, the answer is reported.

The correctness of the algorithm follows from Lemma 1. To solve $H(i, j)$, (i) identifying b_h needs $O(n)$ time, and (ii) we need to compute the minimum of the values of $O(k)$ sub-problems corresponding to the points in b_h. For each point $p_d \in b_h$, $O(m + n)$ time is needed to create the set of red and blue rectangles $(B_{i\ell}, R_{i\ell})$ and $(B_{\ell j}, R_{\ell j})$ for solving $H_{i\ell}$ and $H_{\ell j}$. The space complexity is determined by the size of the DP table. Thus, we conclude the following theorem.

Theorem 3. *Our proposed recursive algorithm for the RBHSAR problem correctly computes the optimum solution in $O(k^3(m + n))$ time using $O(k^2)$ space.*

4.2 Rectangles Stabbing a Horizontal Line

In this section we prove that the *RBHS-RSHL* problem, where the rectangles in $R \cup B$ are stabbed by a horizontal line L, is NP-hard by giving a polynomial-time reduction from the *PMR3SAT* problem (see Sect. 2 for the definition). Let ϕ be an instance of the *PMR3SAT* problem with n variables $x_1, x_2, \ldots, x_n$ and m clauses $C_1, C_2, \ldots, C_m$. Below we construct an instance H_ϕ of the *RBHS-RSHL* problem from ϕ as follows.

Variable Gadget: The gadget for the variable x_i is shown in Fig. 3. Let L be a horizontal line. There are $2m + 1$ points $\{p_1^i, p_2^i, \ldots, p_{2m+1}^i\}$ that one side of L and the remaining $2m+1$ points $\{p_{2m+2}^i, p_{2m+3}^i, \ldots, p_{4m+2}^i\}$ that are on the other side of L. The gadget contains $4m + 2$ blue rectangles $\{b_1^i, b_2^i, \ldots, b_{4m+2}^i\}$ and $4m + 2$ red rectangles $\{r_1^i, r_2^i, \ldots, r_{4m+2}^i\}$. The two rectangles b_1^i and b_{2m+2}^i are intersected by L. The $2m$ blue rectangles $\{b_2^i, b_3^i, \ldots, b_{2m+1}^i\}$ and the $2m + 1$ red rectangles $\{r_1^i, r_2^i, \ldots, r_{2m+1}^i\}$ are anchored on L from the above and the remaining $2m$ blue rectangles $\{b_{2m+3}^i, b_{2m+4}^i, \ldots, b_{4m+2}^i\}$ and the $2m + 1$ red rectangles $\{r_{2m+2}^i, r_{2m+3}^i, \ldots, r_{4m+2}^i\}$ are anchored on L from the below. The blue rectangle b_j^i covers two points p_{j-1}^i and p_j^i, for $1 \leq j \leq 4m+2$ (assuming p_0^i as p_{4m+2}^i) and the red rectangle r_j^i covers the point p_j^i, for $1 \leq j \leq 4m+2$. It is clear that there are exactly two sets; $P_1^i = \{p_1^i, p_3^i, \ldots, p_{4m+1}^i\}$ and $P_2^i = \{p_2^i, p_4^i, \ldots, p_{4m+2}^i\}$; such that each of them hits all the blue rectangles and minimum number $(2m + 1)$ of

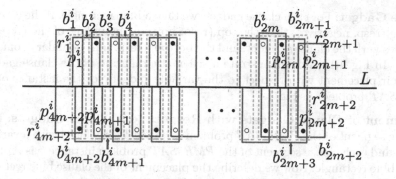

Fig. 3. Structure of a variable gadget.

red rectangles. The set P_1^i represents that the variable x_i is false and the set P_2^i represents that x_i is true.

Modification of Planar Embedding: We modify the planar embedding of the *PMR3SAT* problem in the following way. The variable segments and its placement remain as it is. The clause segments remain the same, however, their placement is changed now. Consider the positive clauses. We reverse the levels of the clause segments i.e., the highest level (largest y-valued) segment are placed in lowest level (smallest y-valued), the second highest level segment are placed in second lowest level (smallest y-valued), and so on. The connection (using legs) between the variable and clauses remain in the same position however their lengths may get increased or decreased. A similar construction is done for negative clauses also. See Fig. 4(a) for this construction.

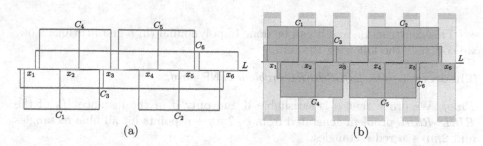

Fig. 4. (a) A modification of the Planar Monotone Rectilinear 3-SAT instance given in Fig. 1. (b) A schematic structure of the position of variable and clause gadgets and their interconnection.

Next, we exchange the positive and negative clauses i.e., the clauses that connect to the variables from above now connect the variables from below and the clauses that connect to the variables from below are now connect the variables from above. See Fig. 4(b) for this new interpretation.

Clause Gadget: For each clause gadget, we take a blue rectangle. If the clause is positive (resp. negative), then the top (resp. bottom) boundary of the rectangle coincides with the clause segment, and the bottom (resp. top) boundary coincides with L. In Fig. 4(b), we demonstrate a schematic diagram of the clause gadgets and their placement with respect to the variable gadgets for the instance of the *PMR3SAT* problem in Fig. 1.

Placement of Clause Gadgets with Respect to Variable Gadgets: Each variable segment of the *PMR3SAT* problem instance ϕ is replaced with a variable gadget and each clause segment of the *PMR3SAT* problem instance ϕ is replaced with a blue rectangle. Now we describe the placement of the clause blue rectangle with respect to the variable gadgets. Let C_ℓ be a positive clause that contains x_i, x_j, and x_k. Further, assume that C_ℓ is interpreted as the t_1-th, t_2-th, and t_3-th clause with respect to the variables x_i, x_j, and x_k, respectively. Then, we shift the three point $p^i_{2t_1}$, $p^j_{2t_2}$, and $p^k_{2t_3}$ vertically downwards from the variables x_i, x_j, and x_k, respectively, along the top boundary of b^ℓ (see Fig. 5).

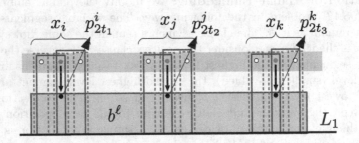

Fig. 5. A clause gadget and its interaction with variable gadgets.

This construction and it can be done in polynomial (in n and m) time. Now, we conclude the following theorem.

Theorem 4. *The RBHS-RSHL problem is* NP-*hard.*

Proof. We prove that ϕ is satisfiable if and only if in the instance H_ϕ of the *RBHS-RSHL* problem generated from ϕ, $2mn + n$ points hit all blue rectangles and $2mn + n$ red rectangles.

Assume that ϕ is satisfiable and let there be a satisfying assignment. If x_i is true, the set P^i_2 is selected, otherwise, the set P^i_1 is selected. So a total $2mn + n$ points are selected across n variables that hit the blue rectangles of the variables and $2mn + n$ red rectangles. Now for a positive clause C_ℓ, the three points p^i_{2t}, p^j_{2t}, and p^k_{2t} are placed inside the blue rectangle b^ℓ. Since C_ℓ is satisfiable at least one of x_i, x_j, and x_k is true and the corresponding point hits b^ℓ.

Assume that $2mn+n$ points hit all blue rectangles and $2mn+n$ red rectangles in H_ϕ. Observe that at least $2m + 1$ points hit the blue rectangles of the gadget of x_i. These points also hit $2m + 1$ red rectangles. Since each variable gadget is disjoint and the solution contains $2mn + n$ points, we say that exactly $2m + 1$ points are required to stab the blue rectangles in the gadget of x_i. Recall that, there are exactly two optimal sets of points either P_1^i and P_2^i that hit minimum number $(2m + 1)$ of red rectangles. So we set x_i to be true if P_1^i is selected, otherwise we set x_i to be false. We now argue that all the clauses are satisfiable. Let C_ℓ be a positive clause contains variables x_i, x_j, and x_k. The rectangle b^ℓ is hit by at least one of p_{2t}^i, p_{2t}^j, and p_{2t}^k. If p_{2t}^κ, for $\kappa \in \{i, j, k\}$, hits b^ℓ, the variable x_κ is true, since $p_{2t}^\kappa \in P_2^\kappa$ and P_2^κ is selected. Resulting C_ℓ is satisfiable. $\qquad\square$

5 APX-Hardness Results for the *RBHS* Problem

We first introduce a restricted variant of the *RBHS* problem in a set system, the *Special-Red-Blue Hitting Set (SPECIAL-RBHS)* problem. We show that this problem is APX-hard. Next, we give encoding of the *RBHS* problem, as *SPECIAL-RBHS* problem, for several classes of objects (see Theorem 6).

Special-Red-Blue Hitting Set (*SPECIAL-RBHS*) Problem:
Let (U, X) be a range space, and $U = S \cup T$, where $S = \{s_i^1, s_i^2, s_i^3 \mid i = 1, 2, \ldots, n\}$ and $T = \{t_q^1, t_q^2, t_q^3, t_q^4 \mid q = 1, 2, \ldots, m\}$ are the sets of elements. Further, $X = R \cup B$, where R and B are the collections of red and blue subsets of U, respectively, such that

1. For each $i = 1, 2, \ldots, n$, the set R contains a subset $\{s_i^1, s_i^2, s_i^3\} \subseteq S$.
2. For each $q = 1, 2, \ldots, m$, there exist two integers i and j, $1 \leq i < j \leq n$, such that X contains five subsets $\{s_i^k, t_q^1\}$, $\{t_q^1, t_q^2\}$, $\{t_q^2, t_q^3\}$, $\{t_q^3, t_q^4\}$, and $\{t_q^4, s_j^{k'}\}$ of U for some $k, k' \in \{1, 2, 3\}$. Further, the sets $\{s_i^k, t_q^1\}$, $\{t_q^2, t_q^3\}$, and $\{t_q^4, s_j^{k'}\}$ are in B and the sets $\{t_q^1, t_q^2\}$ and $\{t_q^3, t_q^4\}$ are in R.
3. Further, each element in U belongs to exactly one set in R and exactly one set in B.

    The goal is to find a subset $U^* \subseteq U$ of points that hits all the sets in B while hitting the minimum number of sets in R.

Theorem 5. *The SPECIAL-RBHS problem is* APX-*hard.*

Proof. We give an L-reduction [20] from an APX-hard problem, the *vertex cover problem on cubic graphs* [1] to the *SPECIAL-RBHS* problem. Let $G = (V, E)$ be a cubic graph where $V = \{v_1, v_2, \ldots, v_n\}$ and $E = \{e_1, e_2, \ldots, e_m\}$. We generate an instance (U, X) of the *SPECIAL-RBHS* problem as follows:

1. For each vertex $v_i \in V$, take three elements s_i^1, s_i^2, s_i^3 in S. Thus, $S = \{s_i^1, s_i^2, s_i^3 \mid i = 1, 2, \ldots, n\}$. Further, for each $i = 1, 2, \ldots, n$, place set $\{s_i^1, s_i^2, s_i^3\}$ in R.
2. For each edge $e_q \in E$, consider four elements t_q^1, t_q^2, t_q^3, and t_q^4 in T. Thus, $T = \{t_q^1, t_q^2, t_q^3, t_q^4 \mid q = 1, 2, \ldots, m\}$.
3. For each edge $e_q = (v_i, v_j) \in E$ where $1 \leq i < j \leq n$ do the following:

(a) Let k be a positive integer such that e_q is the k-th edge incident to v_i in the order $e_1, e_2, \ldots, e_m$. Similarly, let k' be a positive integer such that e_q be the k'-th edge incident on v_j in the order $e_1, e_2, \ldots, e_m$. Since G is a cubic graph, we have $1 \leq k \leq 3$ and $1 \leq k' \leq 3$.

(b) Consider five subsets $\{s_i^k, t_q^1\}$, $\{t_q^1, t_q^2\}$, $\{t_q^2, t_q^3\}$, $\{t_q^3, t_q^4\}$, and $\{t_q^4, s_j^{k'}\}$ of $S \cup T$. Place the sets $\{s_i^k, t_q^1\}$, $\{t_q^2, t_q^3\}$, $\{t_q^4, s_j^{k'}\}$ in B and place the sets $\{t_q^1, t_q^2\}$, $\{t_q^3, t_q^4\}$ in R.

4. Finally, let $U = S \cup T$ and $X = R \cup B$.

The above reduction is an L-reduction with $\alpha = 4$ and $\beta = 1$ and the theorem is proved.   □

Theorem 6. *The RBHS problem is* APX-*hard for the following classes of objects:*

(O1:) *Axis-parallel rectangles where each pair of rectangles intersect exactly either zero or four times.*

(O2:) *Axis-parallel line segments.*

(O3:) *Axis-parallel strips.*

(O4:) *Rectangles containing the origin of the plane.*

(O5:) *Downward shadows of segments.*

Proof. We give an encoding of each class as the *SPECIAL-RBHS* problem.

O1: We place the points $s_1^1, s_1^2, s_1^3, s_2^1, s_2^2, s_2^3, \ldots, s_n^1, s_n^2, s_n^3$ on a horizontal line in the same order from left to right such that for each $i = 1, 2, \ldots, n$, the three points s_i^1, s_i^2, and s_i^3 are very close to each other (see Fig. 6). We place a rectangle which covers only the three points s_i^1, s_i^2, and s_i^3 (see Fig. 7(a)). Further, there is a sufficient gap between the points corresponding to different i's and this gap is used to place the points t_q^2 and t_q^3 for $q = 1, 2, \ldots, m$. Further, for each $q = 1, 2, \ldots, m$, there is a dedicated region in which we place the points t_q^1, t_q^2, t_q^3,

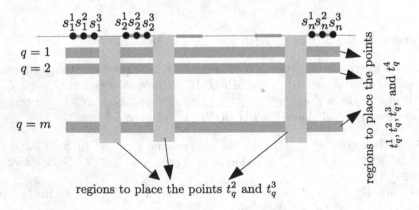

Fig. 6. Outline of placement of points in class **O1**.

and t_q^4 (see Fig. 6). For each $q = 1, 2, \ldots, m$, we place axis-parallel rectangles for the sets $\{s_i^k, t_q^1\}$, $\{t_q^1, t_q^2\}$, $\{t_q^2, t_q^3\}$, $\{t_q^3, t_q^4\}$, and $\{t_q^4, s_j^{k'}\}$ as shown in Fig. 7(a).

O2: The encoding is similar to class O1. Here, every rectangle in O1 is replaced with an appropriate axis-parallel segment (see Fig. 7(b)).

O3: The encoding is similar to class O1 (Fig. 7(c)) except the placement of $s_1^1, s_1^2, s_1^3, \ldots, s_n^1, s_n^2, s_n^3$. For each $i = 1, 2, \ldots, n$, the tuples of points s_i^1, s_i^2, and s_i^3 are placed in an increasing stair-case fashion from bottom to top. The placement

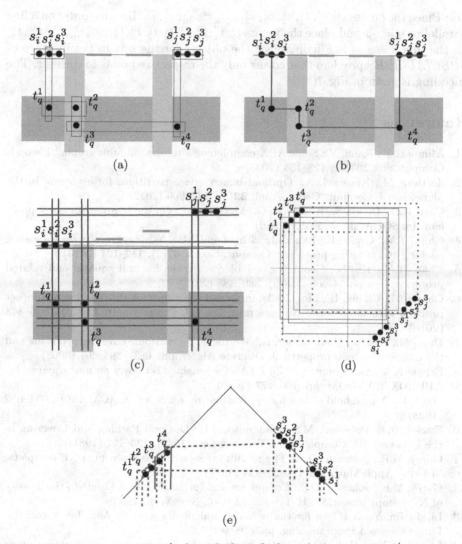

(a)

(b)

(c)

(d)

(e)

Fig. 7. The encoding of sets $\{s_i^k, t_q^1\}$, $\{t_q^1, t_q^2\}$, $\{t_q^2, t_q^3\}$, $\{t_q^3, t_q^4\}$, and $\{t_q^4, s_j^{k'}\}$ for $k = 2$ and $k' = 1$ for classes (a) Class O1 (b) Class O2 (c) Class O3, (d) Class O4, and (e) Class O5.

of other points is the same as in the case of O1. Further, vertical (resp. horizontal) strips are used for blue (resp. red) sets.

O4: The encoding is given in Fig. 7(d). We place the points $s_1^1, s_1^2, s_1^3, s_2^1, s_2^2, s_2^3,$ $\ldots, s_n^1, s_n^2, s_n^3,$ in the order from bottom to top, on a line parallel to $y = x - 1$. Further, place the points $t_1^1, t_1^2, t_1^3, t_1^4, t_2^1, t_2^2, t_2^3, t_2^4, \ldots, t_m^1, t_m^2, t_m^3, t_m^4$ on a line $y = x + 1$ in the same order from bottom to top. Place a rectangle for each set in the instance of *SPECIAL-RBHS* covering only the respective points. These rectangles can be placed such that each one covers the origin of the plane.

O5: Place the points $s_1^1, s_1^2, s_1^3, s_2^1, s_2^2, s_2^3, \ldots, s_n^1, s_n^2, s_n^3,$ in the same order on a line parallel to $y = -x$ and place the points $t_1^1, t_1^2, t_1^3, t_1^4, t_2^1, t_2^2, t_2^3, t_2^4, \ldots, t_m^1, t_m^2, t_m^3, t_m^4$ on the line $y = x - 1$. Finally, place the objects for the sets in the instance of *SPECIAL-RBHS* problem that covers only the respective points in each set. The encoding is given in Fig. 7(e).   □

References

1. Alimonti, P., Kann, V.: Some APX-completeness results for cubic graphs. Theoret. Comput. Sci. **237**(1), 123–134 (2000)
2. de Berg, M., Khosravi, A.: Optimal binary space partitions for segments in the plane. Int. J. Comput. Geom. Appl. **22**(3), 187–206 (2012)
3. Carr, R.D., Doddi, S., Konjevod, G., Marathe, M.: On the red-blue set cover problem. In: SODA, pp. 345–353 (2000)
4. Chan, T.M., Grant, E.: Exact algorithms and APX-hardness results for geometric packing and covering problems. Comput. Geom. **47**(2), 112–124 (2014)
5. Chan, T.M., Hu, N.: Geometric red blue set cover for unit squares and related problems. Comput. Geom. **48**(5), 380–385 (2015)
6. Chang, M., Chung, H., Lin, C.: An improved algorithm for the red-blue hitting set problem with the consecutive ones property. Inf. Process. Lett. **110**(20), 845–848 (2010)
7. Dom, M., Guo, J., Niedermeier, R., Wernicke, S.: Red-blue covering problems and the consecutive ones property. J. Discrete Algorithms **6**(3), 393–407 (2008)
8. Erlebach, T., van Leeuwen, E.J.: PTAS for weighted set cover on unit squares. In: APPROX/RANDOM, pp. 166–177 (2010)
9. Feige, U.: A threshold of $\ln n$ for approximating set cover. J. ACM **45**(4), 634–652 (1998)
10. Fowler, R.J., Paterson, M.S., Tanimoto, S.L.: Optimal Packing and Covering in the Plane are NP-Complete. Inf. Process. Lett. **12**(3), 133–137 (1981)
11. Garey, M.R., Johnson, D.S.: The rectilinear steiner tree problem is NP-complete. SIAM J. Appl. Math. **32**(4), 826–834 (1977)
12. Garey, M.R., Johnson, D.S.: Computers and Intractability; A Guide to the Theory of NP-Completeness. W. H. Freeman & Co., New York (1990)
13. Li, J., Jin, Y.: A PTAS for the weighted unit disk cover problem. In: Automata, Languages, and Programming, pp. 898–909 (2015)
14. Lund, C., Yannakakis, M.: On the hardness of approximating minimization problems. J. ACM **41**(5), 960–981 (1994)
15. Madireddy, R.R., Mudgal, A.: Approximability and hardness of geometric hitting set with axis-parallel rectangles. Inf. Process. Lett. **141**, 9–15 (2019)

16. Madireddy, R.R., Nandy, S.C., Pandit, S.: On the geometric red-blue set cover problem. In: WALCOM, pp. 129–141 (2021)
17. Mudgal, A., Pandit, S.: Geometric hitting set and set cover problem with half-strips. In: CCCG (2014)
18. Mudgal, A., Pandit, S.: Geometric hitting set, set cover and generalized class cover problems with half-strips in opposite directions. Discret. Appl. Math. **211**, 143–162 (2016)
19. Mustafa, N.H., Ray, S.: Improved Results on Geometric Hitting Set Problems. Discrete Comput. Geom. **44**(4), 883–895 (2010). https://doi.org/10.1007/s00454-010-9285-9
20. Papadimitriou, C.H., Yannakakis, M.: Optimization, approximation, and complexity classes. J. Comput. Syst. Sci. **43**(3), 425–440 (1991)
21. Shanjani, S.H.: Hardness of approximation for red-blue covering. In: CCCG (2020)

Bounds for the Oriented Diameter
of Planar Triangulations

Debajyoti Mondal[1][iD], N. Parthiban[2][✉], and Indra Rajasingh[3]

[1] Department of Computer Science, University of Saskatchewan, Saskatoon, Canada
dmondal@cs.usask.ca
[2] Department of Data Science and Business Systems, School of Computing,
SRM Institute of Science and Technology, Kattankulathur, India
parthiban24589@gmail.com
[3] Division of Mathematics, Saveetha School of Engineering, Saveetha Institute of
Medical and Technical Sciences, Chennai, India

Abstract. The diameter of an undirected or a directed graph is defined
to be the maximum shortest path distance over all pairs of vertices in
the graph. Given an undirected graph G, we examine the problem of
assigning directions to each edge of G such that the diameter of the
resulting oriented graph is minimized. The minimum diameter over all
strongly connected orientations is called the oriented diameter of G. The
problem of determining the oriented diameter of a graph is known to be
NP-hard, but the time-complexity question is open for planar graphs.
In this paper we compute the exact value of the oriented diameter for
triangular grid graphs. We then prove an $n/3$ lower bound and an $n/2 +
O(\sqrt{n})$ upper bound on the oriented diameter of planar triangulations. It
is known that given a planar graph G with bounded treewidth and a fixed
positive integer k, one can determine in linear time whether the oriented
diameter of G is at most k. In contrast, we consider a weighted version
of the oriented diameter problem and show it to be weakly NP-complete
for planar graphs with bounded pathwidth.

Keywords: Oriented diameter · Planar graph · Separator · Triangular
grid

1 Introduction

An undirected graph is called *oriented* when each edge of the graph is assigned
an orientation. Computing such orientations often requires the resulting directed
graph to be *strongly connected*, i.e., every vertex in the directed graph must be
reachable from every other vertex. This is useful in transforming two-way traffic or communication networks to one-way networks especially when one-way
communication is preferred or more cost effective over two-way communication

The work of D. Mondal is supported by the Natural Sciences and Engineering Research
Council of Canada (NSERC).

channels [1,28], as well as finds application in the context of network broadcasting and gossiping [10,20].

The *diameter* of a directed or undirected graph is the maximum shortest path distance over all pairs of vertices, where the distance of a path is measured by the number of its edges. The oriented diameter $OD(G)$ of an undirected graph G is the smallest diameter over all the strongly connected orientations of G. In 1978, Chavátal et al. [3] proved that determining whether the oriented diameter of a graph is at most two is NP-complete. They showed that the oriented diameter of a 2-edge-connected graph with diameter 2 is at most 6, and there exist graphs achieving this upper bound. For graphs with diameter 3, the known upper and lower bounds are 9 and 11, respectively [22]. Several studies attempted to provide good upper bounds on the orientated diameter problem of connected and bridgeless graphs [4,7,21]. In 2001, Fomin et al. [7] discovered the relation $OD(G) \leq 9\gamma(G) - 5$ between the oriented diameter and the size $\gamma(G)$ of a minimum dominating set. Dankelmann et. al. [4] showed that every bridgeless graph G of order n and maximum degree Δ has an orientated diameter at most $n - \Delta - 3$.

A rich body of literature examined oriented diameter for interconnection networks [9,16,18,26] and for various interesting graph classes. Gutin et al. [13] studied some well-known classes of strong digraphs where oriented diameter exceeds the diameter of the underlying undirected graph only by a small constant. Fujita et al. [11] considered the problem of finding the minimum oriented diameter of star graphs and proved an upper bound of $\lceil 5n/2 \rceil + 2$ for any $n \geq 3$, which is a significant improvement over the upper bound $2n(n-1)$ derived by Chvátal and Thomassen [3]. Fomin et al. [8] showed that computing oriented diameter remains NP-hard for split graphs and provided approximation algorithms for chordal graphs. Later, they showed that the oriented diameter of an AT-free graph is upper bounded by a linear function in its graph diameter [9].

We consider oriented diameter of planar graphs. Eggemann and Noble [6] showed that given a planar graph G with bounded treewidth and a fixed positive integer k, one can determine in linear time whether the oriented diameter of G is at most k. They also showed how to remove the dependency on the treewidth and gave an algorithm that given a fixed positive integer k, can decide whether the oriented diameter of a planar graph is at most k in linear time. Recently, Wang et al. [30] have showed that the diameter of a maximal outerplanar graph with at least three vertices is upper bounded by $\lceil n/2 \rceil$ with four exceptions, and the upper bound is sharp.

Our Contribution: In this paper we compute the exact value on the oriented diameter for triangular grid graphs. This result relates to the vast literature that attempts to compute diameter preserving or optimal orientation for well-known graph classes (e.g., for two-dimensional torus [19], two-dimensional grid [27,28], hypercube [23], products of graphs [17]). We then generalize the idea of computing oriented diameter of triangular grid graphs to give an algorithm for planar triangulations. Given a planar triangulation with n vertices, we show how to orient its edges such that the diameter of the resulting oriented graph is upper bounded

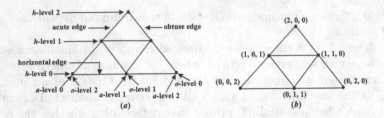

Fig. 1. The triangular-grid network T_2. Illustration for (a) levels and (b) 3-tuples.

by $n/2 + O(\sqrt{n})$. This is interesting since $\lceil n/2 \rceil$ is already known to be an upper bound on the oriented diameter for the maximal outerplanar graphs [30]. We next show that there exist planar triangulations with oriented diameter at least $n/3$. Finally, we show that the weighted version of the oriented diameter problem is weakly NP-complete even for planar graphs of bounded pathwidth, which contrasts the linear-time algorithm of Eggemann and Noble [6] for the unweighted variant.

2   Oriented Diameter of a Triangular Grid

In this section we compute the exact value of the oriented diameter of triangular grid graphs. A triangular tessellation of the plane with equilateral triangles is called a triangular sheet or triangular grid. The vertices are the intersection of lines and the lines between two vertices are the edges.

Definition 1. ([31]). *Consider the ordered 3-tuples of integers (i, j, k) such that $i + j + k = r$. Let these 3-tuples represent the vertices and let two vertices be joined if the sum of the absolute differences of their coordinates is 2. The graph generated is referred to as a triangular grid graph T_r of dimension r.*

There are $r + 1$ levels in a triangular grid T_r and in each level i, there are $i + 1$ vertices, $0 \leq i \leq r$. This implies that T_r has $(r+1)(r+2)/2$ vertices and $3r(r+1)/2$ edges. Diameter of T_r is r. The edges of T_r can be partitioned into horizontal edges, acute edges, and obtuse edges. The shortest path comprising of horizontal (acute, obtuse) edges with end vertices of degree 2 in T_r is said to be at h-level (a-level, o-level) 0. Inductively, the path through the parents of vertices at level i is said to be at h-level (a-level, o-level) $i+1$, $0 \leq i \leq r-1$. The vertex $v = (i, j, k)$ in T_r is the point of intersection of the paths representing its h-level i, a-level j, and o-level k. See Figs. 1(a) and 1(b).

A simple observation on the length of a shortest cycle passing through any two vertices of length 2 in T_r, $r \geq 2$ yields the following result.

Lemma 1. *Let G be the triangular grid T_r, $r \geq 2$. Then $OD(G) \geq r + 1$.*

Proof. The diameter of T_r is r, which is realized between two end vertices of h-*level* 0 line. Any shortest cycle in G passing through these two end vertices has length $r + (r+1) = 2r + 1$. Hence $OD(G) \geq r + 1$.

The following algorithm yields an oriented triangular grid with diameter $r+1$. In the sequel, if $\vec{P}$ and $\vec{Q}$ are directed paths then $\vec{P} \circ \vec{Q}$ denotes the concatenation of the paths $\vec{P}$ and $\vec{Q}$ with end of $\vec{P}$ as the beginning of $\vec{Q}$.

Input: A triangular grid graph T_r of dimension r.
Output: An orientation of T_r with diameter $r + 1$.
Algorithm: Direct the h-*level*0 line from right to left; a-*level*0 line from bottom to top; o-*level*0 line from top to bottom. Direct all other horizontal lines from left to right, acute lines from top to bottom and the obtuse lines from bottom to top.

Proof of Correctness: Given any two vertices u and v in T_r, we have to show that there exist a (u,v)-directed path and a (v,u)-directed path, both of length at most $r + 1$. Let u be (i,j,k) and v be (l,m,n). Without loss of generality assume that $i \leq l$.

Case 1(a): $j < m$ and $i + j \geq l$

Path from u to v: Let $\vec{P}$ be the directed path from u along the o-level k till it reaches vertex w in the h-level l. Let $\vec{Q}$ be the directed path from w along the h-level l till it reaches vertex v. Then $\vec{P} \circ \vec{Q}$ is a directed path from u to v of length $(l - i) + (k - n) \leq j + k - n = r - (i+n) \leq r$. See Fig. 2(a).

Path from v to u: When $(i+j) < m$, let $\vec{P}$ be the directed path from v along the a-level m till it reaches vertex w in h-level 0. Let $\vec{Q}$ be the directed path from w along h-level 0 till it reaches x at o-level k. Let $\vec{R}$ be the directed path from x along o-level k till it reaches vertex u. Then $\vec{P} \circ \vec{Q} \circ \vec{R}$ is a directed path from v to u of length $l + (m - (i + j)) + i = r - n - j < r$. See Fig. 2(b). When $m \leq i + j$, let $\vec{P}$ be the directed path from v along the a-level m till it reaches vertex w in o-level k; let $\vec{Q}$ be the directed path from w along o-level k till it reaches u. Then $\vec{P} \circ \vec{Q}$ is a directed path from v to u of length at most of $l + (m - j) = r - n - j < r$. See Fig. 2(c).

Case 1(b): $j < m$ and $i + j < l$

Path from u to v: Let $\vec{P}$ be the directed path from u along the o-level k till it reaches vertex w on the a-level 0. Let $\vec{Q}$ be the directed path from w along the a-level 0 till it reaches vertex x at h-level l. Let $\vec{R}$ be the directed path from x along h-level l till it reaches v. Then $\vec{P} \circ \vec{Q} \circ \vec{R}$ is a directed path from u to v of length $j + (l - (i + j)) + m = l + m - i = r - i - n < r$. See Fig. 2(d).

Path from v to u: When $(i + j) < m$, let $\vec{P}$ be the directed path from v along the a-level m till it reaches vertex w in h-level 0. Let $\vec{Q}$ be the directed path from w along h-level 0 till it reaches x at o-level k. Let $\vec{R}$ be the directed path from x along o-level k till it reaches vertex u. Then $\vec{P} \circ \vec{Q} \circ \vec{R}$ is a directed path from v to u of length $l + (m - (i + j)) + i = r - n - j < r$. See Fig. 2(e). When $m \leq i + j$, let $\vec{P}$ be the directed path from v along the a-level m till it reaches vertex w in o-level k; let $\vec{Q}$ be the directed path from w along o-level k till it reaches u. Then $\vec{P} \circ \vec{Q}$ is a directed path from v to u of length at most of $l + (m - j) = r - n - j < r$. See Fig. 2(f).

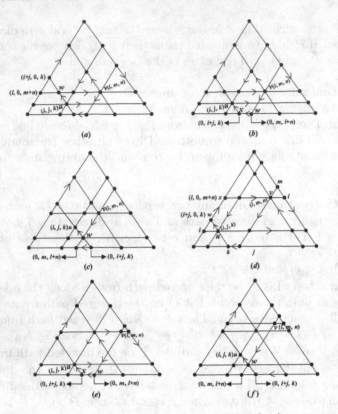

Fig. 2. (a) $\vec{P}$: (u, w)-directed path, $\vec{Q}$:(w, v)-directed path, $\vec{P} \circ \vec{Q}$ is directed path from u to v, (b) $\vec{P}$: (v, w)-directed path, $\vec{Q}$: (w, x)-directed path, $\vec{R}$: (x, u)-directed path, $\vec{P} \circ \vec{Q} \circ \vec{R}$ is directed path from v to u, (c) $\vec{P}$: (v, w)-directed path, $\vec{Q}$: (w, u)-directed path, $\vec{P} \circ \vec{Q}$ is directed path from v to u, (d) $\vec{P}$: (u, w)-directed path, $\vec{Q}$:(w, x)-directed path, $\vec{R}$: (x, v)-directed path, $\vec{P} \circ \vec{Q} \circ \vec{R}$ is (u, v)-directed path, (e) $\vec{P}$: (v, w)-directed path, $\vec{Q}$: (w, x)-directed path, $\vec{R}$: (x, u)-directed path, $\vec{P} \circ \vec{Q} \circ \vec{R}$ is directed path from v to u and (f) $\vec{P}$: (v, w)-directed path, $\vec{Q}$: (w, u)-directed path, $\vec{P} \circ \vec{Q}$ is directed path from v to u.

Case 2: Here we consider the two subcases: (a) $j \geq m$ and $i + k \geq l$, and (b) $j \geq m$ and $i + k < l$. The proof for this is similar to Case 1 and is included in the full version [25].

Case 3: Both the vertices u and v lie in the same h-level, a-level or o-level. As long as u and v are not both vertices of degree 2 in T_r, arguments similar to that of Case 1 show that $\vec{d}(u, v)$ and $\vec{d}(v, u)$ are at most r. Suppose that both u and v are of degree 2 in T_r, then if $\vec{d}(u, v) = r$, then $\vec{d}(v, u) = r + 1$. Thus the oriented graph T_r has diameter $r + 1$.

The above analysis of the algorithm for triangular grid graphs and Lemma 1 yield the following result.

Theorem 1. *Let G be the triangular $T_r, r > 2$. Then $OD(G) = r + 1$.*

3   Upper Bound on the Oriented Diameter of a Planar Triangulation

In this section we provide an $n/2 + O(\sqrt{n})$ upper bound on the oriented diameter of a planar triangulation. The idea is to use edge orientations similar to the one we used for the triangular grid, i.e., three outgoing edges at each internal vertex with interleaving incoming edges and a clockwise orientation on the outer face. In fact, every planar triangulation is known to admit such an orientation for its internal vertices, which is obtained by via a 'Schnyder realizer'. Since we use this concept extensively, we first briefly review some preliminary definitions and notation related to planar graphs and Schnyder realizer.

3.1   Planar Graphs and Schnyder Realizer

A *planar graph* is a graph that can be drawn on the Euclidean plane such that no two edges cross except possibly at their common endpoint. A *plane graph* G is a planar graph with a fixed planar embedding on the plane. G delimits the plane into connected regions called faces. The unbounded face is the outer face of G and all the other faces are the inner faces of G. G is called *triangulated* if every face (including the outerface) of G contains exactly three vertices on its boundary. The vertices on the outer face of G are called the *outer vertices* and all the remaining vertices are called the *inner vertices*. The edges on the outer face are called the *outer edges* of G.

Let $G = (V, E)$ be a triangulated plane graph with the outer vertices v_m, v_r and v_l in clockwise order on the outer face. Then the internal edges of G can be directed in a way such that every inner vertex has three directed paths which are vertex disjoint (except that they start at the common vertex v) and end at the three outer vertices [29]. In other words, there are three directed trees T_m, T_r and T_l rooted at v_m, v_r and v_l, respectively, that span all the internal vertices of G. Let us denote the edges of T_m, T_r and T_l as the m-edges, r-edges and l-edges, respectively. Then the edges at every internal vertex v of G are directed in a certain order, as shown in Fig. 3(c). The trees T_m, T_r and T_l are referred to as *Schnyder realizer*. Figure 3(a) illustrates a Schnyder realizer and Fig. 3(b) shows the tree T_l separately. The trees T_m, T_r and T_l form a partition of all the graph edges except the edges in the outer face.

There exists a Schnyder realizer T_l, T_r, T_m of G with $leaf(T_l) + leaf(T_r) + leaf(T_m) = 2n - 5 - \delta_0$ [2], where $0 \le \delta_0 \le \lfloor (n-1)/2 \rfloor$ is the number of cyclic faces. It is known that there exists a Schnyder realizer where one tree has at least $\lceil (n+1)/2 \rceil$ leaves and such a realizer can be computed in linear time [32].

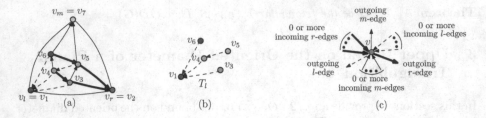

Fig. 3. (a) A Schnyder realizer. (b) Illustration for T_l. (c) Illustration for the edge orientations around a vertex.

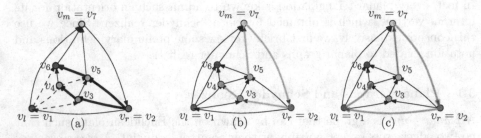

Fig. 4. (a) Illustration for the algorithm of Sect. 3.2. (b–c) The paths $\vec{Q}$ and $\vec{Q'}$, which are shown in red and blue, respectively. Here $v = v_5$ and $w = v_4$. (Color figure online)

3.2   An Initial Upper Bound

We first give an algorithm to compute a $\lceil 3(n-1)/4 \rceil + 2$ upper bound on oriented diameter of G, and in the subsequent section we improve the bound to $n/2 + O(\sqrt{n})$.

Algorithm: We first compute a Schnyder realizer T_l, T_r, T_m such that one tree (assume without loss of generality that T_m) has at least $\lceil \frac{(n+1)}{2} \rceil$ leaves [32]. We now assign the edge orientations as follows (Fig. 4a).

Orient the outer edges of G in clockwise order: $\overrightarrow{(v_m, v_r)}$, $\overrightarrow{(v_r, v_l)}$ and $\overrightarrow{(v_l, v_m)}$. Orient the m-edges such that each vertex of T_m points at its ancestor in T_m. Orient the l-edges such that each vertex of T_l points at its descendent in T_l. Orient the r-edges such that each vertex of T_r points at its descendent in T_r.

Proof of Correctness: We now show that for every pair of vertices u, w in G, their shortest path length is bounded by at most $\lceil 3(n-1)/4 \rceil + 2$.

Case 1 (Both u, w are Internal Vertices of G): Let $\vec{P}_{u,v_m}$ be the path that starts at u and follows the m-edges to reach v_m. Let $\vec{P}_{v_r,w}$ be the path that starts at v_r and follows the r-edges to reach w. Similarly, let $\vec{P}_{v_l,w}$ be the path that starts at v_l and follows the l-edges to reach w.

Consider now two u to w paths (Fig. 4b–c). One is the path $\vec{Q}$ that first travels along $\vec{P}_{u,v_m}$, then along the edge $\overrightarrow{(v_m, v_r)}$ and finally, along $\vec{P}_{v_r,w}$. The other is the path $\vec{Q'}$ that first travels along $\vec{P}_{u,v_m}$, then along the edges $\overrightarrow{(v_m, v_r)}$ and $\overrightarrow{(v_r, v_l)}$, and finally, along $\vec{P}_{v_l,w}$. We now show that at least one of these two paths are of length at most $\lceil 3(n-1)/4 \rceil + 2$.

Since T_m has at least $\lceil (n+1)/2 \rceil$ leaves, the length β of $\vec{P}_{u,v_m}$ is at most $\lceil n - \frac{(n+1)}{2} \rceil = \lceil (n-1)/2 \rceil$. Since the paths $\vec{P}_{v_r,w}$ and $\vec{P}_{v_l,w}$ are vertex disjoint (except that they start at w), the sum of their lengths is at most $(n - \beta - 1)$. Here the -1 term represents that we can safely skip the vertex v_m. We now can assume without loss of generality that $\vec{P}_{v_l,w}$ has at most $\lceil (n - \beta - 1)/2 \rceil$ edges, where $\beta \leq \lceil (n-1)/2 \rceil$. Therefore, the length of $\vec{Q'}$ is bounded by at most $\beta + \lceil (n - \beta - 1)/2 \rceil + 2 = \lceil (n-1)/2 + \beta/2 \rceil + 2 = \lceil 3(n-1)/4 \rceil + 2$.

Similarly, we can find a w to u path of length at most $\lceil 3(n-1)/4 \rceil + 2$ by swapping the roles of u and w in the argument above.

Case 2 (At Least One of u, w is an Outer Vertex of G): If both u, w are outer vertices then they can reach each other following at most two outer edges. If exactly one is an outer vertex, then without loss of generality assume u be the inner vertex and w be the outer vertex. To compute the u to w path we first travel to v_m via m-edges and then reach w by visiting outer edges. To compute the w to u path, we consider two paths (as we did in Case 1): one that goes through v_r and the other that goes through v_l. In both scenarios, we can use the analysis of Case 1 to find a path of length at most $\lceil 3(n-1)/4 \rceil + 2$.

3.3   An Improved Upper Bound

In this section we improve the upper bound to $n/2 + O(\sqrt{n})$ by leveraging the concept of planar separator. Let G be a triangulated plane graph with n vertices. A simple cycle C in G is called a *simple cycle separator* if the interior and the exterior of C each contains at most $2n/3$ vertices. Every planar graph admits a simple cycle separator of size $O(\sqrt{n})$ [5,12,24].

Let C be a simple cycle separator of size $O(\sqrt{n})$ in G. Let G_{in} be the graph induced by the vertices of C and the vertices inside C. We create a triangulated graph G'_{in} by adding a vertex s_{out} to all the vertices of C. Similarly, we define G_{out} to be the graph induced by the vertices of C and the vertices outside C, and create G'_{out} by adding a vertex s_{in} to all the vertices of C. We then take a planar embedding of G'_{out} such that s_{in} lies on the outerface. Figure 5 illustrates a separator of G and the corresponding G'_{in} and G'_{out}.

Let n_{in} and n_{out} be the number of vertices of G'_{in} and G'_{out}, respectively. We now compute an orientation σ for the edges of G'_{in} such that the shortest path distance between every pair of vertices is at most $\lceil 3(n_{in}-1)/4 \rceil + 2$ as we did in Sect. 3.2. We then remove the orientation of the edges on C and reorient them clockwise. Let the resulting orientation be σ'. We now show that using σ' instead of σ and avoiding s_{out} increases the shortest path distance between a pair of vertices in G_{in} by at most $O(\sqrt{n})$.

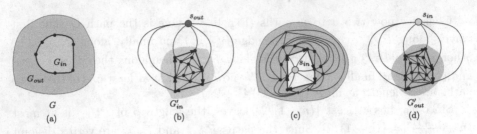

Fig. 5. (a) Illustration for a simple cycle separator. (b–d) The construction of the plane triangulations G'_{in} and G'_{out}. The red vertices in (c) are the outer vertices and they become inner vertices in (d). (Color figure online)

We consider two cases depending on whether the tree T rooted at s_{out} has the maximum number of leaves in the Schnyder realizer of G'_{in}.

Case 1 (T has the Maximum Number of Leaves): In Sect. 3.2 we constructed a path $\vec{P}$ from an inner vertex u to another inner vertex w such that it first moves from u to an outer vertex and then traverses along the outer cycle and finally, travels towards w. Such a path can also be constructed using σ. However, in σ', we need to restrict ourselves within the edges of G_{in}. Let x be a vertex of C and the first such vertex that we encounter while travelling from v to w following $\vec{P}$. Similarly, let y be a vertex of C and the last such vertex that we encounter while travelling from v to w following $\vec{P}$. We now replace the subpath from x to y using a clockwise path on C. This results into a path of length at most $\lceil 3(n_{in} - 1)/4 \rceil + O(\sqrt{n})$ that uses the orientation of σ'. We can find a w to v path of length at most $\lceil 3(n_{in} - 1)/4 \rceil + O(\sqrt{n})$ using the same argument by swapping the roles of u and w. It is straightforward to observe that the argument holds even when one of u and w, or both lie on the cycle separator.

Case 2 (T Does Not Have the Maximum Number of Leaves): Without loss of generality assume that the tree rooted at an outer vertex $v \neq s_{out}$ has the maximum number of leaves in the Schnyder realizer of G'_{in}. The argument here is the same as that of Case 1. To reach from an inner vertex u to another inner vertex w, we take a path $\vec{P}$ from u to v and then traverse along the outer cycle and finally, travel towards w. Since we need to avoid s_{out}, we can leverage the clockwise cycle C. Define the vertices x and y similar to that of Case 1. We then replace the subpath of $\vec{P}$ from x to y using a clockwise path on C. This results into a path of length at most $\lceil 3(n_{in} - 1)/4 \rceil + O(\sqrt{n})$ that uses the orientation of σ'. We can find a w to v path using the same argument by swapping the role of u and w. It is straightforward to observe that the argument holds even when one of u and w or both lie on the cycle separator.

We now compute an orientation σ'' for the edges of G'_{out} in the same way as we did for G'_{in}. In the following we show that the edge orientations of G obtained by taking the edge orientations from σ' and σ'' ensure an oriented diameter of $n/2 + O(\sqrt{n})$.

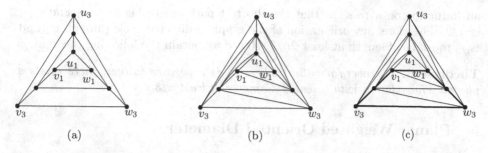

Fig. 6. (a) Illustration for a nested triangles graph. (b) A triangulated nested triangles graph. (c) A shortest cycle of $2n/3$ vertices through w_1 and $v_{n/3}$.

Consider a pair of vertices u and w in G. If both belong to G_{in}, then they can be reached from each other using a path of length $\lceil 3(n_{in} - 1)/4 \rceil + O(\sqrt{n}) = n/2 + O(\sqrt{n})$. The same argument holds if they both belong to G_{out}.

Assume now without generality that u belongs to G_{in} and w belongs to G_{out}. To find a u to w path, we first traverse the path P' that goes from u to a vertex on C (using σ') and then traverse clockwise along C, and finally take the path P'' that goes from a vertex of C to w (using σ''). By the analysis of Sect. 3.2 and then by leveraging σ', we obtain the length of P' to be at most $\lceil (n_{in} - 1)/2 \rceil + O(\sqrt{n})$. By the analysis of Sect. 3.2 we know that there are two vertex disjoint paths in G'_{out} that start at C and reach w. Hence the length of one of these paths is at most $\lceil n_{out}/2 \rceil$. After considering σ'', the corresponding path P'' would have a length of $\lceil n_{out}/2 \rceil + O(\sqrt{n})$. Therefore, the length of the u to w path is upper bounded by at most $n/2 + O(\sqrt{n})$. We can construct a w to u path by swapping the role of u and w.

The computation of the simple cycle separator takes $O(n)$ time [24]. The computation of the required Schnyder realizer also takes $O(n)$ time [32]. Hence it is straightforward to implement the algorithm in linear time. The following theorem summarizes the result of this section.

Theorem 2. *The oriented diameter of a planar triangulation with n vertices is $n/2 + O(\sqrt{n})$ and such an orientation can be computed in $O(n)$ time.*

4 Lower Bound

The lower bound is determined by the nested triangles graph G_n, which is defined as follows.

For $n = 3$, the graph G_3 is a cycle u_1, v_1, w_1 of three vertices. For $n = 3m$, where $m > 1$ is a positive integer, G_n is obtained by enclosing G_{n-1} inside a cycle $u_{n/3}, v_{n/3}, w_{n/3}$ of three vertices and then adding the edges $(u_{n/3-1}, u_{n/3})$, $(v_{n/3-1}, v_{n/3})$, and $(w_{n/3-1}, w_{n/3})$. See Fig. 6(a).

We triangulate the graph G_n by adding the edges (v_i, w_{i+1}), (v_i, u_{i+1}) and (w_i, u_{i+1}), where $1 \leq i < n/3$. See Fig. 6(b). It is now straightforward to employ an induction on n to show that the shortest path length between w_1 and $v_{n/3}$ is $n/3$. Therefore, any orientation of the graph, a directed cycle through w_1 and $v_{n/3}$ must be of length at least $2n/3$. Hence we obtain the following theorem.

Theorem 3. *For every $n = 3m$, where m is a positive integer, there exists a planar triangulation with oriented diameter at least $n/3$.*

5   Planar Weighted Oriented Diameter

In this section we consider the weighted oriented diameter. Formally, given an edge weighted graph, the *weighted oriented diameter* is the minimum weighted diameter over all of its strongly connected orientations.

If the edge weights are small, then one can find an orientation with small weighted diameter. The following corollary is a direct consequence of Theorem 2.

Corollary 1. *Let G be an edge weighted planar graph with n vertices. Assume that the weight of each edge is at most $(1 + \epsilon)/n$, where $0 \le \epsilon < 1$, and the sum of all weights in 1. Then in linear time, one can compute an orientation of G with weighted oriented diameter at most $\frac{(1+\epsilon)}{2} + \frac{1}{O(\sqrt{n})}$.*

In the following section we show that the decision version of the weighted oriented diameter problem is weakly NP-complete.

5.1   Planar Weighted Oriented Diameter Is Weakly NP-complete

Here we show that given a planar graph G and an integer k, it is weakly NP-complete to decide whether G has a strongly connected orientation of diameter at most k. We formally define the problem as follows:

Problem: PLANAR WEIGHTED ORIENTED DIAMETER
Input: An undirected connected planar graph G, where each edge is weighted with a positive real number, and a positive integer D.
Question: Is there a strongly connected orientation of G such that the weighted diameter of the resulting oriented graph is at most D?

The problem is in NP because given an orientation of the edges of G, one can verify whether the weighted oriented diameter is within D in polynomial-time by computing all pair shortest paths.

We will reduce the weakly NP-complete problem PARTITION, which is defined as follows:

Problem: PARTITION
Input: A multiset S of positive integers.
Question: Can S can be partitioned into two subsets with equal sum?

Given an instance $S = \{a_1, a_2, \ldots, a_n\}$ of PARTITION, we construct an instance $I = (G, D)$ of PLANAR WEIGHTED ORIENTED DIAMETER as follows:

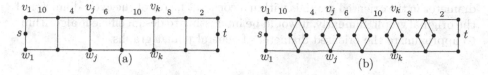

Fig. 7. (a–b) Construction of G from S.

Step 1. Take a $2 \times (n+1)$ grid graph, e.g., Fig. 7(a). Let $v_1, \ldots, v_{n+1}$ be the vertices at the top row and let $w_1, \ldots, w_{n+1}$ be the vertices at the bottom row and set the weights of the top edges $(v_1, v_2), \ldots, (v_n, v_{n+1})$ using the numbers $a_1, \ldots, a_n$ from left to right.

Step 2. Subdivide the leftmost vertical edge (v_1, w_1) with a division vertex s. Similarly, subdivide the rightmost vertical edge with a division vertex t.

Step 3. Replace each internal vertical edge (v_i, w_i), where $1 < i < n+1$, with a cycle $v_i, d_{v_i w_i}, w_i, d'_{v_i w_i}, v_i$, where $d_{v_i w_i}$ and $d'_{v_i w_i}$ are two new vertices, e.g., Fig. 7(b). We will refer to these cycles as *inner cycles*.

Step 4. Set the weight of all the edges except for the top edges to $\epsilon = \frac{1}{m}$, where m is the number of edges in the graph. Set D to be equal to $1 + \frac{1}{2} \sum_{i=1}^{n} a_i - \epsilon$.

We show that S admits a bipartition if and only if G admits a strong orientation with diameter at most D and thus prove the following theorem. The proof is included in the full version [25].

Theorem 4. PLANAR WEIGHTED ORIENTED DIAMETER *is weakly NP-complete.*

Since the pathwidth of G is $O(1)$, we obtain the following corollary.

Corollary 2. PLANAR WEIGHTED ORIENTED DIAMETER *is weakly NP-complete, even when the pathwidth of the input graph is bounded by a constant.*

6 Conclusion

In this paper we computed exact value of the oriented diameter for triangular grid graphs and proved an $n/3$ lower bound and an $n/2 + O(\sqrt{n})$ upper bound on the oriented diameter of planar triangulations. A natural direction for future research would be to close the gap between the lower bound and the upper bound. We also showed that the weighted version of the oriented diameter problem is weakly NP-complete. Although the time complexity of the unweighted version remains open, it would be interesting to examine whether the weighted version is strongly NP-hard.

A related concept here is cycle diameter, which is the maximum shortest cycle length through two vertices, where the maximum is taken over all pair of vertices in the graph. For every planar graph with positive edge weights, Guttmann-Beck and Hassin [14, 15] proved the existence of an orientation where the cycle diameter in the oriented graph is within a constant factor of the cycle diameter of the original graph. Similarly, it would be interesting to design efficient algorithms to approximate the oriented diameter of general planar graphs.

References

1. Ajish Kumar, K.S., Rajendraprasad, D., Sudeep, K.S.: Oriented **Diameter of Star Graphs**. In: Changat, M., Das, S. (eds.) CALDAM 2020. LNCS, vol. 12016, pp. 307–317. Springer, Cham (2020). https://doi.org/10.1007/978-3-030-39219-2_25
2. Bonichon, N., Le Saëc, B., Mosbah, M.: Optimal area algorithm for planar polyline drawings. In: Goos, G., Hartmanis, J., van Leeuwen, J., Kučera, L. (eds.) WG 2002. LNCS, vol. 2573, pp. 35–46. Springer, Heidelberg (2002). https://doi.org/10.1007/3-540-36379-3_4
3. Chvátal, V., Thomassen, C.: Distances in orientations of graphs. J. Combin. Theory, Ser. B **24**(1), 61–75 (1978)
4. Dankelmann, P., Guo, Y., Surmacs, M.: Oriented diameter of graphs with given maximum degree. J. Graph Theory **88**, 5–17 (2018)
5. Djidjev, H.N., Venkatesan, S.M.: Reduced constants for simple cycle graph separation. Acta Inform. **34**(3), 231–243 (1997)
6. Eggemann, N., Noble, S.D.: Minimizing the oriented diameter of a planar graph. Electron. Notes Discrete Math. **34**, 267–271 (2009)
7. Fomin, F.V., Matamala, M., Prisner, E., Rapaport, I.: Bilateral orientations and domination. Electron. Notes Discrete Math. **7** (2001)
8. Fomin, F.V., Matamala, M., Rapaport, I.: The complexity of approximating the oriented diameter of chordal graphs. Int. Workshop Graph-Theoretic Concepts Comput. Sci. **2573**, 211–222 (2002)
9. Fomin, F.V., Matamala, M., Rapaport, I.: AT-free graphs: linear bounds for the oriented diameter. Discret. Appl. Math. **141**, 135–148 (2004)
10. Fraigniaud, P., Lazard, E.: Methods and problems of communication in usual networks. Discret. Appl. Math. **53**(1–3), 79–133 (1994)
11. Fujita, S.: On oriented diameter of star graphs. In: First International Symposium on Computing and Networking, pp. 48–56 (2013)
12. Gazit, H., Miller, G.L.: Planar separators and the Euclidean norm. In: Asano, T., Ibaraki, T., Imai, H., Nishizeki, T. (eds.) SIGAL 1990. LNCS, vol. 450, pp. 338–347. Springer, Heidelberg (1990). https://doi.org/10.1007/3-540-52921-7_83
13. Gutin, G., Yeo, A.: Orientations of digraphs almost preserving diameter. Discret. Appl. Math. **121**, 129–138 (2002)
14. Guttmann-Beck, N., Hassin, R.: Minimum diameter and cycle-diameter orientations on planar graphs. arXiv e-prints pp. arXiv-1105 (2011)
15. Guttmann-Beck, N., Hassin, R.: Series-parallel orientations preserving the cycle-radius. Inf. Process. Lett. **112**(4), 153–160 (2012). https://doi.org/10.1016/j.ipl.2011.10.020
16. Koh, K.M., Tan, B.P., Rapaport, I.: The diameter of an orientation of a complete multipartite graph. Discret. Math. **149**(1–3), 131–139 (1996)
17. Koh, K.M., Tay, E.G.: On optimal orientations of cartesian products of graphs (ii): complete graphs and even cycles. Discret. Math. **211**, 75–102 (2000)
18. Koh, K.M., Tay, E.G.: Optimal orientations of graphs and digraphs: a survey. Graphs Comb. **18**(4), 745–756 (2002)

19. König, J., Krumme, D.W., Lazard, E.: Diameter-preserving orientations of the torus. Networks **32**(1), 1–11 (1998)
20. Krumme, D.W.: Fast gossiping for the hypercube. SIAM J. Comput. **21**(2), 365–380 (1992)
21. Kurzly, S., Latsch, M.: Bounds for the minimum oriented diameter. Discret. Math. Theoret. Computer Sci. **14**(1), 109–142 (2012)
22. Kwok, P.K., Liu, Q., West, D.B.: Oriented diameter of graphs with diameter 3. J. Combinat. Theory **100**(3), 265–273 (2010)
23. McCanna, J.E.: Orientations of the n-cube with minimum diameter. Discret. Math. **68**(2–3), 309–313 (1988)
24. Miller, G.L.: Finding small simple cycle separators for 2-connected planar graphs. J. Comput. Syst. Sci. **32**(3), 265–279 (1986)
25. Mondal, D., Parthiban, N., Rajasingh, I.: Oriented diameter of planar triangulations. CoRR abs/2203.04253 (2022). https://doi.org/10.48550/arXiv.2203.04253
26. Ng, K.L., Koh, K.M.: On optimal orientation of cycle vertex multiplications. Discret. Math. **297**(1–3), 104–118 (2005)
27. Roberts, F.S., Xu, Y.: On the optimal strongly connected orientations of city street graphs I: large grids. SIAM J. Discret. Math. **1**(2), 199–222 (1988)
28. Roberts, F.S., Xu, Y.: On the optimal strongly connected orientations of city street graphs. III. three east-west avenues or north-south streets. Networks **22**(2), 109–143 (1992)
29. Schnyder, W.: Embedding planar graphs on the grid. In: Proceedings of the 1st Annual ACM-SIAM Symposium on Discrete Algorithms (SODA), pp. 138–148. ACM, San Francisco, California, USA (1990)
30. Wang, X., Chen, Y., Dankelmann, P., Guo, Y., Surmacs, M., Volkmann, L.: Oriented diameter of maximal outerplanar graphs. J. Graph Theory **98**(3), 426–444 (2021)
31. West, D.B.: Introduction to Graph Theory. Prentice-Hall (2000)
32. Zhang, H., He, X.: Canonical ordering trees and their applications in graph drawing. Discret. Comput. Geometry **33**(2), 321–344 (2005)

String Rearrangement Inequalities and a Total Order Between Primitive Words

Ruixi Luo[ID], Taikun Zhu[ID], and Kai Jin[(✉)][ID]

School of Intelligent Systems Engineering, Shenzhen Campus of Sun Yat-sen
University, Shenzhen, China
{luorx,zhutk3}@mail2.sysu.edu.cn, jink8@mail.sysu.edu.cn

Abstract. We study the following rearrangement problem: Given n
words, rearrange and concatenate them so that the obtained string is
lexicographically smallest (or largest, respectively). We show that this
problem reduces to sorting the given words so that their repeating strings
are non-decreasing (or non-increasing, respectively), where the repeating
string of a word A refers to the infinite string $AAA\ldots$. Moreover, for fixed
size alphabet Σ, we design an $O(L)$ time sorting algorithm of the words
(in the mentioned orders), where L denotes the total length of the input
words. Hence we obtain an $O(L)$ time algorithm for the rearrangement
problem. Finally, we point out that comparing primitive words via comparing their repeating strings leads to a total order, which can further
be extended to a total order on the finite words (or all words).

Keywords: String rearrangement inequalities · Primitive words ·
Combinatorics on words · String ordering · Greedy algorithm

1 Introduction

Combinatorics on words (MSC: 68R15) have strong connections to many fields
of mathematics and have found significant applications to theoretical computer
science and molecular biology (DNA sequences) [5,8,10,14,17,21]. Particularly,
the primitive words over some alphabet Σ have received special interest, as
they have applications in the formal languages and algebraic theory of codes
[12,13,16,19]. A word is *primitive* if it is not a proper power of a shorter word.

In this paper, we consider the following rearrangement problem of words:
Given n words $A_1, \ldots, A_n$, rearrange and concatenate these words so that the
obtained string S is lexicographically smallest (or largest, respectively). We prove
that the lexicographical smallest outcome of S happens when the words are
arranged so that their repeating strings are increasing, and the largest outcome

Kai Jin is supported by National Natural Science Foundation of China 62002394,
and Shenzhen Science and Technology Program (Grant No. 202206193000001,
20220817175048002).

of S happens when the words are arranged reversely; see Lemma 6. Throughout, the *repeating string* of a word A refers to the infinite string $R(A) = AAA\ldots$.

Based on the above lemma (we suggest to name its results as "string rearrangement inequalities"), the aforementioned rearrangement problem reduces to sorting the words $A_1, \ldots, A_n$ so that $R(A_1) \leq \ldots \leq R(A_n)$. We show how to sort for the special case where $A_1, \ldots, A_n$ are primitive and distinct in $O(\sum_i |A_i|)$ time. The general case can be easily reduced to the special case and can be solved in the same time bound. Note that we assume bounded alphabet Σ and the size of Σ is fixed. Moreover, $|X|$ always denotes the length of word X.

Our algorithm beats the plain algorithm based on sorting (via comparing several pairs $R(A_i), R(A_j)$) by a factor of $\log n$. The algorithm is simple – it only applies basic data structures such as tries and the failure function [10,21]. Nevertheless, its correctness and running time analysis is non-straightforward.

We mention that comparing primitive words via comparing their repeating strings leads to a total order $\leq_\infty$ on primitive words, which can be extended to a total order $\leq_\infty$ on all words (Sect. 5). We show that this order is the same as the lexicographical order over Lyndon words but are different over primitive words and finite words. It is also different from *reflected lexicographic order, co-lexicographic order, shortlex order, Kleene-Brouwer order, V-Order, alternative order* [1–3,9,11]. It seems that order $\leq_\infty$ has not been reported in literature.

1.1 Related Work

It is shown in [6] that the language of Lyndon words is not context-free. Also, many people conjectured that the language of primitive words is not context-free [6,12,13,19]. But this conjecture is unsettled thus far, to the best of our knowledge. It would be interesting to explore whether the results shown in this paper can be helpful for solving this longstanding open problem in the future. See more introductions about primitive words in [16].

Fredricksen and Maiorana [15] showed that if one concatenates, in lexicographic order, all the Lyndon words that have length dividing a given number n, the result is a de Bruijn sequence. Au [4] further showed that if "dividing n" is replaced by "identical to n", the result is a sequence which contains exactly once every primitive word of length n as a factor. Note that concatenating some Lyndon words by lexicographic order is the same as concatenating by $\leq_\infty$ order.

The Lyndon words have many interesting properties and have found plentiful applications, both theoretically and practically. Among others, they are used in constructing de Brujin sequence as mentioned above (which have found applications in cryptography), and they are applied in proving the "runs theorem" [5,8,20]. The famous Chen-Fox-Lyndon Theorem states that any word W can be uniquely factorized into $W = W_1 W_2 \ldots W_m$, such that each W_i is a Lyndon word, and $W_1 \geq \ldots \geq W_m$ [7,14] (Here $\geq$ refers to the opposite of Lexicographical order, but is the same as the opposite of $\leq_\infty$). This factorization is used in the computation of runs in a word [8]. See the Bible of combinatorics on words [17] for more introductions about Lyndon words and primitive words.

2 Preliminaries

Definition 1. The nth **power** of word A is defined as:

$$A^n = \begin{cases} AA^{n-1}, & n > 0; \\ empty\ word, & n = 0. \end{cases}$$

A word A is **non-primitive** if it equals B^k for some word B and integer $k \geq 2$. Otherwise, A is **primitive**. (By this definition the empty word is not primitive.)

The next lemma summarizes three results about the powers proved by Lyndon and Schützenberger [18]; see their Lemmas 3 and 4, and Corollary 4.1. (More introductions of these results can be found in Sect. 1.3 "Conjugacy" of [17].)

Lemma 1 *[18]. Given words A and B, there exist C, k, l such that $A = C^k$ and $B = C^l$ when one of the following conditions holds:*

1. *$AB = BA$.*
2. *Two powers A^{m_1} and B^{m_2} have a common prefix of length $|A| + |B|$.*
3. *$A^{m_1} = B^{m_2}$.*

Definition 2. *The **root** of a word A, denoted by $\mathsf{root}(A)$, is the unique primitive word B such that A is a power of B. The uniqueness of root is obvious, a formal proof can be found in Corollary 4.2 of [18] or in [16].*

Lemma 2. *Assume A is a non-empty word. Find the largest $j < |A|$ such that the prefix of A with length j equals the suffix of A with length j. Let $k = |A| - j > 0$. Then,*

$$|\mathsf{root}(A)| = \begin{cases} k, & |A| = 0 \pmod{k}; \\ |A|, & |A| \neq 0 \pmod{k}. \end{cases} \tag{1}$$

Proof. This result should be well-known. A simple proof is as follows.

Fact 1. If $S = BB' = B'B$ and B, B' are non-empty, S is non-primitive.

This is a trivial fact and is implied by Lemma 1 (condition 1); proof omitted.

Claim 1. $|\mathsf{root}(A)| \geq k$.

Proof: The prefix and suffix of A with length $|A| - |\mathsf{root}(A)|$ are the same, which implies that $j \geq |A| - |\mathsf{root}(A)|$. Consequently, $|\mathsf{root}(A)| \geq |A| - j = k$.

Claim 2. If $|\mathsf{root}(A)| < |A|$ (i.e., A is non-primitive), then $k \geq |\mathsf{root}(A)|$.

Proof: Denote $S = \mathsf{root}(A)$ and assume $|S| < |A|$. Therefore, $A = S^d$ $(d \geq 2)$. Suppose to the opposite that $k < |\mathsf{root}(A)|$. Let B be the prefix of S with length k, and B' be the suffix of S such that $S = BB'$. As $k < |S|$, we have $j > |A| - |S| \geq |S|$. Further since the suffix of A with length j (which starts with $B'B$) equals to the prefix of A with length j (which starts with $S = BB'$), we get $S = B'B$. Applying Fact 1, $\mathsf{root}(A) = S$ is non-primitive. Contradictory.

We are ready to prove the lemma. When $|A|$ is a multiple of k, A is a power of its prefix of length k, which means $|\mathsf{root}(A)| \leq k$. Further by Claim 1, $|\mathsf{root}(A)| = k$. Next, assume $|A|$ is not a multiple of k. Since $|A|$ is a multiple of $|\mathsf{root}(A)|$, we see $|\mathsf{root}(A)| \neq k$. Further by Claims 1 and 2, it follows that $|\mathsf{root}(A)| = |A|$. $\qquad\square$

For a non-empty word A, denote by $R(A)$ the infinite repeating string $AA\ldots$.

The following lemma is fundamental to our algorithm.

Lemma 3. *For non-empty words A and B, the relation between $R(A)$ and $R(B)$ is the same as the relation between AB and BA. In other words,*

$$R(A) < R(B) \quad \Leftrightarrow \quad AB < BA, \tag{2}$$
$$R(A) > R(B) \quad \Leftrightarrow \quad AB > BA. \tag{3}$$
$$R(A) = R(B) \quad \Leftrightarrow \quad AB = BA, \tag{4}$$

Proof. Assume that A, B are words that consist of the decimal symbols '0',...,'9'. The proof can be easily extended to the more general case.

Let α, β denote the number represented by strings A, B. For example, string '89' represents number 89. Denote $a = |A|$ and $b = |B|$. Observe that

$$AB < BA \Leftrightarrow \alpha \cdot 10^b + \beta < \beta \cdot 10^a + \alpha \Leftrightarrow \frac{\alpha}{10^a - 1} < \frac{\beta}{10^b - 1}.$$

Moreover,

$$\frac{\alpha}{10^a - 1} = \alpha \frac{\frac{1}{10^a}}{1 - \frac{1}{10^a}} = \alpha[\frac{1}{10^a} + (\frac{1}{10^a})^2 + (\frac{1}{10^a})^3 + \ldots] = 0.\alpha\alpha\alpha\cdots = 0.\dot{\alpha};$$

$$\frac{\beta}{10^b - 1} = \beta \frac{\frac{1}{10^b}}{1 - \frac{1}{10^b}} = \beta[\frac{1}{10^b} + (\frac{1}{10^b})^2 + (\frac{1}{10^b})^3 + \ldots] = 0.\beta\beta\beta\cdots = 0.\dot{\beta}.$$

So, $AB < BA \Leftrightarrow 0.\dot{\alpha} < 0.\dot{\beta} \Leftrightarrow R(A) < R(B)$. Similarly, (4) and (4) hold. □

A more rigorous but complicated proof of Lemma 3 is given in the appendix.

As an interesting corollary of Lemma 3, we obtain that "if $AB \leq BA$ and $BC \leq CB$, then $AC \leq CA$". This transitivity is not obvious without Lemma 3.

Now, we can formally bring up the main problem.

Problem 1. Given non-empty words $A_1, \ldots, A_n$, find a permutation $\pi_1, \ldots, \pi_n$ of $\{1, \ldots, n\}$ so that

$$R(A_{\pi_1}) \leq \ldots \leq R(A_{\pi_n}).$$

In other words, this is a sorting problem.

Clearly, $R(A) = R(\mathsf{root}(A))$. To solve Problem 1, we can replace A by $\mathsf{root}(A)$ (using a preprocessing algorithm based on Lemma 2), and then it reduces to:

Problem 1'. Given primitive words $A_1, \ldots, A_n$, find a permutation $\pi_1, \ldots, \pi_n$ of $\{1, \ldots, n\}$ so that

$$R(A_{\pi_1}) \leq \ldots \leq R(A_{\pi_n}).$$

3   A Linear Time Algorithm for Sorting the Repeating Words

Assume that $A_1, \ldots, A_n$ are **primitive**. Denote $L = \sum_i |A_i|$ for short. This section presents an $O(L)$ time algorithm for solving Problem 1', that is, sorting $R(A_1), \ldots, R(A_n)$. We start with one definition and two nontrivial observations.

Definition 3. *For any two non-empty words S and A, denote by $\deg_A(S)$ the largest integer d so that S^d is a prefix of A. Moreover, for non-empty word S and set of non-empty words $\mathcal{A} = \{A_1, \ldots, A_n\}$, denote $\deg_{\mathcal{A}}(S) = \max_j \deg_{A_j}(S)$.*

In other words, if we build the trie T of $\mathcal{A}$, $S^{\deg_{\mathcal{A}}(S)}$ is the longest power of S that equals to some path of the trie T starting from its root.

For any i $(1 \leq i \leq n)$, denote

$$N_i = \text{the } \deg_{\mathcal{A}}(A_i)\text{-th power of} A_i \tag{5}$$

$$M_i = N_i A_i^2 = \text{the } (\deg_{\mathcal{A}}(A_i) + 2)\text{-th power of } A_i \tag{6}$$

Lemma 4. *The relation between infinitely repeating strings $R(A_i)$ and $R(A_j)$ is the same as the relation between words M_i and M_j, that is,*

$$R(A_i) = R(A_j) \quad \Leftrightarrow \quad M_i = M_j,$$
$$R(A_i) < R(A_j) \quad \Leftrightarrow \quad M_i < M_j,$$
$$R(A_i) > R(A_j) \quad \Leftrightarrow \quad M_i > M_j.$$

As a corollary, sorting $R(A_1), \ldots, R(A_n)$ reduces to sorting $M_1, \ldots, M_n$.

Proof. Consider the comparison of $R(A_i)$ and $R(A_j)$. Assume $|A_i| \leq |A_j|$. Otherwise it is symmetric.

First, consider the case $R(A_i) = R(A_j)$. Let $m_1 = |A_j|$ and $m_2 = |A_i|$. We know $A_i^{m_1} = A_j^{m_2}$ because $R(A_i) = R(A_j)$. Applying Lemma 1 (condition 3), $A_i = C^k$ and $A_j = C^l$ for some C, k, l. Further since A_i, A_j are primitive, $A_i = C = A_j$. It follows that $M_i = M_j$. Next, assume that $R(A_i) \neq R(A_j)$.

Let $p = \deg_{A_j}(A_i)$. Thus, $A_j = A_i^p S$, where $p \geq 0$ and A_i is not a prefix of S. Be aware that $p \leq \deg_{\mathcal{A}}(A_i)$ by the definition of $\deg_{\mathcal{A}}(A_i)$.

According to Lemma 3, the comparison of $R(A_i)$ and $R(A_j)$ equals to the comparison of $A_i A_j$ and $A_j A_i$. Further since $A_j = A_i^p S$, it equals to the comparison of $A_i^p A_i S$ and $A_i^p S A_i$. In the following, we discuss two subcases.

Subcase 1. $|S| > |A_i|$, or $|S| \leq |A_i|$ and S is not a prefix of A_i.

Recall that A_i is not a prefix of S. In this subcase, we will find an unequal letter if we compare A_i with S (starting from the leftmost letter). Comparing $A_i^p A_i S$ and $A_i^p S A_i$ is thus equivalent to comparing the prefixes $A_i^p A_i$ and $A_i^p S$.

Notice that $A_i^p A_i = A_i^{p+1}$ and $A_i^p S = A_j$ are also prefixes of M_i and M_j, respectively (note that A_i^{p+1} is a prefix of M_i because M_i is the $(\deg_{\mathcal{A}}(A_i) + 2)$-th power of A_i and $\deg_{\mathcal{A}}(A_i) \geq p$ as mentioned above). Therefore, comparing M_i and M_j is also equivalent to comparing the two prefixes $A_i^p A_i$ and $A_i^p S$.

Altogether, comparing $R(A_i), R(A_j)$ is equivalent to comparing M_i, M_j.

Subcase 2. S is a prefix of A_i. (This means S is a proper prefix of A_i as $S \neq A_i$.)

Assume $A_i = ST$. Comparing $A_i^p A_i S$ and $A_i^p SA_i$ is just the same as comparing $A_i^p STS$ and $A_i^p SST$. It reduces to proving that comparing M_i and M_j also reduces to comparing $A_i^p STS$ and $A_i^p SST$.

First, we argue that $ST \neq TS$. Suppose to the opposite that $ST = TS$. Applying Lemma 1 (condition 1), $S = C^k$ and $T = C^l$ for some C, k, l. This implies that A_i and A_j are both powers of C, and hence $R(A_i) = R(A_j)$, contradictory.

Observe that $A_i^p STS$ is a prefix of $A_i^p STST = A_i^{p+2}$, which is a prefix of M_i (because M_i is the $(\deg_A(A_i) + 2)$-th power of A_i and $\deg_A(A_i) + 2 \geq p + 2$).

Observe that $p > 0$. Otherwise $A_j = A_i^0 S$ is shorter than A_i, which contradicts our assumption $|A_i| \leq |A_j|$. As a corollary, A_i is a prefix of $A_j = A_i^p S$. Therefore, $A_i^p SST = A_i^p SA_i = A_j A_i$ is a prefix of A_j^2, which is a prefix of M_j.

To sum up, M_i and M_j admit $A_i^p STS$ and $A_i^p SST$ as prefixes, respectively. Further since $TS \neq ST$, comparing M_i, M_j reduces to comparing $A_i^p STS$ and $A_i^p SST$, which is equivalent to comparing $R(A_i), R(A_j)$ as mentioned above. $\square$

Assume $A_1, \ldots, A_n$ are **distinct** henceforth in this section. To this end, we can use a trie to reduce those duplicate elements in $A_1, \ldots, A_n$, which is trivial.

Lemma 5. *When $A_1, \ldots, A_n$ are primitive and distinct, $\sum_i |N_i| = O(L)$.*

Proof. First, we argue that $N_1, \ldots, N_n$ are distinct. Suppose that $N_i = N_j$ ($i \neq j$). Recall that $N_i = A_i^m$ (for $m = \deg_A(A_i)$) and $N_j = A_j^n$ (for $n = \deg_A(A_j)$). Applying Lemma 1 (condition 3), $A_i = C^k$ and $A_j = C^l$. Further since A_i, A_j are primitive, $A_i = C = A_j$, which contradicts the assumption that $A_i \neq A_j$.

We say N_i *extremal* if it is **not** a prefix of any word in $\{N_1, \ldots, N_n\} \setminus \{N_i\}$. Partition $N_1, \ldots, N_n$ into several groups such that (a) for elements in the same group, one of them is the prefix of the other, and (b) the longest element in each group is extremal. (It is obvious that such a partition exists: we can first distribute the extremal ones to different groups, and then distribute the non-extremal ones to suitable group (each non-extremal one is a prefix of some extremal ones).

Now, consider any such group, e.g., $N_{i_1}, \ldots, N_{i_x}$. It suffices to prove that (X) $|N_{i_1}| + \ldots + |N_{i_x}| = O(|A_{i_1}| + \ldots + |A_{i_x}|)$, and we prove it in the following. Without loss of generality, assume that N_{i_j} is a prefix of $N_{i_{j+1}}$ for $j < x$.

We state two important formulas: (i) $N_{i_x} = A_{i_x}$. (ii) $|N_{i_j}| < |A_{i_j}| + |A_{i_{j+1}}|$ for $j < x$. Equation (X) above follows from formulas (i) and (ii) immediately.

Proof of (i). Suppose to the contrary that $N_{i_x} \neq A_{i_x}$. By the definition of N_{i_x}, there exists some A_j such that N_{i_x} is a prefix of A_j. Clearly, $j \neq i_x$ since N_{i_x} is not a prefix of A_{i_x}. Consequently, N_{i_x} is a prefix of some other N_j, which means N_{i_x} is not extremal, contradicting property (b) of the grouping mentioned above.

Proof of (ii). Suppose to the contrary that $|N_{i_j}| \geq |A_{i_j}| + |A_{i_{j+1}}|$. Because N_{i_j} and $N_{i_{j+1}}$ are powers of $A_{i,j}$ and $A_{i_{j+1}}$ and share a common prefix, N_{i_j}, of length at least $|A_{i_j}| + |A_{i_{j+1}}|$. By Lemma 1 (condition 2), $A_{i_j} = C^k$ and $A_{i_{j+1}} = C^l$ for some C, k, l. Hence $A_{i_j} = A_{i_{j+1}}$, as A_{i_j} and $A_{i_{j+1}}$ are primitive. Contradictory. $\square$

Our algorithm for sorting $R(A_1), \ldots, R(A_n)$ is simply as follows.

First, we build a trie of $A_1, \ldots, A_n$ and use it to compute $N_1, \ldots, N_n$. In particularly, for computing N_i, we walk along the trie from the root and search for maximal powers of A_i, which takes $O(|N_i| + |A_i|) = O(|N_i|)$ time. The total running time for computing $N_1, \ldots, N_n$ is therefore $O(\sum_i |N_i|) = O(L)$.

Second, we compute $M_1, \ldots, M_n$ and build a trie of them. By utilizing this trie, we obtain the lexicographic order of $M_1, \ldots, M_n$, which equals the order of $R(A_1), \ldots, R(A_n)$ according to Lemma 4. The running time of the second step is $\sum_i |M_i| = \sum_i |N_i| + 2\sum_i |A_i| = O(L) + O(L) = O(L)$.

Formally, the algorithm can be described as follows.

Algorithm 1. Algorithm for sorting $R(A_1), \ldots, R(A_n)$

Input: Words $A_1, \ldots, A_n$
Output: The order of $R(A_1), \ldots, R(A_n)$.

1: Build the a trie of $A_1, \ldots, A_n$.
2: **for** each $i \in \{1, \ldots, n\}$ **do**
3:     Walk along the trie from the root and search for maximal powers of A_i.
4: **end for**
5: Compute $M_1, \ldots, M_n$ and build the second trie of them.
6: Obtain the lexicographic order of $M_1, \ldots, M_n$ utilizing the second trie.
7: Output lexicographic order of $M_1, \ldots, M_n$, which equals to the order of $R(A_1), \ldots, R(A_n)$.

Step by step, we can see that step 1 takes $O(\sum_i |A_i|) = O(L)$ time, each step 3 takes $O(|N_i| + |A_i|) = O(|N_i|)$ time and the whole loop takes $O(\sum_i |N_i|) = O(L)$ time, step 5 takes $\sum_i |M_i| = \sum_i |N_i| + 2\sum_i |A_i| = O(L) + O(L) = O(L)$ time, both step 6 and step 7 also takes $O(L)$ time. In total, this algorithm takes $O(L)$ time.

To sum up, we obtain

Theorem 1. *Problem 1' and Problem 1 can be solved in* $O(L) = O(\sum_i |A_i|)$ *time.*

Proof. It remains to be shown that $\mathsf{root}(A_i)$ can be computed in $O(|A_i|)$ time.

Applying Lemma 2, computing $\mathsf{root}(A)$ reduces tao finding the largest $j < |A|$ such that the prefix of A with length j equals the suffix of A with length j. Moreover, the famous KMP algorithm [10] finds this j in $O(|A|)$ time.   □

As a comparison, there exists a less efficient algorithm for solving Problem 1, which is based on a standard sorting algorithm associated with a naïve gadget for comparing $R(A)$ and $R(B)$ – according to Lemma 3, comparing $R(A)$ and $R(B)$ reduces to comparing AB and BA, which takes $O(|A| + |B|)$ time. The time complexity of this alternative algorithm is higher. For example, when A_1="aaaaaa1", A_2="aaaaaa2", etc., the running time would be $\Omega(n \log n |A_1|) = \Omega(L \log n)$.

4   The String Rearrangement Inequalities

We call Eq. (7) right below *the String Rearrangement Inequalities*.

Lemma 6. *For non-empty words $A_1, \ldots, A_n$, where $R(A_1) \leq \ldots \leq R(A_n)$, we claim that*

$$A_1 A_2 \ldots A_n \leq A_{\pi_1} A_{\pi_2} \ldots A_{\pi_n} \leq A_n A_{n-1} \ldots A_1, \tag{7}$$

for any permutation $\pi_1, \ldots, \pi_n$ of $\{1, \ldots, n\}$.

In other words, if several words are to be rearranged and concatenated into a string S, the lexicographical smallest outcome of S occurs when the words are arranged so that their repeating strings are increasing, and the lexicographical largest outcome of S occurs when the words are arranged so that their repeating strings are decreasing. Here, the repeating string of a word A refers to $R(A)$.

Example 1. Suppose there are four given words: "123", "12", "121", "1212". Notice that $R(121) < R(12) = R(1212) < R(123)$. Applying Lemma 6, the lexicographical smallest outcome would be "121121212123", and the lexicographical largest outcome would be "123121212121". The reader can verify this result easily.

Remark 1. If we sort the given words using the lexicographic order instead, the outcome of the concatenation is not optimum. For example, we have "12" < "121" < "1212" <"123", and a concatenation in this order is not the smallest outcome, and a concatenation in its reverse order is neither the largest outcome.

Proof. (of Lemma 6). Consider any concatenation $A_{\pi_1} \ldots A_{\pi_n}$. If A_1 is not at the leftmost position, we swap it with its left neighbor A_x. Note that $R(A_1) \leq R(A_x)$ by assumption. According to Lemma 3, $A_1 A_x \leq A_x A_1$. This means that the entire string becomes smaller or remains unchanged after the swapping. Applying several such swappings, A_1 will be on the leftmost position. Then, we swap A_2 to the second place. So on and so forth. It follows that $A_1 \ldots A_n \leq A_{\pi_1} \ldots A_{\pi_n}$.

The other inequality in (7) can be proved symmetrically; proof omitted.   □

Combining Theorem 1 with Lemma 6, we obtain

Corollary 1. *Given n words $A_1, \ldots, A_n$ that are to be rearranged and concatenated, the smallest and largest concatenation can be found in $O(\sum_i |A_i|)$ time.*

Another corollary of Lemma 6 is the uniqueness of the best concatenation:

Corollary 2. *Given primitive and distinct words $A_1, \ldots, A_n$ that are to be rearranged and concatenated, the smallest (largest, resp.) concatenation is unique.*

Proof. It follows from Lemma 6 and the fact that $R(A_1), \ldots, R(A_n)$ are distinct (see Proposition 1 below).   □

Proposition 1. *For distinct primitive words A and B, we have $R(A) \neq R(B)$.*

Proof. Recall that when A and B are primitive and $R(A) = R(B)$, we can infer that $A = B$ (as proved in the second paragraph of the proof of Lemma 4). Therefore, if A and B are primitive and distinct, $R(A) \neq R(B)$.   □

5   A Total Order $\leq_\infty$ on Words

Definition 4. *Given primitive words A and B, we state that $A \leq_\infty B$ if $R(A) \leq R(B)$. Notice that $\leq_\infty$ is **a total order on primitive words** by Proposition 1. Furthermore, we extend $\leq_\infty$ to the scope of finite nonempty words as follows.*

For non-empty words $A = S^k$ and $B = T^l$, where S, T are primitive, we state that $A \leq_\infty B$ if $(S = T$ and $|S| \leq |T|)$, or $(S \neq T$ and $S \leq_\infty T)$. The symbol $\leq_\infty$ in the equation stands for the relation between primitive words.

For example, $121 \leq_\infty 12 \leq_\infty 1212 \leq_\infty 121212 \leq_\infty 122$.

Obviously, the relation $\leq_\infty$ is **a total order on finite nonempty words**.

The next lemma shows that within the class of Lyndon words, the order $\leq_\infty$ is actually the same as the lexicographical order $\leq_{\text{lex}}$ (denoted by $\leq$ for short). (Note that Lyndon words are primitive, so the unextended $\leq_\infty$ is enough here.)

Lemma 7. *Given Lyndon words A and B such that $A \leq B$, we have $A \leq_\infty B$.*

Proof. Assume that $A \neq B$; otherwise we have $R(A) = R(B)$ and so $A \leq_\infty B$.

By the assumption $A \leq B$, we know $A < B$. Consider two cases:

1. $|A| \geq |B|$, or $|A| < |B|$ and A is not a prefix of B

Combining the assumption $A < B$ with the condition of this case, we can see that the relation between AB, BA is the same as that between A, B: In comparing AB and BA, the result is settled before the $\min\{|A|, |B|\}$-th character.

2. $|A| < |B|$ and A is a prefix of B, i.e., A is a proper prefix of B

Assume that $B = AC$ where C is nonempty. Because B is a Lyndon word by assumption, $AC < CA$. Therefore, $AB = AAC < ACA = BA$.

In both cases, we obtain $AB < BA$. It further implies that $R(A) < R(B)$ by Lemma 3. This means $A \leq_\infty B$.   □

In fact, it is possible to further extend $\leq_\infty$ to all (finite and infinite) words. Define the repeating string of an infinite word A, denoted by $R(A)$, to be A itself. We state that $A \leq_\infty B$ if $R(A) < R(B)$ or $R(A) = R(B)$ and $|A| \leq |B|$.

6   Conclusions

In this paper, we present a simple proof of the "string rearrangement inequalities" (7). These inequalities have not been reported in literature to the best of our knowledge. We also study the algorithmic aspect of these two inequalities, and present a linear time algorithm for rearranging the strings so that $R(A_1) \leq \ldots R(A_n)$. This algorithm beats the trivial sorting algorithm by a factor of $\log n$.

The algorithm itself is direct (indeed, it looks somewhat brute-force) and easy to implement, yet the analysis of its correctness and complexity is built upon nontrivial observations, namely, Lemma 3, Lemma 4, and Lemma 5.

In the future, it is a problem worth attacking that whether we can improve the running time for sorting $R(A_1), \ldots, R(A_n)$ from $O(L)$ to $O(N)$, where N denotes the number of nodes in the trie of $A_1, \ldots, A_n$.

The order $\leq_\infty$ on primitive words has nice connections with repeating decimals as shown in the proof of Lemma 3. It would be interesting to know whether these connections have more applications in the study of primitive words.

A   An alternative proof of Lemma 3

Below we show an alternative proof of Lemma 3. This proof is less clever and much more involved (compared to the other proof in Sect. 2), yet it reflects more insights which helped us in designing our linear time algorithm.

Below we always assume that A, B, X, Y are words.

Definition 5. Word A is **truly less** than word B, if there exists a prefix pair $A_1 A_2 ... A_i$ and $B_1 B_2 ... B_i$, in which $A_1 A_2 ... A_{i-1}$ and $B_1 B_2 ... B_{i-1}$ are equal and A_i is less than B_i. For convenience, let $A <_T B$ denote this case for the rest of this paper. Note that i can be 1 such that A_1 is less than B_1.

For any pair of nonempty words, A and B, we can generalize 3 following properties with Definition 5. Note that any X or Y in the following properties can be any word, including empty word.

Claim (1). Proposition $A <_T B$ is equivalent to $AX < BY$, if A is not prefix of B and B is not prefix of A.

Proof. If A is not prefix of B and B is not prefix of A, the proposition $AX < BY$ implies that A and B fits the case described in Definition 5 and thus $A <_T B$ holds. The proposition $A <_T B$, by Definition 5, also indicates that $AX < BY$. □

Claim (2). If $A <_T B$, it holds that $AX <_T BY$.

Proof. If $A <_T B$, by Definition 5, there exists a prefix pair $A_1, A_2 ... A_i$ and $B_1 B_2 ... B_i$, in which $A_1 A_2 ... A_{i-1}$ and $B_1 B_2 ... B_{i-1}$ are equal and A_i is less than B_i. Since A is the prefix of AX and B is the prefix of BY, AX and BY also have the prefix pair $A_1 A_2 ... A_i$ and $B_1 B_2 ... B_i$ mentioned above, thus it holds that $AX <_T BY$ by Definition 5. □

Claim (3). If $A < B$ and $|A| = |B|$, it holds that $A <_T B$.

Proof. If $A < B$ and $|A| = |B|$, we can find a substring pair $A_1 A_2 ... A_i$ and $B_1 B_2 ... B_i$, in which $A_1 A_2 ... A_{i-1}$ and $B_1 B_2 ... B_{i-1}$ are equal and A_i is less than B_i. This is exactly the case of Definition 5, so naturally $A <_T B$. □

Now, we are ready for proving Lemma 3.

Recall that this lemma states for non-empty words A and B, the relation between $R(A)$ and $R(B)$ is the same as the relation between AB and BA.

We will prove $R(A) = R(B) \Leftrightarrow AB = BA$ and $R(A) < R(B) \Leftrightarrow AB < BA$. Note that $R(A) > R(B) \Leftrightarrow AB > BA$ can be obtained similarly.

Proposition 2. *For nonempty words A and B, $AB = BA \Leftrightarrow R(A) = R(B)$.*

Proof. From $AB = BA$ or $R(A) = R(B)$, we obtain from Lemma 1 that $A = C^k$ and $B = C^l$ for some C, k, l, which implies that $R(A) = R(B)$ and $AB = BA$.□

Proposition 3. *For nonempty words A and B, $AB < BA \Leftrightarrow R(A) < R(B)$.*

We prove the two directions separately in the following.

Proof (of $AB < BA \Rightarrow R(A) < R(B)$).
We discuss two subcases.

Subcase 1. $|A| \leq |B|$
Let $B = A^m S$, in which $m = \deg_B(A)$ by Definition 3.

Note that target $R(A) < R(B)$ equals to $R(A) < R(A^m S)$, which then equals to $R(A) < SR(A^m S)$, by eliminating the leading A^m.

With $AB < BA$, we will get the relation that $AB < BA$ leads to $A^{m+1}S < A^m SA$. And $A^{m+1}S < A^m SA$ leads to $AS < SA$, which eventually leads to $AS <_T SA$.

We will prove that $AA <_T SA$. And since AA is a prefix of $R(A)$ and SA is a prefix of $SR(A^m S)$, proposition $R(A) < SR(A^m S)$ follows by Claim 2, proving the target proposition.

Now we prove $AA <_T SA$ in two cases.

1. If $|A| \leq |S|$, or $|A| > |S|$ but S is not a prefix of A, note that A is not a prefix of S, since $AS < SA$, proposition $A <_T S$ follows by Claim 1. Then $AA <_T SA$ follows by Claim 2.
2. If $|A| > |S|$ and S is a prefix of A, let $A = ST$. Since $AS <_T SA$, we have $STS <_T SST$, then $AA = STST <_T SST = SA$ follows by Claim 2. Thus $AA <_T SA$.

Subcase 2. $|A| > |B|$
Let $A = B^m S$, in which $m = \deg_A(B)$.

Note that target $R(A) < R(B)$ equals to $R(B^m S) < R(B)$, which then equals to $SR(B^m S) < R(B)$, by eliminating the leading B^m.

With $AB < BA$, we will get the relation that $AB < BA$ leads to $B^m SB < B^{m+1}S$. And $B^m SB < B^{m+1}S$ leads to $SB < BS$, which eventually leads to $SB <_T BS$.

We will prove that $SB <_T BB$. And since BB is a prefix of $R(B)$ and SB is a prefix of $SR(B^m S)$, proposition $SR(B^m S) < R(B)$ follows by Claim 2, proving the target proposition.

Now we prove $SB <_T BB$ in 2 cases.

1. If $|B| \leq |S|$, or $|B| > |S|$ but S is not a prefix of B, note that B is not a prefix of S, since $SB < BS$, proposition $S <_T B$ follows by Claim 1. Then $SB <_T BB$ follows by Claim 2.
2. If $|B| > |S|$ and S is a prefix of B, let $B = ST$. Since $SB <_T BS$, we have $SST <_T STS$, then $SB = SST <_T STST = BB$ follows by Claim 2. Thus $SB <_T BB$.

With both cases proved, we have $AB < BA \Rightarrow R(A) < R(B)$.   □

Proof (of $R(A) < R(B) \Rightarrow AB < BA$).

In the following, we discuss two subcases.

Subcase 1. $|A| \leq |B|$.

Let $B = A^m S$, in which $m = \deg_B(A)$.

Note that $AB < BA$ equals to $A^{m+1}S < A^m SA$, which equals to $AS < SA$ by eliminating the leading A^m.

With $R(A) < R(B)$, we will get the relation that $R(A) < R(B)$ equals to $R(A) < R(A^m S)$. And $R(A) < R(A^m S)$ equals to $R(A) < SR(A^m S)$ by eliminating the leading A^m.

Now we will prove $AS < SA$.

1. If $|A| <= |S|$, or $|A| > |S|$ and S is not a prefix of A, note that A is not a prefix of S, since $R(A) < SR(A^m S)$, $A <_T S$ follows by Claim 1. Then $AS < SA$ follows by Claim 2.

2. If $|A| > |S|$ and S is a prefix of A, let $A = ST$. Since $R(A) < SR(A^m S)$, we have $R(ST) < SR((ST)^m S)$, we pay attention to the prefixes with length $2*|S|+|T|$ of these two infinite words: STS and SST. We argue that $STS \neq SST$ otherwise $ST = TS$, then S, T, B, A are powers of a common element by Lemma 1, then $R(A) = R(B)$, which is contradictory. Thus, since $R(ST) < SR((ST)^m S)$, we will have $STS < SST$. It holds that $AS = STS < SST = SA$. Thus, we end up with $AS < SA$.

Subcase 2. $|A| > |B|$.

Let $A = B^m S$, in which $m = \deg_A(B)$.

Note that $AB < BA$ equals to $B^m SB < B^{m+1}S$, which equals to $SB < BS$ by eliminating the leading B^m.

With $R(A) < R(B)$, we will get the relation that $R(A) < R(B)$ equals to $R(B^m S) < R(B)$. And $R(B^m S) < R(B)$ equals to $SR(B^m S) < R(B)$ by eliminating the leading B^m.

Now we will prove $SB < BS$, in two cases.

1. If $|B| <= |S|$, or $|B| > |S|$ and S is not a prefix of B, note that B is not a prefix of S, since $SR(B^m S) < R(B)$, $S <_T B$ follows by Claim 1. Then $SB < BS$ follows by Claim 2.

2. If $|B| > |S|$ and S is a prefix of B, let $B = ST$. Since $SR(B^m S) < R(B)$, we have $SR((ST)^m S) < R(ST)$. we pay attention to the prefixes with length $2 * |S| + |T|$ of these two infinite words: SST and STS. We can argue that $SST \neq STS$ otherwise $ST = TS$, then S, T, B, A are powers of a common element by Lemma 1, then $R(A) = R(B)$, which is contradictory. Thus, since $SR((ST)^m S) < R(ST)$, we will have $SST < STS$. It holds that $SB = SST < STS = BS$. Thus, we end up with $SB < BS$.

With both cases proved, we have $R(A) < R(B) \Rightarrow AB < BA$.   □

Now, with both subcases proved, we have $R(A) < R(B) \Leftrightarrow AB < BA$.

References

1. Orderings - oeiswiki (2022). https://oeis.org/wiki/Orderings
2. Wikipedia: Shortlex order (2022). https://en.wikipedia.org/wiki/Shortlex_order
3. Alatabbi, A., Daykin, J., Rahman, M., Smyth, W.: Simple linear comparison of strings in v-order. In: Pal, S., Sadakane, K. (eds.) WALCOM 2014, LNCS, Vol. 8344, pp. 80–89 (2014)
4. Au, Y.: Generalized de bruijn words for primitive words and powers. Discret. Math. **338**(12), 2320–2331 (2015). https://doi.org/10.1016/j.disc.2015.05.025
5. Bannai, H.I.T., Inenaga, S., Nakashima, Y., Takeda, M., Tsuruta, K.: The "runs" theorem. SIAM J. Comput. **46**(5), 1501–1514 (2017). https://doi.org/10.1137/15M1011032
6. Berstel, J., Boasson, L.: The set of lyndon words is not context-free. Bull. EATCS **63** (1997)
7. Chen, K., Fox, R., Lyndon, R.: Free differential calculus, iv. the quotient groups of the lower central series. Ann. Math. 81–95 (1958)
8. Crochemore, M., Russo, L.: Cartesian and lyndon trees. Theor. Comput. Sci. **806**, 1–9 (2020). https://doi.org/10.1016/j.tcs.2018.08.011
9. Daykin, D., Daykin, J., Smyth, W.: String comparison and lyndon-like factorization using v-order in linear time. In: Giancarlo, R., Manzini, G. (eds.) CPM 2011, LNCS 6661, pp. 65–76 (2011)
10. D.E. Knuth, J.M., Pratt, V.: Fast pattern matching in strings. SIAM J. Comput. **6**(2), 323–350 (1977). https://doi.org/10.1137/0206024
11. Dolce, F., Restivo, A., Reutenauer, C.: On generalized lyndon words. ArXiv:abs/1812.04515 (2019)
12. Dömösi, P., Horváth, S., Ito., M.: On the connection between formal languages and primitive words. In: Proceedings of the First Session on Scientific Communication, pp. 59–67. University of Oradea, Oradea, Romania (1991)
13. Dömösi, P., Ito, M.: Context-free languages and primitive words (2014). https://doi.org/10.1142/7265
14. Duval, J.: Factorizing words over an ordered alphabet. J. Algor. **4**(4), 363–381 (1983). https://doi.org/10.1016/0196-6774(83)90017-2
15. Fredricksen, H., Maiorana, J.: Necklaces of beads in k colors and k-ary de bruijn sequences. Discrete Math. **23**, 207–210 (1979)
16. Lischke, G.: Primitive words and roots of words. Acta Univ. Sapientiae, Informatica **3**(1), 5–34 (2011)
17. Lothaire, M.: Combinatorics on Words. Encyclopedia of Mathematics, vol. 17, Addison-Wesley, MA (1983)
18. Lyndon, R., Schützenberger, M.: The equation $a^m = b^n c^p$ in a free group. Michigan Math. J. **9**(4), 289–298 (1962). https://doi.org/10.1307/mmj/1028998766
19. Petersen, H.: On the language of primitive words. Theor. Comput. Sci. **161**, 141–156 (1996)
20. Smyth, W.: Computing regularities in strings: a survey. Eur. J. Combin. **34**(1), 3–14 (2013). https://doi.org/10.1016/j.ejc.2012.07.010
21. Zhang, D., Jin, K.: Fast algorithms for computing the statistics of pattern matching. IEEE Access **9**, 114965–114976 (2021). https://doi.org/10.1109/ACCESS.2021.3105607

Approximation Algorithms for Prize-Collecting Capacitated Network Design Problems

Lu Han[1], Vincent Chau[2(✉)], and Chi Kit Ken Fong[3]

[1] School of Science, Beijing University of Posts and Telecommunications,
Beijing, China
hl@bupt.edu.cn
[2] School of Computer Science and Engineering, Southeast University, Nanjing, China
vincentchau@seu.edu.cn
[3] Chu Hai College of Higher Education, Hong Kong, China
kenfong@chuhai.edu.hk

Abstract. In this paper, we study the prize-collecting capacitated network design (PCCND) problem. We have to route the demand of each source to some sinks in a network, or eventually pay the prize to not serving the demand. We need to install multiple cables on the edges of the network to support the service. The goal is to find a feasible solution with the minimum cost. We give a 3.482-approximation algorithm for the PCCND. We also consider a special case of the PCCND, in which there is only one given sink, and present a 2.9672-approximation algorithm.

Keywords: Prize-collecting · Network design · Facility location · Steiner tree · Approximation algorithm

1 Introduction

We study the prize-collecting capacitated network design (PCCND) problem, which generalizes both the prize-collecting facility location (PCFL) problem and the prize-collecting Steiner tree (PCST) problem. In the PCCND problem, we are given a set of sinks and a set of sources in a network. Each sink is associated with an opening cost. Each source has a unit of demand and a prize. We could decide to either route the demand of a source to an opened sink or pay its prize for not routing. The amount of demands that can be served by an edge in the network depends on how many cables are installed on the edge. All the cables have the same capacity. Installing l cables on an edge could allow the demands served by the edge up to l times the capacity, and would incur l times the corresponding edge

The research of the first author is supported by the National Natural Science Foundation of China (No. 12001523). The second author is supported by the national key research and development program of China under grant No. 2019YFB2102200, and by the Fundamental Research Funds for the Central Universities No. 2242022R10024.

M. Li and X. Sun (Eds.): IJTCS-FAW 2022, LNCS 13461, pp. 219–232, 2022.
https://doi.org/10.1007/978-3-031-20796-9_16

cost. Our task is to open some sinks, install cables on the edges of the network, and decide to either route the demand of each source to some opened sink or to pay its prize without violating the capacity of each edge, so that the total cost of opening sinks, installing cables on edges, and paying prizes is minimized.

1.1 Related Works

When each cable has a capacity of 1, the PCCND problem simplifies to the PCFL problem. In the PCFL problem, we are given a set of facilities and a set of clients. Each facility has an opening cost. Each client has a unit of demand and is associated with a prize. We could decide to either connect the demand of a client to an opened facility or pay its prize for not connecting. Connecting a client to a facility would incur a connection cost. The objective is to open some facilities, and decide to either connect the demand of each client to some opened facility or to pay its prize, so that the total cost of opening facilities, connecting clients, and paying prizes is minimized. The PCFL problem is also known as the facility location problem with penalties, in which the prize of a client is called its penalty cost. Under the assumption that the connection costs satisfy the triangle inequality, Charikar et al. [3] proposed the PCFL problem and give its first 3-approximation algorithm based on the technique of primal-dual. Subsequently, many researchers began to design approximation algorithms for the PCFL problem [4, 16, 17]. By exploiting the properties of the penalties (i.e., the prizes), Li et al. [11] presented the currently best LP-rounding 1.5148-approximation algorithm for the PCFL problem. Note that if each client has an infinite prize, the PCFL problem becomes the well-known uncapacitated facility location (UFL) problem [12]. Other generalizations of the UFL problem can be found in the literature [9, 10, 15, 18].

When there is only one given sink without opening cost, each cable has an infinite capacity, and all the other vertices in the network are the sources, the PCCND problem simplifies to the PCST problem. In the PCST problem, we are given a graph consisting of a set of vertices and a set of edges, as well as a root vertex. Each vertex is associated with a prize, and each edge is associated with a cost. The objective is to find a tree containing the root vertex so that the total cost of the edge costs of the edges in the tree and the prizes of the vertices not in the tree is minimized. A 3-approximation algorithm could be obtained for the PCST problem by adopting a similar idea of the algorithm for the prize-collecting traveling salesman problem in [2]. Goemans and Williamson presented a 2-approximation primal-dual algorithm in [5]. Archer et al. [1] gave the currently best 1.9672-approximation algorithm for the PCST problem, which overcame the integrality gap barrier. For other generalizations of the PCST problem, we refer to [6–8, 14].

When each source has an infinite prize, the PCCND problem becomes the capacitated-cable facility location (CCFL) problem. In the CCFL problem, we are given a set of facilities and a set of clients in a network. Each facility has an opening cost. Each client has a unit of demand. We have to connect the demand of a client to an opened facility. The amount of demands that can be served by an edge in the network depends on how many cables are installed on the edge. All cables have the same capacity. Installing l cables on an edge could allow the demands served by the edge up to l times the capacity, and would incur l times the corresponding edge cost. The objective is to open some facilities, install cables on the edges of the network, and connect the demand of each client to some opened facilities without violating the capacity of each edge so that the total cost of opening facilities and installing cables on the edges is minimized. Under the assumption that the edge costs satisfy the triangle inequality, Ravi and Sinha [13] proposed the CCFL problem and gave an approximation algorithm for it. It is worth mentioning that the idea of our algorithm in this paper is inspired by the work of [13].

1.2   Our Results

As our main contribution, we propose a 3.482-approximation algorithm for the PCCND problem. The algorithm first solves two constructed instances of the PCFL and PCST problems, and then combines the solutions of these two instances to obtain a feasible solution for the PCCND instance. The intuition behind the algorithm is based on an observation that two well-constructed instances of the PCFL and PCST problems could provide two useful lower bounds for the optimal objective of the PCCND instance. Moreover, we also consider a special case of the PCCND problem, in which there is only one given sink, called the single-sink prize-collecting capacitated network design (single-sink PCCND) problem. We show that the idea of our algorithm for the PCCND problem could yield a 2.9672-approximation algorithm for the single-sink PCCND problem.

The remainder of the paper is structured as follows. In Sect. 2, we give the formal description of the studied problem as well as the description of the PCFL and PCST problems. Then in Sect. 3, we present our main algorithm for the PCCND problem. In Sect. 3.3, we study the single-sink PCCND problem.

2   Preliminaries

As described above, our problem is closely related to the PCFL and PCST problems. We first describe our problem in Sect. 2.1, in what follows, we give the formal description of the two other problems. Finally we show how to construct two relevant instances of the PCFL and PCST in order to provide lower bounds to an optimal solution of the PCCND instance.

2.1   Prize-Collecting Capacitated Network Design Problem

In a PCCND instance, we are given an undirected graph $G = (V, E)$, a set of sinks $T \subseteq V$, a set of sources $S \subseteq V$, and an integer U which corresponds to the cable capacity. Opening a sink $t \in T$ incurs a non-negative opening cost of f_t. Each source $s \in S$ has a unit of demand and a non-negative prize p_s. For a source, we could decide to either route its demand to an opened sink via a path, or to pay its prize for not routing. Each edge $e \in E$ has a non-negative edge cost c_e, and we assume that the edge costs satisfy the triangle inequality. We can install multiple cables on the same edge. If we install l times on an edge

e, the edge e can serve up to lU units of demands go through it, and we should pay a cost of lc_e. The objective is to open a set of sinks $O \subseteq T$, install cables on the edges, and decide to either route the demand of each source to some opened sink or to pay its prize without violating the capacity of any edges, such that the total cost (i.e., the cost of opening sinks, installing cables on edges, and paying prizes) is minimized.

Note that if the edge costs do not follow the triangle inequality, then the problem does not admit a constant approximation ratio since it is a generalization of the PCFL problem.

We use (O, P, τ) to denote a solution of a PCCND instance. Here O is the set of opened sinks, P is the set of sources deciding not to route, and $\tau : E \to \mathbb{Z}$ indicates the times of the cable installed on an edge. For a solution (O, P, τ), denote by $C_O(O, P, \tau)$ its total opening cost, i.e.,

$$C_O(O, P, \tau) = \sum_{t \in O} f_t;$$

and denote by $C_P(O, P, \tau)$ its total prize, i.e.,

$$C_P(O, P, \tau) = \sum_{s \in P} p_s;$$

and denote by $C_E(O, P, \tau)$ its total edge cost of installing, i.e.,

$$C_E(O, P, \tau) = \sum_{e \in E} \tau(e) c_e.$$

Therefore, the total cost of the solution (O, P, τ) is

$$C_O(O, P, \tau) + C_P(O, P, \tau) + C_E(O, P, \tau) = \sum_{t \in O} f_t + \sum_{s \in P} p_s + \sum_{e \in E} \tau(e) c_e.$$

2.2   Prize-Collecting Facility Location Problem

In a PCFL instance, we are given a set of facilities F and a set of clients D. Opening a facility $i \in F$ incurs a non-negative opening cost of f_i^{F}. Each client $j \in D$ has a unit of demand and a non-negative prize p_j^{F}. For a client, we could decide to either connect its demand to an opened facility or to pay its prize for not connecting. Connecting a client $j \in D$ to a facility $i \in F$ incurs a connection cost of c_{ij}^{F}. Assume that the connection costs satisfy the triangle inequality. The objective is to open a set of facilities $O^{\mathrm{F}} \subseteq F$, and decide to either connect the demand of each client to some opened facility or to pay its prize, such that the total cost (i.e., the cost of opening facilities, connecting clients, and paying prizes) is minimized.

We use $(O^{\mathrm{F}}, P^{\mathrm{F}}, \sigma^{\mathrm{F}})$ to denote a solution of a PCFL instance. Here O^{F} is the set of opened facilities, P^{F} is the set of clients deciding not to connect, and

$\sigma^F : D \setminus P^F \to O^F$ indicates the assigned facility of a client. For a solution (O^F, P^F, σ^F), denote by $C_O(O^F, P^F, \sigma^F)$ its total opening cost, i.e.,

$$C_O(O^F, P^F, \sigma^F) = \sum_{i \in O^F} f_i^F;$$

and denote by $C_P(O^F, P^F, \sigma^F)$ its total prize, i.e.,

$$C_P(O^F, P^F, \sigma^F) = \sum_{j \in P^F} p_j^F;$$

and denote by $C_C(O^F, P^F, \sigma^F)$ its total connection cost, i.e.,

$$C_C(O^F, P^F, \sigma^F) = \sum_{j \in D \setminus P^F} c_{\sigma(j)j}^F.$$

Therefore, the total cost of the solution (O^F, P^F, σ^F) is

$$C_O(O^F, P^F, \sigma^F) + C_P(O^F, P^F, \sigma^F) + C_C(O^F, P^F, \sigma^F)$$
$$= \sum_{i \in O^F} f_i^F + \sum_{j \in P^F} p_j^F + \sum_{j \in D \setminus P^F} c_{\sigma(j)j}^F.$$

From any given PCCND instance $\mathcal{I}_{PCCND}$ with inputs of a graph $G = (V, E)$, a sink set $T \subseteq V$, a source set $S \subseteq V$, an integer U, opening costs $\{f_t\}_{t \in T}$, prizes $\{p_s\}_{s \in S}$ and edge costs $\{c_e\}_{e \in E}$, we could construct a corresponding PCFL instance $\mathcal{I}_{PCFL}$ with inputs of a facility set F, a client set D, opening costs $\{f_i^F\}_{i \in F}$, prizes $\{p_j^F\}_{j \in D}$ and connection costs $\{c_{ij}^F\}_{i \in F, j \in D}$ as follows. Define the facility set F and client set D as the sink set T and source set S, respectively. Define the opening cost f_i^F of a facility $i \in F$ to be the same as the opening cost f_t of the corresponding sink $t = i$ in T, and the prize p_j^F of a client $j \in D$ to be the same as the prize p_s of the corresponding source $s = j$ in S, and the connection cost c_{ij}^F of a facility-client pair (i, j) to be $1/U$ times the minimum total edge cost of a path from $i = t \in T$ to $j = s \in S$ in the graph G. An illustration of the above construction can be found in Fig. 1, in which $U = 2$.

The following lemma gives a lower bound to an optimal solution of the PCCND instance.

Lemma 1. *The total cost of an optimal solution of the instructed PCFL instance $\mathcal{I}_{PCFL}$ is no more than the total cost of an optimal solution of the PCCND instance $\mathcal{I}_{PCCND}$.*

Proof. Assume that (O^*, P^*, τ^*) is an optimal solution of the PCCND instance $\mathcal{I}_{PCCND}$. We could construct a feasible solution (O^F, P^F, σ^F) for the PCFL instance $\mathcal{I}_{PCFL}$, where $O^F = O^*$, $P^F = P^*$, $\sigma^F(j) := \arg\min_{i \in O^*} c_{ij}^F$ for each $j \in D \setminus P^F$. It can be seen that the total opened cost and total prize of these two solutions are the same. Note that under the solution (O^*, P^*, τ^*), the demand of a source s may not be served by an opened sink with the minimum total

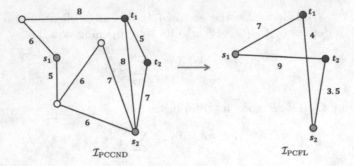

Fig. 1. Construction of the relevant instance of PCFL. Sources are represented in red, and sinks are represented in blue. (Color figure online)

edge cost of a path from s, and j's share of the total edge cost is at least $1/U$ of the total edge cost of the path from j to its assigned sink. Thus, the total edge cost incurred by a source $s \in S \setminus P^*$ under the solution (O^*, P^*, τ^*) is at least the connection cost of $c^F_{\sigma(j)j}$ of the same corresponding client $j = s$ under the solution (O^F, P^F, σ^F). Therefore, the total cost of a feasible solution of the PCFL instance $\mathcal{I}_{PCFL}$ is no more than the total cost of an optimal solution of the PCCND instance $\mathcal{I}_{PCCND}$, implying the lemma. □

2.3 Prize-Collecting Steiner Tree Problem

In a PCST instance, we are given an undirected graph $G^S = (V^S, E^S)$ and a root vertex $r \in V^S$. Each vertex $v \in V^S$ has a non-negative prize p^S_v, and each edge $e \in E^S$ has a non-negative edge cost c^S_e. The objective is to find an r-rooted tree, such that the sum of the edge costs of the edges in the tree and the prizes of the vertices not in the tree is minimized.

We use (R^S, P^S) to denote a solution of a PCST instance. Here R^S is an r-rooted tree, and P^S is the set of vertices not in the tree R^S. Denote by $V(R^S)$ and $E(R^S)$ the vertices and edges in the tree R^S, respectively. Note that $P^S = V^S \setminus V(R^S)$. For a solution (R^S, P^S), denote by $C_P(R^S, P^S)$ its total prize, i.e.,

$$C_P(R^S, P^S) = \sum_{v \in P^S} p^S_v;$$

and denote by $C_E(R^S, P^S)$ its total edge cost, i.e.,

$$C_E(R^S, P^S) = \sum_{e \in E(R^S)} c^S_e.$$

Therefore, the total cost of the solution (R^S, P^S) is

$$C_P(R^S, P^S) + C_E(R^S, P^S) = \sum_{v \in P^S} p^S_v + \sum_{e \in E(R^S)} c^S_e.$$

From any given PCCND instance $\mathcal{I}_{\text{PCCND}}$ with inputs of a graph $G = (V, E)$, a sink set $T \subseteq V$, a source set $S \subseteq V$, an integer U, opening costs $\{f_t\}_{t \in T}$, prizes $\{p_s\}_{s \in S}$ and edge costs $\{c_e\}_{e \in E}$, we could construct a relevant PCST instance $\mathcal{I}_{\text{PCST}}$ with inputs of a graph $G^S = (V^S, E^S)$, a root $r \in V^S$, prizes $\{p_v^S\}_{v \in V^S}$ and edge costs $\{c_e^S\}_{e \in E^S}$ as follows. We add a dummy vertex r as the root and construct an undirected graph $G^S = (V^S, E^S)$, where $V^S = V \cup \{r\}$ and $E^S = E \cup \{(v, r) : v \in T\}$. Define the prize p_v^S of a vertex $v \in S$ to be the same as the prize p_s of the corresponding source $s = v$ in S, and the prize p_v^S of a vertex $v \in V^S \setminus S$ to be zero. Define the edge cost c_e^S of an edge $e \in E$ remains the same the edge cost c_e as in the PCCND instance $\mathcal{I}$, and the edge cost c_e^S of an edge $e \in \{(v, r) : v \in T\}$ to be f_v. An illustration of the above construction can be found in Fig. 2.

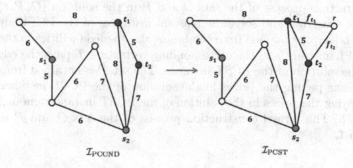

Fig. 2. Construction of the relevant instance of PCST.

The following lemma gives another lower bound to an optimal solution of the PCCND instance.

Lemma 2. *The total cost of an optimal solution of the instructed PCST instance $\mathcal{I}_{\text{PCST}}$ is no more than the total cost of an optimal solution of the PCCND instance $\mathcal{I}_{\text{PCCND}}$.*

Proof. Assume that (O^*, P^*, τ^*) is an optimal solution of the PCCND instance $\mathcal{I}_{\text{PCCND}}$. We could construct a feasible solution (R^S, P^S) for the PCST instance $\mathcal{I}_{\text{PCST}}$, where R^S obtained from deleting all the redundant edges in $\{e \in E : \tau^*(e) \geq 1\} \bigcup \{(v, r) : v \in O^*\}$ but one, and $P^S = V^S \setminus V(R^S)$. It can be seen that the total opening and edge cost of the solution (O^*, P^*, τ^*) is at least the total edge cost of the solution (R^S, P^S), and that the total prize of the solution (O^*, P^*, τ^*) is at least the total prize of the solution (R^S, P^S). Therefore, the total cost of a feasible solution of the PCFL instance $\mathcal{I}_{\text{PCST}}$ is no more than the total cost of an optimal solution of the PCCND instance $\mathcal{I}_{\text{PCCND}}$, implying the lemma. □

3   Algorithm and Analysis

This section presents an approximation algorithm for the PCCND problem. The main idea is based on using the solutions of the relevant constructed instances of the PCFL and PCST to build a feasible solution (O, P, τ) for the PCCND instance. In Sect. 3.1, we show how the set of opened sinks and the set of sources paying the prizes (i.e., the sets O and P in the solution (O, P, τ)) are determined. Then in Sect. 3.2, we show the construction process of the mapping τ in the solution (O, P, τ). Finally, we conclude this section by proving the approximation ratio of our algorithm.

3.1   Bound the Cost of Opening Sinks and Paying Prizes

The construction process of the sets O and P in the solution (O, P, τ) is quite simple. We construct and solve the relevant instances of the PCFL and PCST. The set O is simply obtained from combining the opened facilities in the solution of the PCFL instance and the corresponding vertex $v \in T$ pays the edge cost of (v, r) in the solution of the PCST instance. The set P is obtained from combining the clients paying the prizes in the solution of the PCFL instance and the vertices paying the prizes in the solution of the PCST instance among the given source set S. The formal construction process of the sets O and P is given in Algorithm 1.

Algorithm 1. Selection of the set of opened sinks O and the set of sources P

Input: A PCCND instance $\mathcal{I}_{\mathrm{PCCND}}$ with inputs of $G = (V, E)$, $T \subseteq V$, $S \subseteq V$, U, $\{f_t\}_{t \in T}$, $\{p_s\}_{s \in S}$ and $\{c_e\}_{e \in E}$.

Output: A set O of opened sinks and a set P of the sources paying the prizes for the PCCND instance $\mathcal{I}_{\mathrm{PCCND}}$.

Step 1 Construct and solve a relevant PCFL instance.

    Step 1.1 Construct a PCFL instance $\mathcal{I}_{\mathrm{PCFL}}$ with inputs of F, D, $\{f_i^{\mathrm{F}}\}_{i \in F}$, $\{p_j^{\mathrm{F}}\}_{j \in D}$ and $\{c_{ij}^{\mathrm{F}}\}_{i \in F, j \in D}$, where $F = T$, $D = S$, $f_i^{\mathrm{F}} = f_i$ for any $i \in F = T$, $p_j^{\mathrm{F}} = p_j$ for any $j \in D = S$, and set c_{ij}^{F} to be $1/U$ times the minimum total edge cost of a path from $i \in F = T$ to $j \in D = S$ in the graph G.

    Step 1.2 Solve the PCFL instance $\mathcal{I}_{\mathrm{PCFL}}$ with the currently best α^{F}-approximation algorithm for the PCFL and obtain a solution $(O^{\mathrm{F}}, P^{\mathrm{F}}, \sigma^{\mathrm{F}})$.

Step 2 Construct and solve a relevant PCST instance.

    Step 2.1 Add a dummy vertex r as a root and construct a PCST instance $\mathcal{I}_{\mathrm{PCST}}$ with inputs of $G^{\mathrm{S}} = (V^{\mathrm{S}}, E^{\mathrm{S}})$, $r \in V^{\mathrm{S}}$, $\{p_v^{\mathrm{S}}\}_{v \in V^{\mathrm{S}}}$ and $\{c_e^{\mathrm{S}}\}_{e \in E^{\mathrm{S}}}$, where $V^{\mathrm{S}} = V \cup \{r\}$, $E^{\mathrm{S}} = E \cup \{(v, r) : v \in T\}$, $p_s^{\mathrm{S}} = p_s$ for any $v \in S$, $p_v^{\mathrm{S}} = 0$ for any $v \in V^{\mathrm{S}} \setminus S$, $c_e^{\mathrm{S}} = c_e$ for any $e \in E$, and $c_e^{\mathrm{S}} = f_v$ for any edge $e \in \{(v, r) : v \in T\}$.

    Step 2.2 Solve the PCST instance $\mathcal{I}_{\mathrm{PCST}}$ with the currently best α^{S}-approximation algorithm for the PCST and obtain a solution $(R^{\mathrm{S}}, P^{\mathrm{S}})$.

Step 3 Construct the sets O and P.

    Define $O^{\mathrm{S}} := \{t \in T : (t, r) \in E(R^{\mathrm{S}})\}$, and $P' := P^{\mathrm{S}} \cap S$. Set $O := O^{\mathrm{F}} \cup O^{\mathrm{S}}$, $P := P^{\mathrm{F}} \cup P'$, and output O and P.

The following Lemma 3 bounds the total opening cost of O, while Lemma 4 bounds the total prize of P.

Lemma 3. *The total opening cost under f of the sinks in O is no more than the total opening cost under f^F of the facilities in O^F plus the total edge cost under c^S of the edges in $\{(t,r) \in E(R^S) : t \in T\}$.*

Proof. Since $O := O^F \cup O^S$, therefore,

$$\sum_{t \in O} f_t \leq \sum_{t \in O^F} f_t + \sum_{t \in O^S} f_t$$

$$= \sum_{t \in O^F} f_t^F + \sum_{e \in \{(t,r) \in E(R^S) : t \in T\}} c_e^S.$$

This completes the proof. □

Lemma 4. *The total prize under p of the sources in P is no more than the total prize under p^F of clients in P^F plus the total prize under p^S of the vertices in P^S.*

Proof. Since $P := P^F \cup P'$ and $P' := P^S \cap S$, therefore,

$$\sum_{s \in P} p_s \leq \sum_{s \in P^F} p_s + \sum_{s \in P'} p_s$$

$$\leq \sum_{s \in P^F} p_s^F + \sum_{s \in P'} p_s^S$$

$$\leq \sum_{s \in P^F} p_s^F + \sum_{s \in P^S} p_s^S.$$

This completes the proof. □

3.2   Bound the Cost of Installing Cables

The construction process of the mapping τ in the solution (O, P, τ) is more challenging. Note that all the demands of the sources needing to be routed are on the tree R^S obtained with Algorithm 1. We could divide all these demands according to the subtrees of R^S rooted at the some vertex $v \in O^S$. For the demands needing to be routed in each subtree, we try to install a suitable number of cables on the edges of the graph in order to satisfy their demands. The formal construction process of the mapping τ is given in Algorithm 2.

Lemma 5. *The solution (O, P, τ) is a feasible solution for the PCCND instance $\mathcal{I}_{PCCND}$.*

Proof. If we prove that the number of cables installed on the edges, according to the mapping τ, are sufficient for routing all the demands from the sources in $S \setminus P$ to the opened sinks in O, we prove the feasibility of the solution (O, P, τ).

Algorithm 2. Solving the PCCND problem

Input: The PCCND instance $\mathcal{I}_{\text{PCCND}}$, the solution (O^F, P^F, σ^F) of the PCFL instance $\mathcal{I}_{\text{PCFL}}$, the solution (R^S, P^S) of the PCST instance $\mathcal{I}_{\text{PCST}}$, the sets of O^S, O, and P.

Output: A mapping $\tau : E \to \mathbb{Z}$ for the PCCND instance $\mathcal{I}_{\text{PCCND}}$.

Step 0 Initialization.

Set $\tau(e) := 0$ for each edge $e \in E$. We orient all edges of the tree R^S in a direction towards the root r. For any vertex $v \in V(R^S)$, denote by $R(v)$ the subtree of R^S rooted at v.

Step 1 Classify the subtrees.

Step 1.1 For each subtree $R(v)$, where $v \in O^S$, delete any source $s \in S \cap P$ and its incident edges of the subtree, if s is not an intermediate vertex on a path from any other source in $S \setminus P$ to the root v. After pruning edges, we name each updated subtree of $R(v)$ as $R'(v)$.

Step 1.2 We call the subtree $R'(v)$, where $v \in O^S$, a light subtree, if the number of sources in $S \setminus P$ on it is no more than U. Otherwise, we call the subtree $R'(v)$ a heavy one. Denote by LR all the light subtrees, and HR all the heavy subtrees. Go to Step 2 to deal with the light subtrees, and go to Step 3 to deal with the heavy subtrees.

Step 2 Install cables on the light subtrees.

Step 2.1 Install a cable on each edge of the subtrees in LR, and update $\tau(e) := \tau(e) + 1$ for each edge $e \in \bigcup_{R_l \in LR} E(R)$.

Step 2.2 For each light subtree R_l, route the demand of each source in $(S \setminus P) \cap V(R_l)$ to the root of R_l.

Step 3 Install cables on the heavy subtrees.

Step 3.1 Install a cable on each edge of the subtrees in HR, and update $\tau(e) := \tau(e) + 1$ for each edge $e \in \bigcup_{R_h \in HR} E(R)$.

Step 3.2 For each heavy subtree R_h, let $V'(R_h)$ be the vertices of the subtree, which have at most U units of demands on each of their incoming incident edge and have more than U units of demands on each of their outgoing incident edge. For a vertex $u \in V(R_h)$ of some heavy subtree R_h, denote by $R''(u)$ the subtree of R_h rooted at u.

Step 3.3 For all v' in $V'(R_h)$ of some heavy subtree R_h do

Step 3.3.1 Find all the vertices that directly orient to v' via an edge of the subtree R_h. We call these vertices the children of v', and denote them by $V_{\text{ch}}(v')$. We call v' the parent of each vertex in $V_{\text{ch}}(v')$. For each subtree $R''(u)$, where $u \in V_{\text{ch}}(v')$, let $S(u)$ be the sources needing to be routed on the subtree, i.e., the sources in $(S \setminus P) \cap V(R''(u))$. Then for each $R''(u)$, find a source-sink pair (s_u, t_u), where $s_u \in S(u)$ and $t_u \in O^F$, which has the minimum connection cost $c_{s_u t_u}^F$. Select $\lfloor \sum_{u \in V_{\text{ch}}(v')} S(u)/U \rfloor$ such pairs with lowest connection cost. Update $\tau(e) := \tau(e) + 1$ for each edge on the path connecting the selected pairs with a minimum total edge cost.

Step 3.3.2 If the source-sink pair (s_u, t_u) corresponding to a subtree $R''(u)$ is selected, then route the demand of each source in $S(u)$ to the sink t_u. If the source-sink pair (s_u, t_u) of a subtree $R''(u)$ is not selected, then route some or none of the demands of the sources in $S(u)$ to some selected sinks via a path on the heavy subtree and the corresponding selected pair, such that the amount of demands being routed to each sink is exactly U.

Step 3.3.3 Remove all the routed demands and update each heavy subtree. Go to Step 2.2 to deal with the updated light subtrees, and go to Step 3.2 to deal with the updated heavy subtrees.

Step 4 Output the mapping.

Once all the demands needing to be routed have been processed, we stop the algorithm and output the mapping of τ.

Since the number of sources needing to be routed on any light subtree is no more than U, installing one cable on each edge of the subtree is sufficient for routing all the demands to the root. Recall that the root of any light subtree is in O^S. Therefore, all the demands on any light subtree could be routed to an opened sink in O without violating the capacities of the corresponding edges.

For a heavy subtree, recall that we first find all the vertices with more than U units outgoing demands and at most U incoming demands on their incident edges. Then, we try to deal with the demands on the subtrees of their children by selecting a certain number of source-sink pairs, so that most of the demands on the subtrees of the children can be routed to the sinks in O^F and there are no more than U units of demands left. It is worth mentioning that the routing idea for dealing with most of the demands in Step 3.3.2 of Algorithm 2 may cause a violation of the capacity requirements, but that kind of situation can be handled. Consider the situation that the demand of a source s_1 on a children subtree R'' is routed to some selected source-sink pairs (s_2, t_2) related to a sibling subtree of R'', and then in later iterations, one of the un-routed source s_3 on R'' is selected as a source of a source-sink pair. Since Step 3.3.2 of Algorithm 2 guarantees that the amount of demands being routed to each selected source-sink pair is exactly U, there must be some demands of the sources on a sibling subtree of R'' which need to be routed to s_3. Note that routing the demand of s_1 to the pair (s_2, t_2) occupies a unit of capacity on the edge from the root of R'' to its parent, and routing the demand of some source s_4 to s_3 also occupy a unit of capacity on the edge from the root of R'' to its parent. When the source s_1 in this situation is a set of sources on R'', and the source s_4 is also a set of sources, it is very likely that the capacity of the edge from the root of R'' to its parent is violated. To handle the violation, we could reroute the demands of the sources like s_1 from s_2 to s_3, and reroute the demands of the sources like s_4 from s_3 to s_2. Therefore, it can be seen that installing a cable on each edge of an initially heavy subtree and on each edge related to the selected source-sink pairs is sufficient for serving all the demands on the heavy subtree.

From the above analysis, we complete the proof of this lemma.   □

Lemma 6. *The total edge cost under c of the edges according to the mapping τ is no more than the total connection cost under c^F of the clients according to the mapping σ^F plus the total edge cost under c^S of the edges in $V(R^S)$.*

Proof. The total edge cost incurred by installing a cable on each edge related to the selected source-sink pairs is no more than the corresponding total connection cost of the solution (O^F, P^F, σ^F), since there are exactly U units of demands being routed to each selected source-sink pair and the connection cost of each source $s \in S \setminus P$ routing its demand through the pair under the mapping of σ^F is no less than the connection cost of the selected pair. It can be seen that the total edge cost incurred by installing a cable on each edge of the light and heavy subtrees are no more than the total edge cost of the edges in $E(R^S) \setminus \{(t, r) \in E(R^S) : t \in T\}$, therefore, we have that

$$\sum_{e \in E} \tau(e)c_e \le \sum_{j \in D \backslash P} c^F_{\sigma^F(j)j} + \sum_{e \in E(R^S) \backslash \{(t,r) \in E(R^S) : t \in T\}} c_e$$

$$= \sum_{j \in D \backslash P} c^F_{\sigma^F(j)j} + \sum_{e \in E(R^S) \backslash \{(t,r) \in E(R^S) : t \in T\}} c^S_e.$$

This completes the proof.   □

We are now ready to present our main result for the PCCND problem.

Theorem 1. *Algorithm 2 is a 3.482-approximation algorithm for the PCCND problem.*

Proof. Let OPT_{PCCND}, OPT_{PCFL} and OPT_{PCST} be the total costs of the optimal solutions of the PCCND instance $\mathcal{I}_{\text{PCCND}}$, the constructed PCFL instance $\mathcal{I}_{\text{PCFL}}$ and the constructed PCST instance $\mathcal{I}_{\text{PCST}}$, respectively.

Recall that $P := P^F \cup P'$ and $P' := P^S \cap S$, that the solutions (O^F, P^F, σ^F) and (R^S, P^S) are obtained from using the currently best approximation algorithms with ratios of α^F and α^S for the PCFL and PCST to solve the instances $\mathcal{I}_{\text{PCFL}}$ and $\mathcal{I}_{\text{PCST}}$, respectively. Combining Lemmas 1-4 and 6, we obtain that

$$\sum_{t \in O} f_t + \sum_{s \in P} p_s + \sum_{e \in E} \tau(e)c_e$$

$$\le \left(\sum_{t \in O^F} f^F_t + \sum_{s \in P^F} p^F_s + \sum_{j \in D \backslash P} c^F_{\sigma^F(j)j} \right) + \left(\sum_{s \in P^S} p^S_s + \sum_{e \in E(R^S)} c^S_e \right)$$

$$\le \left(\sum_{t \in O^F} f^F_t + \sum_{s \in P^F} p^F_s + \sum_{j \in D \backslash P^F} c^F_{\sigma^F(j)j} \right) + \left(\sum_{s \in P^S} p^S_s + \sum_{e \in E(R^S)} c^S_e \right)$$

$$\le \alpha^F OPT_{\text{PCFL}} + \alpha^S OPT_{\text{PCST}}$$

$$\le (\alpha^F + \alpha^S) OPT_{\text{PCCND}}.$$

Using the values of $\alpha^F = 1.5148$ and $\alpha^S = 1.9672$ yields the approximation ratio of 3.482 of our algorithm.   □

3.3   Case of Single Sink

In a single-sink PCCND instance, we are given an undirected graph $G = (V, E)$, a sink $t \in V$, a set of sources $S \subseteq V$, and an integer U of cable capacity. Each source $s \in S$ has a unit of demand and a non-negative prize p_s. For a source, we could decide to either route its demand to the sink t via a path, or to pay its prize for not routing. Each edge $e \in E$ has a non-negative edge cost c_e, and assume that the edge costs satisfy the triangle inequality. We can install multiple times the cable on an edge. If we install l times on an edge e, the edge e can serve lU units of demand, and we should pay a cost of lc_e. The objective is to install cables on the edges, and decide to either route the demand of each source to the sink t or to pay its prize without violating the capacity of any edge, such that the total cost (i.e., the cost of installing cables on edges and paying prizes) is minimized.

Based on the approximation algorithm for the PCCND, we provide a better approximation algorithm for the single-sink PCCND. This comes from the fact that the optimal solution of the corresponding constructed PCFL instance could be found. Here is our main result for the single-sink PCCND.

Corollary 1. *Algorithm 2 is a 2.9672-approximation algorithm for the PCCND problem with a single sink.*

4   Conclusion

In this paper, we assume that each source has a unit of demand in the PCCND and single-sink PCCND problems. It is of great interest to study a more general case where demands can be arbitrary and are not splittable. A natural question is whether our algorithm can be adapted to the case of non-uniform demands.

References

1. Archer, A., Bateni, M., Hajiaghayi, M., Karloff, H.: Improved approximation algorithms for prize-collecting steiner tree and tsp. SIAM J. Comput. **40**(2), 309–332 (2011)
2. Bienstock, D., Goemans, M.X., Simchi-Levi, D., Williamson, D.: A note on the prize collecting traveling salesman problem. Math. Program. **59**(1), 413–420 (1993)
3. Charikar, M., Khuller, S., Mount, D.M., Narasimhan, G.: Algorithms for facility location problems with outliers. In: SODA, vol. 1, pp. 642–651 (2001)
4. Geunes, J., Levi, R., Romeijn, H.E., Shmoys, D.B.: Approximation algorithms for supply chain planning and logistics problems with market choice. Math. Program. **130**(1), 85–106 (2011)
5. Goemans, M.X., Williamson, D.P.: A general approximation technique for constrained forest problems. SIAM J. Comput. **24**(2), 296–317 (1995)
6. Hajiaghayi, M.T., Jain, K.: The prize-collecting generalized steiner tree problem via a new approach of primal-dual schema. In: SODA, vol. 6, pp. 631–640. Citeseer (2006)
7. Hajiaghayi, M., Nasri, A.A.: Prize-Collecting Steiner Networks via Iterative Rounding. In: López-Ortiz, A. (ed.) LATIN 2010. LNCS, vol. 6034, pp. 515–526. Springer, Heidelberg (2010). https://doi.org/10.1007/978-3-642-12200-2_45
8. Han, L., Xu, D., Du, D., Wu, C.: A 5-approximation algorithm for the k-prize-collecting steiner tree problem. Optim. Let. **13**(3), 573–585 (2019)
9. Han, L., Xu, D., Du, D., Zhang, D.: A local search approximation algorithm for the uniform capacitated k-facility location problem. J. Combin. Opt. **35**(2), 409–423 (2018)
10. Han, L., Xu, D., Li, M., Zhang, D.: Approximation algorithms for the robust facility leasing problem. Opt. Let. **12**(3), 625–637 (2018). https://doi.org/10.1007/s11590-018-1238-x

11. Li, Y., Du, D., Xiu, N., Xu, D.: Improved approximation algorithms for the facility location problems with linear/submodular penalties. Algorithmica **73**(2), 460–482 (2015)
12. Mahdian, M., Pál, M.: Universal facility location. In: Di Battista, G., Zwick, U. (eds.) ESA 2003. LNCS, vol. 2832, pp. 409–421. Springer, Heidelberg (2003). https://doi.org/10.1007/978-3-540-39658-1_38
13. Ravi, R., Sinha, A.: Approximation algorithms for problems combining facility location and network design. Oper. Res. **54**(1), 73–81 (2006)
14. Sharma, Y., Swamy, C., Williamson, D.P.: Approximation algorithms for prize collecting forest problems with submodular penalty functions. In: Proceedings of the Eighteenth Annual ACM-SIAM Symposium on Discrete Algorithms, pp. 1275–1284. Citeseer (2007)
15. Swamy, C., Kumar, A.: Primal-dual algorithms for connected facility location problems. Algorithmica **40**(4), 245–269 (2004)
16. Xu, G., Xu, J.: An LP rounding algorithm for approximating uncapacitated facility location problem with penalties. Inform. Proces. Let. **94**(3), 119–123 (2005)
17. Xu, G., Xu, J.: An improved approximation algorithm for uncapacitated facility location problem with penalties. J. Combin. Opt. **17**(4), 424–436 (2009)
18. Xu, Y., Xu, D., Du, D., Wu, C.: Improved approximation algorithm for universal facility location problem with linear penalties. Theor. Comput. Sci. **774**, 143–151 (2019)

Computational and Network Economics

Possible and Necessary Winner Problems in Iterative Elections with Multiple Rules

Peihua Li[✉] and Jiong Guo[✉]

Department of Computer Science, Shandong University, Qingdao, China
lipeihua@mail.sdu.edu.cn, jguo@sdu.edu.cn

Abstract. An iterative election eliminates some candidates in each round until the remaining candidates have the same score according to a given voting rule. Prominent iterative voting rules include Hare, Coombs, Baldwin, and Nanson. The Hare/Coombs/Baldwin rules eliminate in each round the candidates with the least plurality/veto/Borda scores, while the Nanson rule eliminates the candidates with below-average Borda scores. Recently, it has been demonstrated that iterative elections admit some desirable properties such as polynomial-time winner determination and NP-hard control/manipulation/bribery.

We study new aspects of iterative elections. We suppose that a set R of iterative voting rules is given and each round of the iterative election can choose one rule in R to apply. The question is whether there is a combination of rules, such that a specific candidate p becomes the unique winner (the Possible Winner problem), or whether a specific candidate p wins under all rule combinations (the Necessary Winner problem). The Possible Winner problem can be considered as a special control problem for iterative elections. We prove that for all subsets R of {Hare, Coombs, Baldwin, Nanson} with $|R| \geq 2$, both Possible and Necessary Winner problems are hard to solve, with the only exception of $R = \{$Baldwin, Nanson$\}$. We further provide special cases of the Necessary Winner problem with $R = \{$Baldwin, Nanson$\}$, which are polynomial-time solvable. We also discuss the parameterized complexity of the Possible Winner problems with respect to the number of candidates and the number of votes, and achieve fixed-parameter tractable (FPT) results.

Keywords: Computational social choice · Iterative elections · Combinatorial opitimization · Computational complexity.

1 Introduction

The theory of social choice aims at the aggregation of preferences of individuals to achieve a collective decision. Over the past two decades, computational social choice has become an interdisciplinary area to study the computational

The authors are supported by the National Natural Science Foundation of China (No.62072275 and 61772314).

perspectives of voting problems [BCE+16]. An instance of a voting problem consists of an election and a voting rule (or correspondence). An election is denoted as $E = (C, V)$, where C is a set of candidates $C = \{c_1, ..., c_m\}$ and V is a multiset of votes $V = \{v_1, ..., v_n\}$, where each vote is a total order of C, $v_i = c_{i_1} > c_{i_2} > ... > c_{i_m}$, for each $1 \leq i \leq n$ and $i_j \in \{1, ..., m\}$ with $1 \leq j \leq m$. In v_i, the first candidate c_{i_1} is the most favorite candidate of v_i and c_{i_m} is the least favorite candidate. A voting rule maps the election to a set of winning candidates (the winners). Hereby, the unique winner case refers to the scenario with only one winner. We also use co-winners to refer to the multiple winners.

Most voting rules directly output the winners based on the computation of some scores for the candidates or pairwise comparisons of the candidates. For instance, the most extensively studied rules are from the class of positional scoring rules, each of which is associated with a scoring vector $< \alpha_1, ..., \alpha_m >$ with $\alpha_1 \geq \alpha_2 \geq ... \geq \alpha_m$. From a vote v, the candidate at the i-th position of v receives a score of α_i. The candidates with the highest overall score from all votes win the election. The most prominent positional scoring rules are plurality $< \alpha_1 = 1, \alpha_2 = ... = \alpha_m = 0 >$, veto $< \alpha_1 = ... = \alpha_{m-1} = 1, \alpha_m = 0 >$, Borda $< \alpha_1 = m - 1, \alpha_2 = m - 2, ..., \alpha_m = 0 >$, and k-approval for $1 \leq k \leq m - 1$ $< \alpha_1 = ... = \alpha_k = 1, \alpha_{k+1} = ... = \alpha_m = 0 >$.

Recently, iterative elections have attracted more and more attention [MNRS18, BSW20, ZG20]. Given an election, iterative voting rules eliminate the candidates in rounds. The winners are the candidates remaining in the end. Until now, the most prominent iterative voting rules include Hare, Coombs, Baldwin and Nanson. The Hare rule [TP08] eliminates in each round the candidates with the minimum plurality score, the Coombs rule [LN95] the ones with the minimum veto score, the Baldwin rule [Bal26] the ones with minimum Borda score, and the Nanson rule [Nan82] the ones with below-average Borda scores. The iterative process terminates, when all remaining candidates have the same score. These candidates are then the winners. It is easy to observe that the winner determination problem with respect to these four rules is easy to solve. The resistance of these rules against strategic behaviors such as control, manipulation, and bribery has been proved [ENR+21, FMS21]. We study a new aspect of iterative elections. Hereby, we assume that a set of iterative voting rules is given and each round is allowed to choose a rule from the set to apply. The question here is whether there is a combination of the rules, such that a specific candidate p becomes the winner of the election (the Possible Winner problem), or whether a specific candidate p wins the election under all possible rule combinations (the Necessary Winner problem). A rule combination refers to a vector $< r_1, r_2, ... >$ where each r_i is from the given set, and is applied to the i-th round. For the Possible Winner problem, we require that p becomes the unique winner, while p can be a co-winner in the Necessary Winner problem. The other unique winner/co-winner settings can be handled in similar ways. Possible and Necessary Winner problems have been studied for traditional single-round rules under different circumstances such as with incomplete votes [XC08, CDK+21] and for sequential voting [GNNW14]. The version studied in this paper can also be considered as a constructive control problem for iterative elections. The election chair has the power to choose the rule applied to each round and his aim is to make a specific candidate win the election.

We achieve hardness results for both Possible/Necessary problems with the rules from a subset R of {Hare, Coombs, Baldwin, Nanson}. Clearly, if $|R| = 1$, the Possible and Necessary Winner problems become the winner determination problem for the rule in R and are solvable in polynomial time. With $|R| \geq 2$, the Possible Winner problem is NP-hard and the Necessary Winner problem is CoNP-hard. The only exception is with $R = $ {Baldwin, Nanson}, for which both Possible and Necessary problems remain open. We achieve these results by complex reductions from 3SAT, which require careful construction to balance the scores of the candidates and to guarantee the correspondence between the choice of rules and the assignment of the variables. Further, we discuss the parameterized complexity of the Possible Winner problems with respect to the number of candidates and the number of votes, and achieve fixed-parameter tractable (FPT) results. Moreover, we consider a special case with Necessary Winner for R = {Balwin, Nanson} and show that a Condorcet winner of the initial election is always a necessary winner and thus this special case is polynomial-time solvable. Some proofs are deferred to a long version.

2   Preliminaries

An election is denoted as $E = (C, V)$, where C is a set of candidates $C = \{c_1, ..., c_m\}$ and V is a multiset of votes $V = \{v_1, ..., v_n\}$, where each vote is a total order of C, $v_i = c_{i_1} > c_{i_2} > ... > c_{i_m}$, for each $1 \leq i \leq n$ and $i_j \in \{1, ..., m\}$ with $1 \leq j \leq m$. The rules studied in this paper are from the class of positional scoring rules, each of which is associated with a scoring vector $< \alpha_1, ..., \alpha_m >$ with $\alpha_1 \geq \alpha_2 \geq ... \geq \alpha_m$. From a vote v, the candidate at the i-th position of v receives a score of α_i. The overall score of a candidate is the sum of the scores that the candidate gets from all votes under the corresponding rule.

The most prominent positional scoring rules are plurality $< \alpha_1 = 1, \alpha_2 = ... = \alpha_m = 0 >$, veto $< \alpha_1 = ... = \alpha_{m-1} = 1, \alpha_m = 0 >$, Borda $< \alpha_1 = m - 1, \alpha_2 = m - 2, ..., \alpha_m = 0 >$.

2.1   Iterative Voting Rules

An iterative election eliminates some candidates in each round until the remaining candidates have the same score according to a given voting rule. Prominent iterative voting rules include Hare, Coombs, Baldwin, and Nanson.

- The Hare rule eliminates in each round the candidates with the least plurality score.
- The Coombs rule eliminates in each round the candidates with the least veto score.
- The Baldwin rule eliminates in each round the candidates with the least Borda score.

- The Nanson rule eliminates in each round the candidates with below-average Borda scores.

2.2 Possible and Necessary Winner Problems

Possbile Winner
Input: An election $E = (C, V)$, a specific candidate p, a voting rules set R.
Question: Is there a combination of the rules from R, such that p becomes the winner of the election?

Necessary Winner
Input: An election $E = (C, V)$, a specific candidate p, a voting rules set R.
Question: Can p win the election under all possible rule combinations?

3   Possible Winner

In this section, we prove the NP-hardness and the parameterized complexity of the Possible Winner problem.

Theorem 1. *Possible Winner is NP-hard for all subsets R of {Hare, Coombs, Baldwin, Nanson} satisfying $|R| \geq 2$ and $R \neq \{Baldwin, Nanson\}$.*

Proof. The NP-hardness is achieved by reductions from 3SAT. Here, we present the reduction for $R = \{Hare, Baldwin\}$. The other cases follow from similar reductions. The 3SAT problem asks whether, for a given set of Boolean variables $x_1, .., x_s$ and a set of clauses $y_1, ..., y_t$ with each clause consisting of three literals, there exists an assignment of the variables satisfying all clauses.

The basic idea of the reduction is as follows. We create four variable candidates $x_i^1, x_i^2, x_i^T, x_i^F$ for each variable x_i, and one clause candidate z_i for each clause y_i. The votes are constructed in the way that 1) the eliminations of these candidates occur in s main rounds $r_1, ..., r_s$ and each main round r_i eliminates the variable candidates created for x_i, and 2) each main round has two secondary rounds, where the first secondary round r_i^1 of r_i eliminates x_i^1, x_i^T or x_i^1, x_i^F, representing that x_i is assigned to TRUE or FALSE, and the second secondary round r_i^2 eliminates the remaining two. To eliminate x_i^1 and x_i^T in r_i^1, we have to apply Hare, while Baldwin is applied to eliminate x_i^1 and x_i^F. The clause candidate z_j is eliminated in the second secondary round of the main round r_i, depending on whether variable x_i occurs in the clause y_j and whether the assignment of x_i satisfies y_j, that is, which variable candidate in $\{x_i^T, x_i^F\}$ is eliminated in the first secondary round of r_i. Finally, the specific candidate p remains in the end only if all clause candidates are eliminated in the s main rounds, meaning that all clauses are satisfied by the assignment of the variables. Then by combining the rules applied in all main rounds, we can get a rule combination such that p becomes the unique winner.

Next, we present the details of the reduction. As stated above, we have four variable candidates $x_i^1, x_i^2, x_i^T, x_i^F$ for each variable x_i and one clause candidate z_j for each clause y_j. Then, the candidate set is set as $C :=$ $\left(\bigcup_{i=1}^s \{x_i^1, x_i^2, x_i^T, x_i^F\} \right) \cup \left(\bigcup_{j=1}^t \{z_j\} \right) \cup \widehat{E} \cup \{p\}$ with the set of auxiliary candidates $\widehat{E} := \left(\bigcup_{j=1}^t \{u_{j,1}, u_{j,2}\} \right) \cup \{\alpha, \beta, q\} \cup \left(\bigcup_{i=1}^s \{w_i^{1,1}, w_i^{2,1}, w_i^{T,1}, w_i^{F,1}, \right.$ $\left. w_i^{1,2}, w_i^{2,2}, w_i^{T,2}, w_i^{F,2}\} \right)$. Particularly, the pair of candidates $u_{j,1}$ and $u_{j,2}$ is used to balance the plurality score of clause candidate z_j. Analogously, the pair of candidates $w_i^{g,1}$ and $w_i^{g,2}$ with g in $\{1, 2, T, F\}$ is used for the plurality score setting of variable candidate x_i^g. Moreover, candidate q plays the role of comparison with p with respect to the plurality score after s main rounds. The candidates α and β are used to construct votes, where the Borda scores of other candidates are influenced but their plurality scores remain the same. For a subset $S \subseteq C$, we use $\overrightarrow{S}$ to denote an arbitrary but fixed ordering of S and $\overleftarrow{S}$ to denote the reversed ordering of $\overrightarrow{S}$. Further, we use $D(c_1, c_2)$ with $c_1, c_2 \in C$ to denote a pair of votes: $\alpha > c_1 > c_2 > \overrightarrow{S} > \beta$ and $\beta > \overleftarrow{S} > c_1 > c_2 > \alpha$, where $S = C \setminus \{\alpha, \beta, c_1, c_2\}$. If we have two votes being two completely reversed orders, then all candidates have the same Borda score from these two votes. However, in the two votes in $D(c_1, c_2)$, all candidates with the only exception of c_1 and c_2 have the same Borda score $m - 1$ with $m = |C|$, while the Borda score of c_1 is m and the Borda score of c_2 is $m - 2$. If we use γ to denote the "standard" Borda score $m - 1$, then the score of c_1 is $\gamma + 1$ and the score of c_2 is $\gamma - 1$. Further, if we remove one candidate c from C, then all candidates in $C \setminus \{c, c_1, c_2\}$ have again the standard Borda score $m - 2$ from these two votes. By abusing γ to denote the new standard Borda score, we can say that the Borda scores of most candidates remain the same, γ. If $c \notin \{c_1, c_2\}$, then the scores of c_1 and c_2 also remain the same, $\gamma + 1$ and $\gamma - 1$, respectively. If $c = c_1$, then the Borda score of c_2 becomes γ. We then say that the removal of c_1 increases the Borda score of c_2. In the case of removing $c = c_2$, the Borda score of c_1 becomes γ and we say the removal of c_2 decreases the Borda score of c_1. Note that in the following, we often construct the votes in a pairwise manner as $D(c_1, c_2)$, and the above change of Borda score can also be observed there. Define $W = (2s + 1 + t)t$.

The set $\mathcal{V}$ of votes consists of six subsets which are created in the order of their indices. The first subset $\mathcal{V}_1$ contains the votes for the clause candidates. For a clause candidate z_i with $1 \leq i \leq t$ whose corresponding clause y_i contains variables x_j, x_k, x_l with $1 \leq j < k < l \leq s$, the votes are constructed according to the positive/negative occurrences of the variables. Hereby, we define four vote forms $\chi_1, \chi_2, \chi_3, \chi_4$:

- $\chi_1 := z_i > u_{i,1} > p > \overrightarrow{S} > u_{i,2}$ with $S = C \setminus \{p, z_i, u_{i,1}, u_{i,2}\}$;

- $\chi_2 :=$

$$\begin{cases} x_j^F > z_i > u_{i,1} > p > \overrightarrow{S} > u_{i,2} \text{ with } S = C \setminus \{p, z_i, x_j^F, u_{i,1}, u_{i,2}\}, \text{ if} \\ x_j \text{ occurs positively in } y_i; \\ x_j^T > z_i > u_{i,1} > p > \overrightarrow{S} > u_{i,2} \text{ with } S = C \setminus \{p, z_i, x_j^T, u_{i,1}, u_{i,2}\}, \text{ if} \\ x_j \text{ occurs negatively in } y_i; \end{cases}$$

- $\chi_3 :=$

$$\begin{cases} x_k^F > z_i > u_{i,1} > p > \overrightarrow{S} > u_{i,2} \text{ with } S = C \setminus \{p, z_i, x_k^F, u_{i,1}, u_{i,2}\}, \text{ if} \\ x_k \text{ occurs positively in } y_i; \\ x_k^T > z_i > u_{i,1} > p > \overrightarrow{S} > u_{i,2} \text{ with } S = C \setminus \{p, z_i, x_k^T, u_{i,1}, u_{i,2}\}, \text{ if} \\ x_k \text{ occurs negatively in } y_i; \end{cases}$$

- $\chi_4 :=$

$$\begin{cases} x_l^F > z_i > u_{i,1} > p > \overrightarrow{S} > u_{i,2} \text{ with } S = C \setminus \{p, z_i, x_l^F, u_{i,1}, u_{i,2}\}, \text{ if} \\ x_l \text{ occurs positively in } y_i; \\ x_l^T > z_i > u_{i,1} > p > \overrightarrow{S} > u_{i,2} \text{ with } S = C \setminus \{p, z_i, x_l^T, u_{i,1}, u_{i,2}\}, \text{ if} \\ x_l \text{ occurs negatively in } y_i. \end{cases}$$

Then, the set $\mathcal{V}_1^i$ contains the votes created for the clause candidate z_i, which are as follows:

- one vote $v = z_i > p > u_{i,1} > \overrightarrow{S} > u_{i,2}$ with $S = C \setminus \{p, z_i, u_{i,1}, u_{i,2}\}$ and one vote $\overleftarrow{v}$;
- $W + 2j - 1$ many of the form χ_1 and $W + 2j - 1$ many of the form $\overleftarrow{\chi_1}$;
- $2k - 2j$ many of the form χ_2 and $2k - 2j$ many of the form $\overleftarrow{\chi_2}$;
- $2l - 2k$ many of the form χ_3 and $2l - 2k$ many of the form $\overleftarrow{\chi_3}$;
- $2s + 1 + t - 2l$ many of the form χ_4 and $2s + 1 + t - 2l$ many of the form $\overleftarrow{\chi_4}$.

Note that $|\mathcal{V}_1^i| = 2(W + 2s + 1 + t)$ for all $1 \le i \le t$, and $\mathcal{V}_1 = \bigcup_{i=1}^{t} \mathcal{V}_1^i$. Next, in $\mathcal{V}_2$, we construct a vote set $\mathcal{V}_2^i$ for each variable x_i, consisting of two subsets, P^i and B^i, which serve mainly the purpose of controlling the plurality and Borda scores of variable candidates, respectively, and thus, determine the order of eliminations. Hereby, we use $\mathcal{V}_1(x)$ to denote the set of votes in $\mathcal{V}_1$, where x is ranked at the first position. The subset P^i consists of the following votes:

- one vote $v = x_i^1 > x_i^T > w_i^{1,1} > p > \overrightarrow{S} > w_i^{1,2}$ with $S = C \setminus \{p, x_i^1, x_i^T, w_i^{1,1}, w_i^{1,2}\}$ and one vote $\overleftarrow{v}$;
- $W + 2i - 2$ votes $v = x_i^1 > w_i^{1,1} > p > \overrightarrow{S} > w_i^{1,2}$ with $S = C \setminus \{p, x_i^1, w_i^{1,1}, w_i^{1,2}\}$ and $W + 2i - 2$ votes $\overleftarrow{v}$;
- $W + 2i$ votes $v = x_i^2 > w_i^{2,1} > p > \overrightarrow{S} > w_i^{2,2}$ with $S = C \setminus \{p, x_i^2, w_i^{2,1}, w_i^{2,2}\}$ and $W + 2i$ votes $\overleftarrow{v}$.

- $W + 2i - 1 - |\mathcal{V}_1(x_i^T)|$ votes $v = x_i^T > w_i^{T,1} > p > \vec{S} > w_i^{T,2}$ with $S = C \setminus \{p, x_i^T, w_i^{T,1}, w_i^{T,2}\}$ and $W + 2i - 1 - |\mathcal{V}_1(x_i^T)|$ votes $\overleftarrow{v}$;
- $W + 2i - |\mathcal{V}_1(x_i^F)|$ votes $v = x_i^F > w_i^{F,1} > p > \vec{S} > w_i^{F,2}$ with $S = C \setminus \{p, x_i^F, w_i^{F,1}, w_i^{F,2}\}$ and $W + 2i - |\mathcal{V}_1(x_i^F)|$ votes $\overleftarrow{v}$.

The subset B^1 contains six votes: two votes in $D(p, x_1^1)$, two votes in $D(x_1^2, p)$, and two votes in $D(x_1^2, x_1^F)$. For each $2 \le i \le s$, B^i contains ten votes: two votes in $D(p, x_i^1)$, two votes in $D(x_i^2, p)$, two votes in $D(x_i^2, x_i^F)$, two votes in $D(x_i^1, x_{i-1}^2)$, and two votes in $D(x_i^F, x_{i-1}^2)$. Then, $\mathcal{V}_2 := \bigcup_{i=1}^{s} \mathcal{V}_2^i = \bigcup_{i=1}^{s} \left(P^i \cup B^i \right)$.

The third subset $\mathcal{V}_3$ contains $2(W + 2s + 1 + t - 3) + 6$ votes to set the scores of p and q:

- $W + 2s + 1$ votes $v = p > \vec{S} > q$ with $S = C \setminus \{p, q\}$ and $W + 2s + 1$ votes $\overleftarrow{v}$;
- $t - 3$ votes $v = q > \vec{S} > \alpha$ with $S = C \setminus \{q, \alpha\}$ and $t - 3$ votes $\overleftarrow{v}$;
- two votes in $D(q, p)$;
- two votes in $D(x_s^2, q)$;
- two votes in $D(p, x_s^2)$.

The fourth subset $\mathcal{V}_4$ contains the following votes for setting the plurality scores of the candidates in $\widehat{E} \setminus \{\alpha, \beta, q\}$:

- for each $1 \le i \le t$, $W + 2s + 1 + t$ votes $v = u_{i,1} > p > \vec{S} > u_{i,2}$ with $S = C \setminus \{p, u_{i,1}, u_{i,2}\}$ and $W + 2s + 1 + t$ votes $\overleftarrow{v}$;
- for each $1 \le i \le s$, $2(W + 2s + 1 + t) - (W + 2i - 1)$ votes $v = w_i^{1,1} > p > \vec{S} > w_i^{1,2}$ with $S = C \setminus \{p, w_i^{1,1}, w_i^{1,2}\}$ and $2(W + 2s + 1 + t) - (W + 2i - 1)$ votes $\overleftarrow{v}$;
- for each $1 \le i \le s$, $2(W + 2s + 1 + t) - (W + 2i)$ votes $v = w_i^{2,1} > p > \vec{S} > w_i^{2,2}$ with $S = C \setminus \{p, w_i^{2,1}, w_i^{2,2}\}$ and $2(W + 2s + 1 + t) - (W + 2i)$ votes $\overleftarrow{v}$.
- for each $1 \le i \le s$, $2(W + 2s + 1 + t) - (W + 2i - 1 - |\mathcal{V}_1(x_i^T)|)$ votes $v = w_i^{T,1} > p > \vec{S} > w_i^{T,2}$ with $S = C \setminus \{p, w_i^{T,1}, w_i^{T,2}\}$ and $2(W + 2s + 1 + t) - (W + 2i - 1 - |\mathcal{V}_1(x_i^T)|)$ votes $\overleftarrow{v}$;
- for each $1 \le i \le s$, $2(W + 2s + 1 + t) - (W + 2i - |\mathcal{V}_1(x_i^F)|)$ votes $v = w_i^{F,1} > p > \vec{S} > w_i^{F,2}$ with $S = C \setminus \{p, w_i^{F,1}, w_i^{F,2}\}$ and $2(W + 2s + 1 + t) - (W + 2i - |\mathcal{V}_1(x_i^F)|)$ votes $\overleftarrow{v}$.

The fifth subset $\mathcal{V}_5$ contains $6(2t + 8s)$ votes for setting the Borda scores of auxiliary candidates:

- two votes in $D(p, e)$ for each $e \in \widehat{E} \setminus \{\alpha, \beta, q\}$;
- two votes in $D(q, p)$ for each $e \in \widehat{E} \setminus \{\alpha, \beta, q\}$;
- two votes in $D(e, q)$ for each $e \in \widehat{E} \setminus \{\alpha, \beta, q\}$.

The sixth subset $\mathcal{V}_6$ contains $2(W + 2s + 1 + t) + 12$ votes for adjusting the Borda scores of α, β:

- one vote $\beta > p > \alpha > \overrightarrow{S} > q$ and one vote $q > \overleftarrow{S} > p > \alpha > \beta$ with $S = C \setminus \{p, q, \alpha, \beta\}$;
- one vote $\alpha > q > \overrightarrow{S} > \beta$ and one vote $\beta > \overleftarrow{S} > \alpha > q$ with $S = C \setminus \{q, \alpha, \beta\}$;
- two votes in $D(q, p)$;
- $W + 2s + 1 + t$ votes of the same form $\alpha > \overrightarrow{S} > \beta$ and $W + 2s + 1 + t$ votes of the same form $\beta > \overleftarrow{S} > \alpha$ with $S = C \setminus \{\alpha, \beta\}$;
- one vote $\alpha > p > \beta > \overrightarrow{S} > q$ and one vote $q > \overleftarrow{S} > p > \beta > \alpha$ with $S = C \setminus \{p, q, \alpha, \beta\}$;
- one vote $\beta > q > \overrightarrow{S} > \alpha$ and one vote $\alpha > \overleftarrow{S} > \beta > q$ with $S = C \setminus \{q, \alpha, \beta\}$;
- two votes in $D(q, p)$.

The vote set is $\mathcal{V} := \mathcal{V}_1 \cup \mathcal{V}_2 \cup \mathcal{V}_3 \cup \mathcal{V}_4 \cup \mathcal{V}_5 \cup \mathcal{V}_6$. The reduction is clearly doable in polynomial time.

In order to prove the correctness of the reduction, we give a detailed illustration of the first main round. The other main rounds excute in the same way. For the ease of presentation, we adopt a variation of the computation of Borda scores. Note that all votes are created in a pairwise manner. If two votes are completely reversed orders, as the votes in $\mathcal{V}_1 \cup (\bigcup_{i=1}^{s} P^i) \cup \mathcal{V}_4$, then all candidates receive the same Borda score from these two votes. After deleting arbitrarily many candidates from these votes, the remaining candidates have always the same Borda score. These votes can therefore be ignored from the computation of Borda scores. If two votes are created as $D(c_1, c_2)$, then all candidates have the same score except for c_1 and c_2, which we call the "standard score". Then, we calculate only the deviation from this standard score for each candidate.

Claim 1. *Before the first main round, x_1^1 and x_1^F have the minimum Borda score, while x_1^1 and x_1^T have the minimum plurality score.*

By this claim, we know that the candidates to be eliminated in the first secondary round can only be x_1^1 and x_1^T (by Hare) or x_1^1 and x_1^F (by Baldwin). This round is the first secondary round of the first main round dealing with the candidates for x_1.

Claim 2. *1) If we apply the Hare rule to eliminate x_1^1 and x_1^T in the first secondary round of the first main round, then p has the minimum Borda score and x_1^2, x_1^F and the clause candidates z_i, whose corresponding clauses contain positive occurrences of x_1, have the minimum plurality score.*

2) If we apply the Baldwin rule to eliminate x_1^1 and x_1^F in the first secondary round of the first main round, then p has the minimum Borda score and x_1^2, x_1^T and the clause candidates z_i, whose corresponding clauses contain negative occurrences of x_1, have the minimum plurality score.

By this claim, we cannot apply the Baldwin rule in the second secondary round, since, otherwise, p would be eliminated. By applying the Hare rule, x_1^2, x_1^F and the clause candidates, whose corresponding clauses contain positive occurrences of x_i, are eliminated. Thus, we can interprete the application of the Hare rule in the first secondary round as assigning TRUE to x_1. The application of the Baldwin rule in the first secondary round corresponds to the assignment of FALSE to x_1.

Claim 3. *After the first main round, x_2^1 and x_2^T have the minimum plurality score, while x_2^1 and x_2^F have the minimum Borda score.*

From Claims 1–3, there are only two possible rule combinations for the two secondary rounds of the first main round, that is, <Hare, Hare> and <Baldwin, Hare>. The former one corresponds to the assignment of TRUE to x_1, while the latter means the assignment of FALSE to x_1. Moreover, the second main round faces the same situation as the first main round, dealing with x_2 instead of x_1. The same argument applies. Next, we consider the situation after the s main rounds, that is, after all variable candidates are eliminated. Clearly, all elements e in $\widehat{E}$ and p remain in the election. Some clause candidates z_i might remain.

Claim 4. *If only the candidates in $\widehat{E}$ and p remain after the s main rounds, then p is the unique winner.*

By Claim 4, we can conclude that if the given 3SAT instance is a yes-instance, then we can apply the rules according to Claim 2 and all clause candidates are eliminated in the s main rounds. In the $(2s+1)$-th round, only the candidates in $\widehat{E} \cup \{p\}$ remain and p becomes the unique winner by applying first the Hare rule and then the Baldwin rule. Thus, the corresponding Possible Winner instance is a yes-instance. Next, we prove the reversed direction.

Claim 5. *If there is a clause candidate z_i remaining in the $(2s+1)$-th round together with $\widehat{E} \cup \{p\}$, then p cannot be the unique winner.*

By Claim 2, we can conclude that if the given 3SAT instance is a no-instance, we cannot eliminate all clause candidates in the s main rounds. There is at least one clause candidate z_i remaining in the $(2s+1)$-th round. Then by Claim 5, p cannot be the unique winner. Thus, the corresponding Possible Winner instance is a no-instance. Combining Claims 4 and 5, we establish the equivalence of the instances. The NP-hardness for $R = \{\text{Hare, Baldwin}\}$ then follows.   □

Now, we explore the parameterized complexity of the Possbile Winner problem with respect to the parameter of the number of the candidates or the number of the votes.

Theorem 2. *The Possible Winner problem is FPT with respect to the number of candidates.*

Proof. Let $C = \{c_1, ..., c_m\}$ be the candidate set, $V = \{v_1, ..., v_n\}$ be the vote set and R be the given voting rule set. We first enumerate all possible orders of the candidates, where p is at the last position. Each order specifies the order of candidate eliminations. There are at most $m!$ different elimination orders. For each elimination order, we partition the candidates such that p is in a singleton subset. There are at most $f(m) = \sum_{d=1}^{m} \left(\frac{\sum_{i=0}^{d-1} (-1)^i \cdot \binom{d}{i} \cdot (d-i)^m}{d!} \right)$ different partitions. Each subset in the partition specifies the candidates eliminated in a single round. Thus, there are at most $m! \cdot f(m)$ different partitions in total. For each partition, we iterate over all subsets in their order in the permutation and find out which rules in R can eliminate all candidates in each subset. If there exists a partition, for which there is a sequence of rules, which eliminates the candidates in the corresponding subset, then we get a combination of rules making p the unique winner. The running time is $F(m,n) = m! \cdot f(m) \cdot |R| \cdot poly(m,n)$, which is FPT with respect to the number of candidates. $\qquad \square$

Next, we investigate the parameterized complexity of the Possbile Winner problem with respect to the parameter of the number of votes.

Theorem 3. *Given the voting rule set R with $|R| = 2$ and R containing Hare, the Possible Winner problem is FPT with respect to the number of votes.*

Proof. Let $C = \{c_1, ..., c_m\}$ be the candidate set, $V = \{v_1, ..., v_n\}$ be the vote set and $R = \{\text{Hare}, r\}$. We observe that, once we apply Hare, then the number of the remaining candidates is at most n. It means that we can apply r several times before applying Hare and then in the remaining rounds the numbers of the candidates and the votes are both bounded by n and then we can use the algorithm in the proof of Theorem 2 to examinate if the specific candidate p can be a unique winner. The rule r can be applied at most m times before applying Hare. Thus, the running time is $m \cdot F(n,n) = n! \cdot f(n) \cdot poly(m,n)$, where $f(n) = \sum_{d=1}^{n} \left(\frac{\sum_{i=0}^{d-1} (-1)^i \cdot \binom{d}{i} \cdot (d-i)^n}{d!} \right)$. $\qquad \square$

4   Necessary Winner

In this section, we prove the CoNP-hardness of the Necessary Winner problem and provide special cases, for which Necessary Winner are easy to solve.

Theorem 4. *Necessary Winner is CoNP-hard for $R \subseteq \{\text{Hare, Coombs, Baldwin, Nanson}\}$ with $|R| \geq 2$ and $R \neq \{\text{Baldwin, Nanson}\}$.*

Proof. We consider only the subset $R = \{\text{Hare, Baldwin}\}$. The other cases follow from similar reductions. Again we reduce from 3SAT and the reduction shares some common features with the reduction in the proof of Theorem 1. We give here only a brief discussion of the main differences and leave the details to the full version.

We also construct the same set of candidates and apply the first s main rounds to eliminate the variable candidates. The "satisfied" clause candidates are also eliminated in the second secondary rounds of the main rounds. However, if no clause candidate remains after the s main rounds, then there is a rule combination for the remaining rounds such that the specific candidate p loses the election; otherwise, p wins with all possible rule combinations. This means that, the given 3SAT-instance is a yes-instance, if and only if the Necessary Winner instance is a no-instance. The candidate q also plays the same role but we slightly modify the construction of votes. More precisely, we have to exchange the positions of p and q in some votes, such that p's plurality score becomes the minimum after all clause candidates have been eliminated in the s main rounds. In this way, we can apply the Hare rule in the $(2s + 1)$-th round to eliminate p, resulting in that p loses the election. Moreover, in the case that some clause candidates remain after the s main rounds, q's plurality score is the minimum, guaranteeing that applying the Hare rule in the $(2s + 1)$-th round does not eliminate p. We also assure that the Borda score of p in the $(2s + 1)$-th round is not the minimum and thus, applying the Baldwin rule does not eliminate p.

As in the proof of Theorem 1, the vote set $\mathcal{V}$ consists of six subsets. The votes in $\mathcal{V}_2 \cup \mathcal{V}_4$ remain the same. As for $\mathcal{V}_1$, in $\mathcal{V}_1^i$ for each $1 \leq i \leq t$, we replace the first pair of votes $z_i > p > u_{i,1} > \overrightarrow{S} > u_{i,2}$ and $u_{i,2} > \overleftarrow{S} > u_{i,1} > p > z_i$ with $S = C \setminus \{p, z_i, u_{i,1}, u_{i,2}\}$ by $z_i > q > p > u_{i,1} > \overrightarrow{S'} > u_{i,2}$ and $u_{i,2} > \overleftarrow{S'} > u_{i,1} > p > q > z_i$ with $S' = C \setminus \{p, q, z_i, u_{i,1}, u_{i,2}\}$ and $\mathcal{V}_1 = \bigcup_{i=1}^{t} \mathcal{V}_1^i$.

The set $\mathcal{V}_3$ is set completely different and contains the following votes:

- $W + 2s + 1$ votes $q > p > \overrightarrow{S} > \alpha$ and $W + 2s + 1$ votes $\alpha > \overleftarrow{S} > p > q$ with $S = C \setminus \{p, q, \alpha\}$;
- $W + 2s + 1 + t - 2$ votes $p > \overrightarrow{S} > \beta$ and $W + 2s + 1 + t - 2$ votes $\beta > \overleftarrow{S} > p$ with $S = C \setminus \{p, \beta\}$.

The fifth subset $\mathcal{V}_5$ contains the following votes for setting the Borda scores of auxiliary candidates:

- two votes in $D(p, e)$ for each $e \in \widehat{E} \setminus \{\alpha, \beta\}$;
- two votes in $D(x_s^2, p)$ for each $e \in \widehat{E} \setminus \{\alpha, \beta\}$;
- two votes in $D(e, x_s^2)$ for each $e \in \widehat{E} \setminus \{\alpha, \beta\}$.

The sixth subset $\mathcal{V}_6$ contains now the following votes:

- one vote $p > \alpha > \overrightarrow{S} > \beta$ and one vote $\beta > \overleftarrow{S} > p > \alpha$ with $S = C \setminus \{p, \alpha, \beta\}$;
- one vote $\alpha > x_s^2 > \overrightarrow{S} > \beta$ and one vote $\beta > \overleftarrow{S} > \alpha > x_s^2$ with $S = C \setminus \{x_s^2, \alpha, \beta\}$;
- two votes in $D(x_s^2, p)$;
- $W + 2s + 1 + t$ votes of the same form $\alpha > \overrightarrow{S} > \beta$ and $W + 2s + 1 + t$ votes of the same form $\beta > \overleftarrow{S} > \alpha$ with $S = C \setminus \{\alpha, \beta\}$;
- one vote $p > \beta > \overrightarrow{S} > \alpha$ and one vote $\alpha > \overleftarrow{S} > p > \beta$ with $S = C \setminus \{p, \alpha, \beta\}$;
- one vote $\beta > x_s^2 > \overrightarrow{S} > \alpha$ and one vote $\alpha > \overleftarrow{S} > \beta > x_s^2$ with $S = C \setminus \{x_s^2, \alpha, \beta\}$;
- two votes in $D(x_s^2, p)$.

The correctness proof works similarly as in the proof of Theorem 1.   □

Next, we provide special cases, for which Necessary Winner always returns yes. Given a set of n votes, each being a linear order of m candidates, we say that candidate c_1 "beats" candidate c_2, if in more than half of the votes, c_1 is ranked in front of c_2. A candidate c is called a Condorcet winner, if c beats all other candidates. A rule r is called "Condorcet-consistent", if applying r to the current election does not eliminate the Condorcet winner c. Note that after the elimination, c remains the Condorcet winner. The following theorem follows trivially from the definition of Condorcet-consistency.

Theorem 5. *If p is the Condorcet winner of the current election and each rule in R has the property of Condorcet consistency, then p is the necessary winner.*

Next, we consider only the rules, which are based on Borda scores.

Corollary 1. *Let n and m be the numbers of votes and candidates, respectively. If p is a Condorcet winner of the current election and each rule r in R satisfies the following conditions, then p is a necessary winner:*

1) *r is based on Borda scores;*
2) *In the case n being even, if r is applied to the current election, then all candidates eliminated in this round have Borda scores less than $\frac{(n+2)(m-1)}{2}$;*
3) *In the case n being odd, if r is applied to the current election, then all candidates eliminated in this round have Borda scores less than $\frac{(n+1)(m-1)}{2}$.*

Proof. It is easy to verify that every Condorcet winner of an election with n votes and m candidates has a minimum Borda score of $\frac{(n+2)(m-1)}{2}$ for the case n being even or $\frac{(n+1)(m-1)}{2}$ for the case n being odd. Then, by the conditions, all rules in R are Condorcet-consistent. The claim follows from Theorem 5. □

Since both Baldwin and Nanson satisfy the conditions in Corollary 1, we have the following result.

Corollary 2. *If p is a Condorcet winner of the current election and $R = \{Baldwin, Nanson\}$, then p is the necessary winner.*

5   Conclusion

In this paper, we introduce a new version of Possible and Necessary Winner problems with respect to iterative elections and achieve NP-hard and CoNP-hard results for most subsets of the four prominent iterative election rules. The complexity status of both Possible and Necessary Winner problems for $R = \{Baldwin, Nanson\}$ remains open. Since for the same election, the candidates eliminated by the Baldwin rule form a subset of the set of candidates eliminated by the Nanson rule, we conjecture that both problems are polynomial-time solvable with these two rules. Moreover, it would be a challenging research topic to give a general classification of the rules for which Possible and Necessary Winner are solvable in polynomial time or hard to solve.

References

[Bal26] Baldwin, J.: The technique of the Nanson preferential majority system of election. Proc. Royal Soc. Victoria **39**, 42–52 (1926)

[BCE+16] Brandt, F., Conitzer, V., Endriss, U., Lang, J., Procaccia, A.D. (eds.) Handbook of Computational Social Choice. Cambridge University Press (2016)

[BSW20] Baumeister, D., Selker, A.-K., Wilczynski, A.: Manipulation of opinion polls to influence iterative elections. In: Proceedings of the 19th International Conference on Autonomous Agents and MultiAgent Systems, AAMAS 2020, pp. 132–140 (2020)

[CDK+21] Chakraborty, V., Delemazure, T., Kimelfeld, B., Kolaitis, P.G., Relia, K., Stoyanovich, J.: Algorithmic techniques for necessary and possible winners. Trans. Data Sci. **2**(3):22:1–22:23 (2021)

[ENR+21] Erdélyi, G., Neveling, M., Reger, C., Rothe, J., Yang, Y., Zorn, R.: Towards completing the puzzle: complexity of control by replacing, adding, and deleting candidates or voters. Auton. Agents Multi-Agent Syst. **35**(2), 1–48 (2021). https://doi.org/10.1007/s10458-021-09523-9

[FMS21] Faliszewski, P., Manurangsi, P., Sornat, K.: Approximation and hardness of shift-bribery. Artif. Intell. **298**, 103520 (2021)

[GNNW14] Gaspers, S., Naroditskiy, V., Narodytska, N., Walsh, T.: Possible and necessary winner problem in social polls. In: Proceedings of the 13th International Conference on Autonomous Agents and MultiAgent Systems, AAMAS 2014, pp. 613–620 (2014)

[LN95] Levin, J., Nalebuff, B.: An introduction to voteacounting schemes. J. Econ. Perspect. 3–26 (1995)

[MNRS18] Maushagen, C., Neveling, M., Rothe, J., Selker, A.-K.: Complexity of shift bribery in iterative elections. In: Proceedings of the 17th International Conference on Autonomous Agents and MultiAgent Systems, AAMAS 2018, pp. 1567–1575, 2018

[Nan82] Nanson, E.: Methords of election. In: Transactions and Proceedings of the Royal Society of Victoria, pp. 197–240 (1882)

[TP08] Taylor, A., Pacelli, A.: Mathematics and Politics: Strategy. Power, and Proof. Springer Science and Business Media, Voting (2008)

[XC08] Xia, L., Conitzer, V.: Determining possible and necessary winners under common voting rules given partial orders. In: Proceedings of the 23th International Conference on Artificial Intelligence, AAAI 2008, pp. 196–201 (2008)

[ZG20] Zhou, A., Guo, J.: Parameterized complexity of shift bribery in iterative elections. In: Proceedings of the 19th International Conference on Autonomous Agents and MultiAgent Systems, AAMAS 2020, pp. 1665–1673 (2020)

A Mechanism Design Approach
for Multi-party Machine Learning

Mengjing Chen[1]([⊠]), Yang Liu[2], Weiran Shen[3], Yiheng Shen[4],
Pingzhong Tang[1], and Qiang Yang[2,5]

[1] Tsinghua University, Beijing, China
ccchmj@qq.com
[2] WeBank Co., Ltd., Shenzhen, China
yangliu@webank.com , qyang@cse.ust.hk
[3] Renmin University, Beijing, China
shenweiran@ruc.edu.cn
[4] Duke University, Durham, USA
ys341@duke.edu
[5] Hong Kong University of Science and Technology, Clear Water Bay, Hong Kong

Abstract. In a multi-party machine learning system, different parties
cooperate on optimizing towards better models by sharing data in a
privacy-preserving way. A major challenge in learning is the incentive
issue. For example, if there is competition among the parties, one may
strategically hide his data to prevent other parties from getting better
models.

In this paper, we study the problem through the lens of mechanism
design and incorporate the features of multi-party learning in our setting.
First, each agent's valuation has externalities that depend on others'
types and actions. Second, each agent can only misreport a type lower
than his true type, but not the other way round. We call this setting
interdependent value with type-dependent action spaces. We provide the
optimal truthful mechanism in the quasi-monotone utility setting. We
also provide necessary and sufficient conditions for truthful mechanisms
in the most general case. We show the existence of such mechanisms is
highly affected by the market growth rate. Finally, we devise an algorithm
to find the desirable mechanism that is truthful, individually rational,
efficient and weakly budget-balance.

Keywords: Mechanism design · Federated learning · Incentive design

1 Introduction

In multi-party machine learning, a group of parties cooperates on optimizing
towards better models. This concept has attracted much attention recently [12,
22,23]. The advantage of this approach is that, it can make use of the distributed
datasets and computational power to learn a powerful model that anyone in the
group cannot achieve alone.

M. Li and X. Sun (Eds.): IJTCS-FAW 2022, LNCS 13461, pp. 248–268, 2022.
https://doi.org/10.1007/978-3-031-20796-9_18

To make multi-party machine learning practical, a large body of works focus on preserving data privacy in the learning process [1, 22, 26]. However, the incentive issues in the multi-party learning have largely been ignored in most previous studies, which results in a significant reduction in the effectiveness when putting their techniques into practice. Previous works usually let all the parties share the same global model with the best quality regardless of their contributions. This allocation works well when there are no conflicts of interest among the parties. For example, an app developer wants to use the users' usage data to improve the user experience. All users are happy to contribute data since they can all benefit from such improvements [16].

When the parties are competing with one another, they may be unwilling to participate in the learning process since their competitors can also benefit from their contributions. Consider the case where companies from the same industry are trying to adopt federated learning to level up the industry's service qualities. Improving other companies' services can possibly harm their own market share, especially when there are several monopolists that own most of the data.

Such a cooperative and competitive relation poses an interesting challenge that prevents the multi-party learning approach from being applied to a wider range of environments. In this paper, we view this problem from the multi-agent system perspective, and address the incentive issues mentioned above with the mechanism design theory.

Our setting is a variant of the so-called interdependent value setting [18]. A key difference between our setting and the standard interdependent value setting is that each agent cannot "make up" a dataset that is of higher quality than his actual one. Thus the reported type of an agent is capped by his true type. We call our setting *interdependent value with type-dependent action spaces*. The setting that agents can never over-report is common in practice. One straightforward example is that the sports competitions where athletes can show lower performance than their actual abilities but not over-perform. The restriction on the action space poses more constraints on agents' behaviors, and allows more flexibility in the design space.

We first formulate the problem mathematically, and then apply techniques from the mechanism design theory to analyze it. Our model is more general than the standard mechanism design framework, and is also able to describe other similar problems involving both cooperation and competition.

We make the following contributions in this paper:

- We model and formulate the mechanism design problem in multi-party machine learning, and identify the differences between our setting and the other mechanism design settings.
- For the quasi-monotone externalities setting, we provide the revenue-optimal and truthful mechanism. For the general valuation functions, we provide both the necessary and the sufficient conditions for all truthful and individually rational mechanisms.

- We analyze the influence of the market size on mechanisms. When the market grows slowly, there may not exist a mechanism that achieves all the desirable properties we focus on.
- We design an algorithm to find the mechanisms that guarantee individual rationality, truthfulness, efficiency and weak budget balance simultaneously when the valuation functions are given.

1.1   Related Works

A large body of literature studies mechanisms with interdependent values [18], where agents' valuations depend on the types of all agents and the intrinsic qualities of the allocated object. Roughgarden et al. [21] extend Myerson's auction to specific interdependent value settings and characterize truthful and rational mechanisms. They consider a bayesian setting while we do not know any prior information. Chawla et al. [5] propose a variant of the VCG auction with reserve prices that can achieve high revenues. They consider value functions that are single-crossing and concave while we consider environments with more general value functions. Mezzetti [17] gives a two-stage Groves mechanism that guarantees truthfulness and efficiency. He requires agents to report their types and valuations before the final monetary transfer are made while in our model, agents can only report their types.

In our setting, agents have restricted action spaces, i.e., they can never report types exceeding their actual types. There is a series of works that focus on mechanism design with a restricted action space [2–4]. The discrete-bid ascending auctions [2,6,7] specify that all bidders' action spaces are the same bid level set. Several works restrict the number of actions, such as bounded communications [4]. Previous works focus on mechanisms with independent values and discrete restricted action spaces, while we study the interdependent values and continuous restricted action spaces setting.

The learned model can be copied and distributed to as many agents as possible, so the supply is unlimited. A line of literature focuses on selling items in unlimited supply such as digital goods [9–11]. However, the seller sells the same item to buyers while in our setting we can allocate models with different qualities to different agents.

Redko et al. [20] study the optimal strategies of agents for collaborative machine learning problems. Both their work and ours capture the cooperation and competition among the agents, but they only consider the case where agents reveal their total datasets to participate while agents can choose to contribute only a fraction in our setting. Kang et al. [14] study the incentive design problem for federated learning, but all their results are about a non-competitive environment, which may not hold in real-world applications.

Our work contributes to the growing body of literature on incentive mechanism design for federated learning [15,27]. Jia et al. and Song et al. [13,24] design mechanisms based on the Shapley value and Ding et al. [8] apply the contract theory. However, the existing works do not consider the interdependent values of participants and type-dependent action space as our model does.

2   Preliminaries

In this section, we introduce the general concepts of mechanism design and formulate the multi-party machine learning as a mechanism design problem. A multi-party learning consists of a central platform and several parties (called agents hereafter). The agents serve their customers with their models trained using their private data. Each agent can choose whether to enter the platform. If an agent does not participate, then he trains his model with only his own data. The platform requires all the participating agents to contribute their data in a privacy-preserving way and trains a model for each participant using a (weighted) combination of all the contributions. Then the platform returns the trained models to the agents.

We assume that all agents use the same model structure. Therefore, each participating agent may be able to train a better model by making use of his private data and the model allocated to him. One important problem in this process is the incentive issue. For example, if the participants have conflicts of interest, then they may only want to make use of others' contributions but are not willing to contribute with all their own data. To align their incentives, we allow the platform to charge the participants according to some predefined rules.

Our goal is to design allocation and payment rules that encourage all agents to join the multi-party learning as well as to contribute all their data.

2.1   Valid Data Size (Type)

Suppose there are n agents, denoted by $N = (1, 2, \ldots, n)$, and each of them has a private dataset D_i where $D_i \cap D_j = \emptyset, \forall i \neq j$. For ease of presentation, we assume that a model is fully characterized by its quality Q (e.g., the prediction accuracy), and the quality only depends on the data used to train it. We have the following observation:

Observation 1. *If the agents could fake a dataset with a higher quality, any truthful mechanism would make agents gain equal final utility.*

Suppose that two agents have different true datasets D_1 and D_2. We assume that all other agents truthfully report datasets D_{-i}. If truthfully reporting dataset D_1 and D_2 finally leads to different utility, w.l.o.g, we let $u(D_1, D_{-i}) < u(D_2, D_{-i})$, then if an agent has true dataset D_1, he would report D_2 and use the dataset allocated by the platform in the market. All his behavior is the same as that of an agent with real dataset D_2. Thus if a mechanism is truthful, any reported dataset would lead to the same final utility and it is pointless to discuss the problem. Hence, we make the assumption that the mechanism is able to identify the quality of any dataset. All agents can only report a dataset with a lower quality.

For simplicity, we measure the contribution of a dataset to a trained model by its *valid data size*. Thus we have the following assumption:

Assumption 1. *The model quality Q is bounded and monotone increasing with respect to the valid data size $s \geq 0$ of the training data:*

1. $Q(0) = 0$ and $Q(s) \leq 1$, $\forall s$;
2. $Q(s') > Q(s)$, $\forall s' > s$.

The valid data size of every contributor's data is validated by the platform in a secure protocol (which we propose in the full version). Let $t_i \in \mathbb{R}_+$ be the valid data size of agent i 's private dataset D_i. We call t_i the agent's *type*. The agent can only falsify his type by using a dataset of lower quality (for example, using a subset of D_i, or adding fake data), which decreases the contribution to the trained model as well as the size of valid data. As a result, the agent with type t_i cannot contribute to the platform with a dataset higher than his type:

Assumption 2. *Each agent i can only report a type lower than his true type t_i, i.e., the action space of agent i is $[0, t_i]$.*

2.2   Learning Protocol

In this section, we describe the learning protocol that could enable the implementation of our mechanism. We assume that the platform has a validation dataset. The platform requires the agents to report their valid data size t_i. This could be done by asking each agent to submit the best model that he can possibly obtain by using his own dataset. Then the platform computes the model quality q_i using the validation dataset and get the agent's valid data size t_i by $t_i = Q^{-1}(q_i)$. The agent type t_i will be used in the training process (e.g., aggregate weighted model updates), as well as to determine the final allocation, which is a model with quality x_i.

The platform should also guarantee to deliver to each agent the promised model. However, it is possible that an agent reports t_i in the beginning but only contribute $t'_i < t_i$ in the actual training process. In the extreme case where all agents contribute nothing to the training process, the platform will fail to allocate a model to each agent with the quality determined by the mechanism. To address this issue, the platform can train n additional models simultaneously, with the i-th model trained only using the data from agent i. During the training process, the platform can apply secure multi-party computation techniques, such as homomorphic encryption [25,26], to prevent the agent from knowing which model is sent to him to compute the update. And after the training, the platform can compute the quality t'_i of the i-th model again using the validation dataset. If the qualities t'_i and t_i match, we know with high probability that the dataset contributed by the agent is consistent with the type he reports. Otherwise, the platform can just exclude the agent and start over the training process again.

The above protocol only ensures that the type reported by each agent is the same as the type he uses in the actual training process with. To encourage all agents to join and contribute all their data, we still need to design mechanisms with desirable properties, to which we devote the rest of the paper.

2.3   Mechanism

Let $t = (t_1, t_2, \ldots, t_n)$ and $t_{-i} = (t_1, \ldots, t_{i-1}, t_{i+1}, \ldots, t_n)$ be the type profile of all agents and all agents without i, respectively. Given the reported types of agents, a mechanism specifies a numerical allocation and payment for each agent, where the allocation is a model in the multi-party learning. Formally, we have:

Definition 1 (Mechanism). *A mechanism* $\mathcal{M} = (x, p)$ *is a tuple, where*

- $x = (x_1, x_2, \cdots, x_n)$, *where* $x_i \colon \mathbb{R}_+^n \mapsto \mathbb{R}$ *is the allocation function for agent* i, *which takes the agents' reported types as input and decides the model quality for agent* i *as output;*
- $p = (p_1, p_2, \cdots, p_n)$, *where* $p_i \colon \mathbb{R}_+^n \mapsto \mathbb{R}$ *is the payment function for agent* i, *which takes the agents' reported types as input and specifies how much agent* i *should pay to the mechanism.*

In a competitive environment, a strategic agent may hide some of data and does not use the model he receives from the platform. Thus the final model quality depends on both the allocation and his actual type. We use valuation function $v_i(x(t'), t)$ to measure the profit of agent i.

Definition 2 (Valuation). *We consider valuation functions* $v_i(x(t'), t)$ *that depend not only on the allocation outcome* $x(t')$ *where* t' *is the reported type profile, but also on the actual type profile* t.

We assume the model agent i uses to serve customers is:

$$q_i = \max\{x_i(t'), Q(t_i)\},$$

where $Q(t_i)$ is the model trained with his own data. The valuation of agent i depends on the final model qualities of all agents due to their competition. Hence v_i can also be expressed as $v_i(q_1, \ldots, q_n)$.

We make the following assumption on agent i's valuation:

Assumption 3. *Agent* i*'s valuation is monotone increasing with respect to true type* t_i *when the outcome* x *is fixed.*

$$v_i(x, t_i, t_{-i}) \geq v_i(x, \hat{t}_i, t_{-i}), \forall x, \forall t_i \geq \hat{t}_i, \forall t_{-i}, \forall i.$$

This is because possessing more valid data will not lower one's valuation. Otherwise, an agent is always able to discard part of his dataset to make his true type t_i'. Suppose that each agent i's utility $u_i(t, t')$ has the form:

$$u_i(t, t') = v_i(x(t'), t) - p_i(t'),$$

where t and t' are true types and reported types of all agents respectively. As we mentioned above, an agent may lie about his type in order to benefit from the mechanism. The mechanism should incentivize truthful reports to keep agents from lying.

Definition 3 (Incentive Compatibility (IC)). *A mechanism is said to be incentive compatible, or truthful, if reporting truthfully is always the best response for each agent when the other agents report truthfully:*

$$u_i(x(t_i, t_{-i}), t) \geq u_i(x_i(t'_i, t_{-i}), t), \forall t_i \geq t'_i, \forall t_{-i}, \forall i.$$

For ease of presentation, we say agent i reports $\emptyset$ if he chooses not to participate (so we have $x_i(\emptyset, t_{-i}) = 0$ and $p_i(\emptyset, t_{-i}) = 0$). To encourage the agents to participate in the mechanism, the following property should be satisfied:

Definition 4 (Individual Rationality (IR)). *A mechanism is said to be individually rational, if no agent loses by participation when the other agents report truthfully:*

$$u_i(x(t_i, t_{-i}), t) \geq u_i(x(\emptyset, t_{-i}), t), \forall t_i, t_{-i}, \forall i.$$

The revenue and welfare of a mechanism are defined to be all the payments collected from the agents and all the valuations of the agents.

Definition 5. *The revenue and welfare of a mechanism* (x, p) *are:*

$$\text{REV}(x, p) = \sum_{i=1}^{n} p_i(t'), \ \text{WEL}(x, p) = \sum_{i=1}^{n} v_i(x, t).$$

We say that a mechanism is efficient *if*

$$(x, p) = \arg\max_{(x,p)} \text{WEL}(x, p),$$

A mechanism is *weakly budget-balance* if it never loses money.

Definition 6 (Weak Budget Balance). *A mechanism is weakly budget-balance if:*

$$\text{REV}(x, p) \geq 0, \forall t.$$

Definition 7 (Desirable Mechanism). *We say a mechanism is desirable if it is IC, IR, efficient and weakly budget-balance.*

2.4   Comparison with the Standard Interdependent Value Setting

Although each agent's valuation depend on both the outcome of the mechanism and all agent's true types, our interdependent value with type-dependent action spaces setting, however, fundamentally different from standard interdependent value settings:

- In our setting, the type of each agent is the "quality" of his dataset, thus each agent cannot report a higher type than his true type. While in the standard interdependent value setting, an agent can possibly report any type.

- In our setting, the agents do not have the "exit choice" (not participating in the mechanism and getting 0 utility) as they do in the standard setting. This is due to the motivation of this paper: companies from the same industry are trying to improve their service quality, and they are always in the game regardless of their choices. A non-participating company may even have a negative utility if all other companies improved their services.
- To capture the cooperation among the agents, the item being sold, i.e., the learned model, also depends on all agents types. The best model learned by the multi-party learning platform will have high quality if all agents contribute high-quality datasets. However, the objects for allocation are usually fixed in standard mechanisms instead.

3   Quasi-Monotone Externality Setting

In the interdependent value with type-dependent action spaces setting, each agent's utility may also depend on the models that other agents actually use. Such externalities lead to interesting and complicated interactions between the agents. For example, by contributing more data, one may improve the others' model quality, and end up harming his own market share. In this section, we study the setting where agents have *quasi-monotone* externalities.

Definition 8 (Quasi-Monotone Valuation). *Let q_i be the final selected model quality of the agent and q_{-i} be the profile of model qualities of all the agents except i. A valuation function is quasi-monotone if it is in the form:*

$$v_i(q_i, q_{-i}) = F_i(q_i) + \theta_i(q_{-i}),$$

where F_i is monotone and θ_i is an arbitrary function.

Example 1. Let's consider a special quasi-monotone valuation: the linear externality setting, where the valuation for each agent is defined as $v_i = \sum_j \alpha_{ij} q_j$ with q_j being the model that agent j uses. The externality coefficient α_{ij} means the influence of agent j to agent i and captures either the competitive or cooperative relations among agents. If the increase of agent j's model quality imposes a negative (positive) effect on agent i's utility (e.g. major opponents or collaborators in the market), α_{ij} would be negative (positive). Additionally, α_{ii} should always be positive, naturally.

In the linear externality setting, the efficient allocation is straightforward. For each agent i, we give i the training model with best possible quality if $\sum_j \alpha_{ji} \geq 0$. Otherwise, agent i are not allocated any model if $\sum_j \alpha_{ji} < 0$.

We introduce a payment function called *maximal exploitation payment*, and show that the mechanism with efficient allocation and the maximal exploitation payment guarantees IR, IC, efficiency and revenue optimum.

Definition 9 (Maximal Exploitation Payment (MEP)). *For a given allocation function x, if the agent i reports a type t'_i and the other agents report t'_{-i}, the maximal exploitation payment is to charge agent i*

$$p_i(t'_i, t'_{-i}) = v_i(x(t'_i, t'_{-i}), t'_i, t'_{-i}) - v_i(x(\emptyset, t'_{-i}), t'_i, t'_{-i}).$$

We emphasize that our MEP mechanism and the VCG are quite different. The VCG charges each agent for the harm he causes to others due to his participation while the MEP charges each agent the profit he gets from the mechanism due to his participation. We will show that the MEP is truthful in the quasi-monotone valuation setting in the following theorem, while it is already known that VCG cannot guarantee truthfulness in the interdependent setting [17].

Theorem 1. *Under the quasi-monotone valuation setting, any mechanism with MEP is the mechanism with the maximal revenue among all IR mechanisms, and it is IC.*

Corollary 1. *Any efficient allocation mechanism with MEP under the linear externality setting with all the linear coefficients $\alpha_{ji} \geq 0$ should be IR, IC, weakly budget-balance and efficient.*

In the standard mechanism design setting, the Myerson-Satterthwaite Theorem [19] is a well-known classic result, which says that no mechanism is simultaneously IC, IR, efficient and weakly budget-balance. The above Corollary 1 shows that in our setting, the Myerson-Satterthwaite Theorem fails to hold.

4   General Externality Setting

In this section, we consider the general externality setting where the valuations of agents can have any forms of externalities. The restrictions on the action space and the value functions make the IC and IR mechanisms hard to characterize. It is possible that given a allocation rule, there exist several mechanisms with different payments that satisfy both IC and IR constraints. To understand what makes a mechanism IC and IR, we analyze some properties of truthful mechanisms in this section. For ease of presentation, we assume that the functions $v(\cdot)$, $x(\cdot)$ and $p(\cdot)$ are differentiable.

Theorem 2 (Necessary Condition). *If a mechanism (x, p) is both IR and IC, for all possible valuation functions satisfying Assumption 3, then the payment function satisfies $\forall t_i \geq t_i', \forall t_i, \forall t_{-i}, \forall i$,*

$$p_i(0, t_{-i}) \leq v_i(x(0, t_{-i}), 0, t_{-i}) - v_i(x(\emptyset, t_{-i}), 0, t_{-i}), \tag{1}$$

$$p_i(t_i, t_{-i}) - p_i(t_i', t_{-i}) \leq \int_{t_i'}^{t_i} \left. \frac{\partial v_i(x(s', t_{-i}), s, t_{-i})}{\partial s'} \right|_{s=s'} ds', \tag{2}$$

where we view $v_i(x(t_i', t_{-i}), t_i, t_{-i})$ as a function of t_i, t_i' and t_{-i} for simplicity. The partial derivative in Eq. (2) is computed using the chain rule, i.e.,

$$\frac{\partial v_i(x(t_i', t_{-i}'), t_i, t_{-i})}{\partial t_i'} = \sum_{j=1}^{n} \frac{\partial v_i(x(t_i', t_{-i}'), t_i, t_{-i})}{\partial x_j(t_i', t_{-i}')} \frac{\partial x_j(t_i', t_{-i}')}{\partial t_i'}$$

Theorem 2 describes what the payment p is like in all IC and IR mechanisms. In fact, the conditions in Theorem 2 are also crucial in making a mechanism truthful. However, to ensure IC and IR, we still need to restrict the allocation.

Theorem 3 (Sufficient Condition). *A mechanism (x, p) satisfies both IR and IC, for all possible valuation functions satisfying Assumption 3, if for each agent i, for all $t_i \geq t'_i$, and all t_{-i}, Eqs. (1) and the following two hold*

$$t'_i = \arg\min_{t_i:t_i>t'_i} \frac{\partial v_i(x(t'_i, t_{-i}), t_i, t_{-i})}{\partial t'_i} \tag{3}$$

$$p_i(t_i, t_{-i}) - p_i(t'_i, t_{-i})$$
$$\leq \int_{t'_i}^{t_i} \frac{\partial v_i(x(s', t_{-i}), s, t_{-i})}{\partial s'}\bigg|_{s=s'} \mathrm{d}s' - \int_{t'_i}^{t_i} \frac{\partial v_i(x(\emptyset, t_{-i}), s, t_{-i})}{\partial s} \mathrm{d}s. \tag{4}$$

5 Market Growth Rate

In this section, we will analyze a factor, the market growth rate, for the existence of the desirable mechanism. Expanding the market size would reduce competition among the agents, meaning that the damage to an agent's existing market caused by joining the mechanism is more likely to be covered by the market growth. Thus our intuition is that if the market grows quickly, a desirable mechanism is more likely to exist.

As mentioned above, each agent's valuation is the profit made from the market, so formally we define the market size to be the sum of the valuations of all the agents. Let $M(q)$ be the agents' total valuations where $q = (q_1, q_2, \ldots, q_n)$ is the set of actual model qualities they use. We have:

$$M(q) = \sum_{i=1}^{n} v_i(x, t).$$

In general, the multi-party learning process improves all agents' models. So we do not consider the case where the market shrinks due to the agents' participation, and assume that the market is growing.

Assumption 4 (Growing Market). $q \succeq q'$ *implies* $M(q) \geq M(q')$.

A special case of the growing market is the non-competitive market where agent's values are not affected by others' model qualities, formally:

Definition 10 (Non-competitive Market). *A market is non-competitive iff* $\frac{\partial v_i(q)}{\partial q_j} \geq 0, \forall i, j$.

Theorem 4. *In a non-competitive market, there always exists a desirable mechanism, that gives the best possible model to each agent and charges nothing.*

Since the efficient mechanism both redistributes existing markets and enlarges the market size by giving the best learned model when the market is growing, it is difficult to determine whether a desirable mechanism exists if the competition exists. We will give the empirical analysis of the influence of the growth rate of competitive growing markets for desirable mechanisms in Appendix G.

6   Finding a Desirable Mechanism

In the linear externality setting, we provide a mechanism that satisfies all the desirable properties. But this mechanism is not applicable to all valuation functions in the general setting, since the existence of a desirable mechanism depends on the agents' actual valuation functions. We provide an algorithm, that given the agents' valuations, computes whether such a mechanism exists, and outputs the one that optimizes revenue, if any.

Since each agent can only under-report, according to the IR property, we must have:

$$u_i(x(t_i, t_{-i}), t) \geq u_i(x(\emptyset, t_{-i}), t), \forall t, \forall i.$$

Equivalently, we get $\forall t, \forall i$,

$$u_i(x(\emptyset, t_{-i}), t) \leq v_i(x(t_i, t_{-i}), t) - p_i(t_i, t_{-i}),$$
$$p_i(t_i, t_{-i}) \leq v_i(x(t_i, t_{-i}), t) - u_i(x(\emptyset, t_{-i}), t),$$
$$p_i(t) \leq v_i(x(t_i, t_{-i}), t) - u_i(x(\emptyset, t_{-i}), t).$$

For simplicity, we define the upper bound of $p(t')$ as

$$\overline{p(t)} \triangleq \{v_i(x(t_i, t_{-i}), t) - u_i(x(\emptyset, t_{-i}), t).$$

The IC property requires that $\forall t_i \geq t'_i, \forall t_{-i}, \forall i$,

$$u_i(x(t_i, t_{-i}), t) \geq u_i(x(t'_i, t_{-i}), t).$$

A little rearrangement gives:

$$p_i(t_i, t_{-i}) - p_i(t'_i, t_{-i}) \leq v_i(x(t_i, t_{-i}), t) - v_i(x(t'_i, t_{-i}), t) \triangleq Gap_i(t'_i, t_i, t_{-i}).$$

Note that the inequality correlations between the payments form a system of difference constraints. The form of update of the payments is almost identical to that of the shortest path problem. Therefore, we make use of this observation to design the algorithm.

We assume that all the value functions are common knowledge, the efficient allocation is then determined because the mechanism always chooses the one that maximizes the social welfare. Thus it suffices to figure out whether there is a payment rule $p(t')$ which makes the mechanism IR, IC and weakly budget-balance. Since the valid data size for each agent is bounded in practice, we assume the mechanism only decides the payment functions on the data range

$[0, D]$, and discretize the type space into intervals of length ϵ, which is also the minimal size of the data. Thus each agent's type is a multiple of ϵ. Note that since the utility function is general, all the points in the action space would influence the properties and existence of the mechanism, thus it is necessary to enumerate all the points in the space. The exponential value function space, i.e., the exponential input space, determines that the complexity of our algorithm is exponential in D.

We give the following algorithmic characterization for the existence of a desirable mechanism.

Algorithm 1: Finding desirable mechanisms

input: Agents' valuation functions v.

Use the function v_i to calculate all the $Gap_i(t'_i, t_i, t_{-i})$ and $\overline{p_i(t_i, t_{-i})}$ for each i;

Initialize all $p_i^{max}(t_i, t_{-i})$ to be $\overline{p_i(t_i, t_{-i})}$ for each i;

for $i = 1$ *to* n **do**

  for $t_{-i} = (\emptyset, \emptyset, \cdots, \emptyset)$ *to* $(D, D, \cdots, D)$ *(increment = ϵ on each dimension)* **do**

    Build an empty graph;

    For each $p_i(t_i, t_{-i})$, construct a vertex $V_{t_i t_{-i}}$ and insert it into the graph;

    Construct a base vertex $VB_{t_{-i}}$ which denotes the payment zero into the graph;

    for $t_i = 0$ *to* D *(increment = ϵ)* **do**

      Add an edge from $VB_{t_{-i}}$ to $V_{t_i t_{-i}}$ with weight $\overline{p(t_i, t_{-i})}$;

      for $t'_i = 0$ *to* t_i *(increment = ϵ)* **do**

        Add an edge with weight $Gap_i(t'_i, t_i, t_{-i})$ from $V_{t'_i t_{-i}}$ to $V_{t_i t_{-i}}$;

    Use the Single-Source Shortest-Path algorithm to find the shortest path from $VB_{t_{-i}}$ to all the other vertices. These are the maximum solutions $p_i^{max}(t_i, t_{-i})$ for each payment case;

    if $\sum_{j=1}^{n} p_j^{max}(t) < 0$ **then**

      return There is no desirable mechanism.

return p_i^{max} as the payment functions.

The following theorem proves the correctness of Algorithm 1.

Theorem 5. *Taking agents' valuation functions as input, Algorithm 1 outputs the answer of the decision problem of whether there exists a mechanism that guarantees IR, IC, efficiency and weak budget balance simultaneously, and specifies the payments that achieve maximal revenue if the answer is yes.*

7 Conclusion

In this paper, we study the mechanism design problem for multi-party machine learning. We restrict the action space of each agent where he can only misreport a lower type than his actual type and consider the valuation function that is about the allocation outcome and the true types of all agents. The VCG mechanism does not guarantee IR and DSIC and the Myerson-Satterthwaite Theorem in the standard mechanism design setting does not hold in our setting, implying the desirable mechanisms that are both IR, DSIC, efficient and weakly budget-balance exist in our setting. We propose a maximal exploitation payment mechanism and show that this mechanism is truthful and revenue-optimal in the quasi-monotone externalities setting. Then we give sufficient and necessary conditions for designing a truthful mechanism for the general setting. These conditions restrict both the allocation function and the payment function. We show that the data size disparity between agents and the market growth rate highly affect the existence of the desirable mechanism. If the market grows fast and the disparity is small, a desirable mechanism is more likely to exist. Finally, we devise an algorithm to find desirable mechanisms that are truthful, individually rational, efficient and weakly budget-balance simultaneously.

Appendix

A   Proof of Theorem 1

Proof. Intuitively, the MEP rule charges agent i the profit he gets from an model that the mechanism allocates to him. If the mechanism charges higher than the MEP, an agent would have negative utility after taking part in. The IR constraint would then be violated. So it's easy to see that the MEP is the maximal payment among all IR mechanisms.

Then we prove that this payment rule also guarantees the IC condition. It suffices to show that if an agent hides some data, no matter which model he chooses to use, he would never get more utility than that of truthful reporting. We suppose that agent i's type is t_i' and he untruthfully reports t_i'.

Suppose that the agent i truthfully reports the type $t_i' = t_i$, since the payment function is defined to charge this agent until he reaches the valuation when he does not take part in the mechanism, the utility of this honest agent would be

$$u_i^0(t') = F_i(Q(t_i)) + \theta_i(q_{-i}(\emptyset, t'_{-i})).$$

If the agent does not report truthfully, we suppose that the agent reports t_i' where $t_i' \leq t_i$. According to the MEP, the payment function for agent i would be

$$p_i(t_i', t'_{-i}) = F_i(q_i(t_i', t'_{-i})) + \theta_i(q_{-i}(t_i', t'_{-i})) - F_i(Q(t_i')) - \theta_i(q_{-i}(\emptyset, t'_{-i})).$$

It can be seen that the mechanism would never give an agent a worse model than the model trained by its reported data, otherwise the agents would surely select their private data to train models. Hence it is without loss of generality to assume that the allocation $x_i(t_i', t'_{-i}) \geq Q(t_i')$, $\forall t_i', t'_{-i}, \forall i$. Thus we have $q_{-i}(t_i', t'_{-i}) = x_{-i}(t_i', t'_{-i})$. We discuss the utility of agent i by two cases of choosing models.

Case 1: the agent chooses the allocation x_i. Since agent i selects the allocated model, we have $q_i = x_i(t'_i, t'_{-i})$. Then the utility of agent i would be

$$
\begin{aligned}
u_i^1 =& v_i(t'_i, t'_{-i}) - p_i(t'_i, t'_{-i}) \\
=& F_i(x_i(t'_i, t'_{-i})) + \theta_i(x_{-i}(t'_i, t'_{-i})) + F_i(Q(t'_i)) \\
& + \theta_i(x_{-i}(\emptyset, t'_{-i})) - F_i(x_i(t'_i, t'_{-i})) - \theta_i(x_{-i}(t'_i, t'_{-i})) \\
=& F_i(Q(t'_i)) + \theta_i(x_{-i}(\emptyset, t'_{-i})).
\end{aligned}
$$

Because both F_i and Q are monotone increasing functions and $t_i \geq t'_i$, we have

$$
u_i^1 \leq F_i(Q(t_i)) + \theta_i(x_{-i}(\emptyset, t'_{-i})) = u_i^0.
$$

Case 2: the agent chooses $Q(t_i)$. Since agent i selects the model trained by his private data, we have $q_i = Q(t_i)$. The final utility of agent i would be

$$
\begin{aligned}
u_i^2 =& v_i(t'_i, t'_{-i}) - p_i(t'_i, t'_{-i}) \\
=& F_i(Q(t_i)) + \theta_i(x_{-i}(t'_i, t'_{-i})) + F_i(Q(t'_i)) \\
& + \theta_i(x_{-i}(\emptyset, t'_{-i})) - F_i(x_i(t'_i, t'_{-i})) - \theta_i(x_{-i}(t'_i, t'_{-i})) \\
=& F_i(Q(t_i)) + F_i(Q(t'_i)) + \theta_i(x_{-i}(\emptyset, t'_{-i})) - F_i(x_i(t'_i, t'_{-i})).
\end{aligned}
$$

Subtract the original utility from the both sides, then we have

$$
\begin{aligned}
u_i^2 - u_i^0 =& F_i(Q(t_i)) + F_i(Q(t'_i)) + \theta_i(x_{-i}(\emptyset, t'_{-i})) \\
& - F_i(x_i(t'_i, t'_{-i})) - F_i(Q(t_i)) - \theta_i(x_{-i}(\emptyset, t'_{-i})) \\
=& F_i(Q(t'_i)) - F_i(x_i(t'_i, t'_{-i})).
\end{aligned}
$$

Because $x_i(t'_i, t'_{-i}) \geq Q(t'_i), \forall t'_i, t'_{-i}, \forall i$ and because F_i is a monotonically increasing function, we can get $u_i^2 - u_i^0 \leq 0$. Therefore $\max\{u_i^1, u_i^2\} \leq u_i^0$, lying would not bring more benefits to any agent, and the mechanism is IC.

B   Proof of Corollary 1

Proof. In Theorem 1 we know that the MEP mechanism is IR and IC. Since the linear coefficients are all positive and the externality setting is linear, any efficient mechanism would allocate the best model to all the agents. Since each agent gets a model with no less quality than his reported one and the payment is equal to the value difference between the case an agent truthfully report and the case he exit the mechanism. The agent's value is always larger than the value when he exits the mechanism. Then the payment is always positive and the mechanism should satisfy all of the four properties.

C   Proof of Theorem 2

Proof. We first prove that Eq. (1) holds. Observe that

$$
\begin{aligned}
&u_i(x(t_i, t'_{-i}), t_i, t_{-i}) - u_i(x(t'_i, t'_{-i}), t'_i, t_{-i}) \\
=&[v_i(x(t_i, t'_{-i}), t_i, t_{-i}) - p_i(t_i, t'_{-i})] - [v_i(x(t'_i, t'_{-i}), t'_i, t_{-i}) - p_i(t'_i, t'_{-i})] \\
\geq&[v_i(x(t_i, t'_{-i}), t_i, t_{-i}) - p_i(t_i, t'_{-i})] - [v_i(x(t'_i, t'_{-i}), t_i, t_{-i}) - p_i(t'_i, t'_{-i})] \\
=&u_i(x(t_i, t'_{-i}), t_i, t_{-i}) - u_i(x(t'_i, t'_{-i}), t_i, t_{-i}) \\
\geq&0,
\end{aligned}
\tag{5}
$$

where the first inequality is because of Assumption 3, and the last inequality is because of the DSIC property.

Let $t'_i = 0$ in Eq. (5). We have

$$
u_i(x(t_i, t'_{-i}), t_i, t_{-i}) \geq u_i(x(0, t'_{-i}), 0, t_{-i}).
$$

The IR property further requires that $u_i(x(0, t'_{-i}), 0, t_{-i}) \geq u_i(x(\emptyset, t'_{-i}), 0, t_{-i})$, which Eq. (1) follows.

To show Eq. (2) must hold, we rewrite Eq. (5):

$$
\begin{aligned}
p_i(t_i, t'_{-i}) - p_i(t'_i, t'_{-i}) &\leq v_i(x(t_i, t'_{-i}), t_i, t_{-i}) - v_i(x(t'_i, t'_{-i}), t'_i, t_{-i}) \\
&= \int_{t'_i}^{t_i} \frac{dv_i(x(s', t'_{-i}), s(s'), t_{-i})}{ds'} \, ds'.
\end{aligned}
\tag{6}
$$

Fixing t_{-i} and t'_{-i}, the total derivative of $v_i(x(s', t'_{-i}), s, t_{-i})$ is:

$$
\begin{aligned}
&dv_i(x(s', t'_{-i}), s, t_{-i}) \\
=&\frac{\partial v_i(x(s', t'_{-i}), s, t_{-i})}{\partial s'} \, ds' + \frac{\partial v_i(x(s', t'_{-i}), s, t_{-i})}{\partial s} \, ds.
\end{aligned}
$$

View s as a function of s' and let $s(s') = s'$:

$$
\begin{aligned}
&\frac{dv_i(x(s', t'_{-i}), s(s'), t_{-i})}{ds'} \\
=&\frac{\partial v_i(x(s', t'_{-i}), s, t_{-i})}{\partial s'}\bigg|_{s=s'} + \frac{\partial v_i(x(s', t'_{-i}), s(s'), t_{-i})}{\partial s(s')} \frac{ds(s')}{ds'}.
\end{aligned}
$$

Plug into Eq. (6), and we obtain:

$$
\begin{aligned}
&p_i(t_i, t'_{-i}) - p_i(t'_i, t'_{-i}) \\
\leq&\int_{t'_i}^{t_i} \frac{\partial v_i(x(s', t'_{-i}), s, t_{-i})}{\partial s'}\bigg|_{s=s'} + \int_{t'_i}^{t_i} \frac{\partial v_i(x(s', t'_{-i}), s(s'), t_{-i})}{\partial s(s')} \, ds'.
\end{aligned}
$$

Since the above inequality holds for any valuation function with $v_i(x, t_i, t_{-i}) \geq v_i(x, t'_i, t_{-i}), \forall x, \forall t_{-i}, \forall t_i \geq t'_i$, we have:

$$
p_i(t_i, t'_{-i}) - p_i(t'_i, t'_{-i}) \leq \int_{t'_i}^{t_i} \frac{\partial v_i(x(s', t'_{-i}), s, t_{-i})}{\partial s}\bigg|_{s=s'} \, ds'.
$$

D   Proof of Theorem 3

Proof. Equation (3) indicates that the function $\frac{\partial v_i(x(t'_i,t'_{-i}),t_i,t_{-i})}{\partial t'_i}$ is minimized at t'_i:

$$\frac{\partial v_i(x(t'_i,t'_{-i}),s,t_{-i})}{\partial t'_i}\bigg|_{s=t'_i} \leq \frac{\partial v_i(x(t'_i,t'_{-i}),t_i,t_{-i})}{\partial t'_i}. \tag{7}$$

Therefore, we have

$$u_i(x(t_i,t'_{-i}),t_i,t_{-i}) - u_i(x(t'_i,t'_{-i}),t_i,t_{-i})$$
$$= \int_{t'_i}^{t_i} \frac{\partial v_i(x(s',t'_{-i}),t_i,t_{-i})}{\partial s'}\,ds' - p_i(t_i,t'_{-i}) + p_i(t'_i,t'_{-i})$$
$$\geq \int_{t'_i}^{t_i} \frac{\partial v_i(x(s',t'_{-i}),s,t_{-i})}{\partial s'}\bigg|_{s=s'}\,ds' - p_i(t_i,t'_{-i}) + p_i(t'_i,t'_{-i})$$
$$\geq \int_{t'_i}^{t_i} \frac{\partial v_i(x(\emptyset,t'_{-i}),s,t_{-i})}{\partial s}\,ds, \tag{8}$$

where the two inequalities are due to Eq. (7) and (4), respectively. Since $v_i(x,t_i,t_{-i}) \geq v_i(x,t'_i,t_{-i})$, $\forall x, \forall t_{-i}, \forall t_i \geq t'_i$ indicates $\frac{\partial v_i(x(\emptyset,t'_{-i}),s,t_{-i})}{\partial s} \geq 0$, the above inequality shows that the mechanism guarantees the DSIC property.

To prove that the mechanism is IR, we first observe that

$$[u_i(x(t_i,t'_{-i}),t_i,t_{-i}) - v_i(x(\emptyset,t'_{-i}),t_i,t_{-i})] - [u_i(x(t'_i,t'_{-i}),t'_i,t_{-i})$$
$$- v_i(\emptyset,x(t_{-i}),t'_i,t_{-i})]$$

$$= u_i(x(t_i,t'_{-i}),t_i,t_{-i}) - u_i(x(t'_i,t'_{-i}),t'_i,t_{-i}) - \int_{t'_i}^{t_i} \frac{\partial v_i(x(\emptyset,t'_{-i}),s,t_{-i})}{\partial s}\,ds$$

$$\geq u_i(x(t_i,t'_{-i}),t_i,t_{-i}) - u_i(x(t'_i,t'_{-i}),t_i,t_{-i}) - \int_{t'_i}^{t_i} \frac{\partial v_i(x(\emptyset,t'_{-i}),s,t_{-i})}{\partial s}\,ds$$

$$\geq 0,$$

where the two inequalities are Assumption 3 and Eq. (8). Letting $t'_i = 0$ using Eq. (2), we get:

$$u_i(x(t_i,t'_{-i}),t_i,t_{-i}) - v_i(x(\emptyset,t'_{-i}),t_i,t_{-i})$$
$$\geq u_i(x(0,t'_{-i}),0,t_{-i}) - v_i(x(\emptyset,t'_{-i}),0,t_{-i})$$
$$= v_i(x(0,t'_{-i}),0,t_{-i}) - p_i(0,t'_{-i}) - v_i(x(\emptyset,t'_{-i}),0,t_{-i})$$
$$\geq 0.$$

E   Proof of Theorem 4

Proof. Suppose that the platform uses the mechanism mentioned in the theorem. Then for each agent, contributing with more data increases all participants' model qualities. By definition, in a non-competitive market, improving others' models does not decrease one's profit. Therefore, the optimal strategy for each participant is to contribute with all his valid data, making the mechanism truthful. Also because of the definition, entering the platform always weakly increases one's model quality. Thus the mechanism is IR. With the IC and IR properties, it is easy to see that the mechanism is also efficient and weakly budget-balance.

F   Proof of Theorem 5

Proof. Suppose that there is a larger payment for agent i such that $p_i(t') > p_i^{\max}(t')$ where t' is the profile of reported types. In the process of our algorithm, the $p_i^{\max}(t')$ is the minimal path length from VB_{-i} to $V_{t_i}t_{-i}$, denoted by $(VB_{-i}, V_{t_{i1}}t_{-i}, V_{t_{i2}}t_{-i}, \cdots, V_{t_{ik}=t_i'}t_{-i})$. By the definition of edge weight, we have the following inequalities:

$$p_i(t_{i1}, t_{-i}) \le \overline{p_i(t_{i1}, t_{-i})},$$
$$p_i(t_{i2}, t_{-i}) - p_i(t_{i1}, t_{-i}) \le Gap_i(t_{i1}, t_{i2}, t_{-i}),$$

$$\vdots$$

$$p_i(t_{ik}, t_{-i}) - p_i(t_{i(k-1)}, t_{-i}) \le Gap_i(t_{i1}, t_{i2}, t_{-i}).$$

Adding these inequalities together, we get

$$p_i(t') \le \overline{p_i(t_{i1}, t_{-i})} + \sum_{j=1}^{k-1} Gap_i(t_{ij}, t_{i(j+1)}, t_{-i}) = p_i^{\max}(t').$$

If $p_i(t') < p_i^{\max}(t')$ holds, this would violate at least 1 of the k inequalities above. If the first inequality is violated, the mechanism would not be IR, by the definition of $\overline{p_i(t_{i1}, t_{-i})}$. If any other inequality is violated, the mechanism would not be IC, by the definition of $Gap_i(t_{ij}, t_{i(j+1)}, t_{-i})$.

On the other hand, if we select $p_i^{\max}(t')$ to be payment of agent i, all the inequalities should be satisfied, otherwise the shortest path would be updated to a smaller length.

Therefore the $p_i^{\max}(t')$ must be the maximum payment for agent i. If the maximal payment sum up to less than 0, there would obviously be no mechanism that is IR, IC and weakly budget-balance under the efficient allocation function.

G   Experiments

We design experiments to demonstrate the performance of our mechanism for practical use. We first show the mechanism with the maximal exploitation payments can guarantee a good quality of trained model and high revenues under the linear externality cases. Then we conduct simulations to exhibit the relation of the market growth of competitive markets to the existence of desirable mechanisms.

G.1   The MEP Mechanism

We consider the valuation with linear externalities setting where α_{ij}'s (defined in Example 1) are generated uniformly in $[-1, 1]$. Each agent's type is drawn uniformly from $[0, 1]$ independently and the $Q(t)$ is $\frac{1-e^{-t}}{1+e^{-t}}$. The performance of a mechanism is measured by the platform's revenue and its best quality of trained model under the mechanism. All the values of each instance are averaged over 50 samples. We both show the performance changes as the number of agents increases and as the agents' type changes.

When the number of agents becomes larger, the platform can obtain more revenues and train better models (see Fig. 1). Particularly, the model quality is close to be optimal when the number of agents over 12. An interesting phenomenon is that the revenue may surpass the social welfare. This is because the average external effect of other agents on one agent i tends to be negative when agent i does not join in the mechanism, thus the second term in the MEP payment is averagely negative and revenue is larger than the welfare.

To see the influence of type on performance, we fix one agent's type to be 1 and set the other agent's type from 0 to 10. It can be seen in Fig. 2 that the welfare and opponent agent's utility (uti_2) increase as the opponent's type increases but the platform's revenue and the utility of the static agent (uti_1) are almost not affected by the type. So we draw the conclusion that the most efficient way for the platform to earn more revenue is to attract more small companies to join the mechanism, since in the Fig. 1 the revenue obviously increases as the number of agents increases.

G.2   Existence of Desirable Mechanisms

We assume all the agents' types lie in $[0, D]$, and the type space can be discretized into intervals of length ϵ, which can be viewed as the minimal size of a dataset. Thus each agent's type is a multiple of ϵ. The data disparity is defined as the ratio of the largest possible data size to the smallest possible data size, namely, D/ϵ. We measure the condition for existence of desirable mechanisms by the maximal data disparity when the market growth rate is given.

To describe the market growth, we use the following form of valuation function and model quality function:

$$Q(t) = t \text{ and } v_i(q) = \left(\sum_{j=1}^{n} Q(q_j)\right)^{\gamma} \cdot Q(q_i), \forall i,$$

where γ indicates the market growth rate. We consider the competitive growing market case where $-1 \leq \gamma < 0$.[1]

The algorithm we use to find desirable mechanisms under different valuation functions is described in Sect. 6. We enumerate the value of γ from -1 to -0.668 with step length 0.002 and run the algorithm to figure out the boundary of D/ϵ

---

[1] When $\gamma < -1$, the market is not a growing market; when $\gamma \geq 0$, the market becomes non-competitive, therefore by Theorem 4, a desirable mechanism trivially exists.

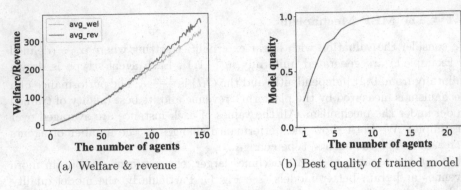

(a) Welfare & revenue

(b) Best quality of trained model

Fig. 1. Performance of MEP under different numbers of agents

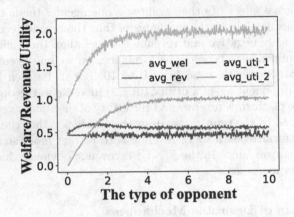

Fig. 2. Performance of MEP under different types

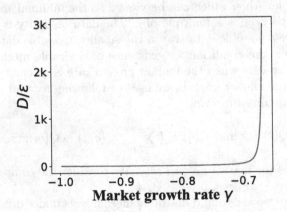

Fig. 3. Data disparity vs. Market growth (Color figure online)

under different γ in a market with 2 agents. The range of γ is determined by our computing capability, and the disparity boundary has been over 10000 when γ is near -0.66.

Figure 3 shows the boundary of data disparity for existence of desirable mechanisms under different market growth rates. For every fixed γ, there does not exist any desirable mechanism when the data disparity is larger than the point on the red line. It can be seen an obvious trend that when γ becomes larger, the constraint on data size disparity would become looser. A desirable mechanism is more likely to exist in a market that grows faster. When the market is not growing, there would not be such a desirable mechanism at all. On the other hand, if the market grows so fast such that there does not exist any competition between the agents, the desirable mechanism always exists.

References

1. Abadi, M., et al.: Deep learning with differential privacy. In: Proceedings of the 2016 ACM SIGSAC Conference on Computer and Communications Security, pp. 308–318. ACM (2016)
2. Ausubel, L.M.: An efficient ascending-bid auction for multiple objects. Am. Econ. Rev. **94**(5), 1452–1475 (2004)
3. Blumrosen, L., Feldman, M.: Implementation with a bounded action space. In: Proceedings of the 7th ACM Conference on Electronic Commerce, pp. 62–71. ACM (2006)
4. Blumrosen, L., Nisan, N.: Auctions with severely bounded communication. In: 2002 Proceedings of the 43rd Annual IEEE Symposium on Foundations of Computer Science, pp. 406–415. IEEE (2002)
5. Chawla, S., Fu, H., Karlin, A.R.: Approximate revenue maximization in interdependent value settings. In: Babaioff, M., Conitzer, V., Easley, D.A. (eds.) ACM Conference on Economics and Computation, EC 2014, Stanford, CA, USA, 8–12 June 2014, pp. 277–294. ACM (2014)
6. Chwe, M.S.Y.: The discrete bid first auction. Econ. Lett. **31**(4), 303–306 (1989)
7. David, E., Rogers, A., Jennings, N.R., Schiff, J., Kraus, S., Rothkopf, M.H.: Optimal design of English auctions with discrete bid levels. ACM Trans. Internet Technol. (TOIT) **7**(2), 12 (2007)
8. Ding, N., Fang, Z., Huang, J.: Optimal contract design for efficient federated learning with multi-dimensional private information. IEEE J. Sel. Areas Commun. **39**(1), 186–200 (2020)
9. Fang, W., Tang, P., Zuo, S.: Digital good exchange. In: Proceedings of the 2016 International Conference on Autonomous Agents & Multiagent Systems, pp. 1277–1278. International Foundation for Autonomous Agents and Multiagent Systems (2016)
10. Goldberg, A.V., Hartline, J.D.: Envy-free auctions for digital goods. In: Proceedings of the 4th ACM Conference on Electronic Commerce, pp. 29–35. ACM (2003)
11. Goldberg, A.V., Hartline, J.D., Wright, A.: Competitive auctions and digital goods. In: Proceedings of the Twelfth Annual ACM-SIAM Symposium on Discrete Algorithms, pp. 735–744. Society for Industrial and Applied Mathematics (2001)

12. Hu, Y., Niu, D., Yang, J., Zhou, S.: FDML: a collaborative machine learning framework for distributed features. In: Proceedings of the 25th ACM SIGKDD International Conference on Knowledge Discovery & Data Mining, pp. 2232–2240. ACM (2019)

13. Jia, R., et al.: Towards efficient data valuation based on the Shapley value. In: The 22nd International Conference on Artificial Intelligence and Statistics, pp. 1167–1176. PMLR (2019)

14. Kang, J., Xiong, Z., Niyato, D., Yu, H., Liang, Y.C., Kim, D.I.: Incentive design for efficient federated learning in mobile networks: a contract theory approach. arXiv preprint arXiv:1905.07479 (2019)

15. Lim, W.Y.B., et al.: Federated learning in mobile edge networks: a comprehensive survey. IEEE Commun. Surv. Tutor. **22**(3), 2031–2063 (2020)

16. McMahan, B., Moore, E., Ramage, D., Hampson, S., y Arcas, B.A.: Communication-efficient learning of deep networks from decentralized data. In: Artificial Intelligence and Statistics, pp. 1273–1282 (2017)

17. Mezzetti, C.: Mechanism design with interdependent valuations: efficiency. Econometrica **72**(5), 1617–1626 (2004)

18. Milgrom, P.R., Weber, R.J.: A theory of auctions and competitive bidding. Econometrica **50**(5), 1089–1122 (1982)

19. Myerson, R.B., Satterthwaite, M.A.: Efficient mechanisms for bilateral trading. J. Econ. Theory **29**(2), 265–281 (1983)

20. Redko, I., Laclau, C.: On fair cost sharing games in machine learning. In: Thirty-Third AAAI Conference on Artificial Intelligence (2019)

21. Roughgarden, T., Talgam-Cohen, I.: Optimal and robust mechanism design with interdependent values. ACM Trans. Econ. Comput. **4**(3), 18:1–18:34 (2016)

22. Shokri, R., Shmatikov, V.: Privacy-preserving deep learning. In: Proceedings of the 22nd ACM SIGSAC Conference on Computer and Communications Security, pp. 1310–1321. ACM (2015)

23. Smith, V., Chiang, C.K., Sanjabi, M., Talwalkar, A.S.: Federated multi-task learning. In: Advances in Neural Information Processing Systems, pp. 4424–4434 (2017)

24. Song, T., Tong, Y., Wei, S.: Profit allocation for federated learning. In: 2019 IEEE International Conference on Big Data (Big Data), pp. 2577–2586. IEEE (2019)

25. Takabi, H., Hesamifard, E., Ghasemi, M.: Privacy preserving multi-party machine learning with homomorphic encryption. In: 29th Annual Conference on Neural Information Processing Systems (NIPS) (2016)

26. Yonetani, R., Naresh Boddeti, V., Kitani, K.M., Sato, Y.: Privacy-preserving visual learning using doubly permuted homomorphic encryption. In: Proceedings of the IEEE International Conference on Computer Vision, pp. 2040–2050 (2017)

27. Zhan, Y., Zhang, J., Hong, Z., Wu, L., Li, P., Guo, S.: A survey of incentive mechanism design for federated learning. IEEE Trans. Emerg. Top. Comput. **10**, 1035–1044 (2021)

Budget-Feasible Sybil-Proof Mechanisms for Crowdsensing

Xiang Liu[1], Weiwei Wu[1(✉)], Wanyuan Wang[1], Yuhang Xu[1], Xiumin Wang[2], and Helei Cui[3]

[1] School of Computer Science and Engineering, Southeast University, Nanjing, China
{xiangliu,weiweiwu,wywang,yuhang_xu}@seu.edu.cn
[2] School of Computer Science and Engineering, South China University of Technology, Guangzhou, China
xmwang@scut.edu.cn
[3] School of Computer Science, Northwestern Polytechnical University, Xi'an 710129, China
chl@nwpu.edu.cn

Abstract. The rapid use of smartphones and devices leads to the development of crowdsensing (CS) systems where a large crowd of participants can take part in performing data collecting tasks in large-scale distributed networks. Participants/users in such systems are usually selfish and have private information, such as costs and identities. Budget-feasible mechanism design, as a sub-field of auction theory, is a useful paradigm for crowdsensing, which naturally formulates the procurement scenario with buyers' budgets being considered and allows the users to bid their private costs. Although the bidding behavior is well-regulated, budget-feasible mechanisms are still vulnerable to the Sybil attack where users may generate multiple fake identities to manipulate the system. Thus, it is vital to provide Sybil-proof budget-feasible mechanisms for crowdsensing. In this paper, we design a budget-feasible incentive mechanism which can guarantee truthfulness and deter Sybil attack. We prove that the proposed mechanism achieves individual rationality, truthfulness, budget feasibility, and Sybil-proofness. Extensive simulation results further validate the efficiency of the proposed mechanism.

Keywords: Crowdsensing · Budget feasibility · Sybil-proofness · Mechanism design · Auction

1 Introduction

The proliferation of smart mobile devices, such as phones, tablets and smartwatch, which are installed with rich sensors (*e.g.*, camera, light sensor, and GPS), has made crowdsensing a new popular economic paradigm which provides the crowd of users with mobile devices chances accomplishing large-scale distributed tasks, like collecting and sharing environmental information. Crowdsensing (CS) systems usually consists of a platform and a collection of users. The platform

© The Author(s), under exclusive license to Springer Nature Switzerland AG 2022
M. Li and X. Sun (Eds.): IJTCS-FAW 2022, LNCS 13461, pp. 269–288, 2022.
https://doi.org/10.1007/978-3-031-20796-9_19

acts as a data requester who posts a set of tasks need to be finished and smartphone users provide services by performing assigned tasks. Applications like reCAPTCHA [18], Amazon Mechanical Turks (AMT) and oDesk have made it possible to exploit human resources solving crowdsensing problems.

Most of smartphone users are not voluntary to work on the tasks since they consume their own resources, *e.g.*, battery, computing power, time, cellular data traffic, and expose private information with potential privacy. Furthermore, the system can be more effective with more users' participation. Thus a good incentive mechanism is vitally important to stimulate users to contribute to the platform. Many works [5,9,21,25] model the crowdsourcing/crowdsensing problems as reverse auctions where the requester works as a buyer and the users act as service sellers who bid for performing tasks. Users achieve monetary reward after submitting results of assigned tasks. Auction-based systems often face the strategic scenario where the participants may take strategic behaviors to obtain more utilities, *e.g.*, bidding false private information. Sufficient works thus make the effort to design truthful mechanisms so that users have no incentive to bid dishonestly [4–6,8,12,15,20,21,25]. Apart from false bidding behaviors, there is another kind of strategic behavior called Sybil attack, also known as false-name attack, that users may generate fake identities to manipulate the system for more utilities. The detection methods for Sybil attack have been considered in various research areas such as combinatorial auctions [16,17,24], spectrum auctions [19], and social networks [2]. Unfortunately, Lin *et al.* [10] show that many existing truthful mechanisms in crowdsourcing are vulnerable to Sybil attack, *e.g.*, by taking Sybil attack, users in [5,25] can increase her payment by reporting false information, and the user in [27] can change from a loser to a winner with a positive utility. Lin *et al.* [10] and Zhang *et al.* [26] are the first to propose incentive mechanisms guaranteeing truthfulness and Sybil-proofness in the auction-based crowdsourcing systems.

However, in the procurement scenario, the requester often comes with budget and the designed procurement mechanism should satisfy the budget constraint that the total payment from the requester cannot exceed a given budget. The goal of requester in this scenario is to maximize total value of assigned tasks finished by users within the budget constraint. This problem falls into research of budget-feasible mechanism design problem first studied in [13] which proposes the first budget-feasible truthful mechanism in the procurement scenarios. After that, many works [1,7,14,28] extend the budget-feasible mechanism design into the crowdsourcing systems. Although the truthfulness/bidding behavior is well regulated in these mechanisms, budget-feasible mechanisms in crowdsourcing systems yet consider the Sybil-proofness and are still vulnerable to Sybil attack of users. And existing Sybil-proof mechanisms proposed for the auction-based systems [10,26] also cannot be applied to the procurement scenario as an unlimited payment is even allowed if necessary to elicit the incentive behaviors.

Therefore, in this paper, we focus on designing a budget-feasible Sybil-proof mechanism for CS systems to deter the untruthful bidding behaviors and Sybil attack. The designed mechanism should guarantee various desired properties like, *individual rationality* that the payment to each seller covers at least (but not necessarily equals) her private cost, *budget feasibility* that the total payment of the requester does not exceed her budget, *truthfulness* that no sellers have incentive to bid dishonestly, and *Sybil-proofness* that users cannot increase their utilities by launching Sybil attack. The main contributions of this paper are as follows:

(1) We are the first to address Sybil attack in procurement scenarios for CSs, and propose a corresponding budget-feasible Sybil-proof mechanism, which moves a step forward to robust budget-feasible mechanisms in crowdsensing.
(2) We design a Mechanism TBS (**T**ruthful **B**udget-feasible and **S**ybil-proof mechanism) and prove that the proposed mechanism achieves computational efficiency, individual rationality, truthfulness, budget feasibility and Sybil-proofness.
(3) We evaluate the performance and validate the desired properties by extensive simulations. Furthermore, it shows that the proposed mechanism spends less when procuring fixed value from users than previous Sybil-proof mechanisms, while ensuring the budget feasibility.

The rest of paper is organized as follows. In Sect. 2, we briefly review the works in truthful auctions, budget-feasible mechanisms in crowdsensing and Sybil-proof mechanisms. In Sect. 3, we introduce the system model and problem formulation. We discuss the vulnerability to Sybil attack in traditional budget-feasible mechanisms in Sect. 4. In Sect. 5, we propose a mechanism TBS and prove the desired properties. The performance evaluation is presented in Sect. 6. Finally, we conclude this paper in Sect. 7.

2   Related Work

Many works consider incentive mechanisms in crowdsensing/crowdsourcing systems. Yang et al. [21] compute the unique Stackelberg Equilibrium for the platform-centric crowdsensing model and designed truthful mechanism for the user-centric crowdsourcing model. Feng et al. [5] further take the location information into consideration when assigning sensing tasks to smartphones. Zhang et al. [25] study three models of crowdsourcing which consider the cooperation and competition among the service and propose incentive mechanisms for each of them. Zhu et al. [29] design incentive mechanisms based on the combination of a reverse auction and a Vickrey auction to address malicious competition behavior in price bidding. Huang et al. [8] design a truthful double auction mechanism which takes max-min fairness into consideration. Cui et al. [4] propose an incentive mechanism for task allocation problem in crowdsourcing systems by designing a bid-independent payment calculation scheme.

Budget-feasible mechanism was first studied in [13] which addresses the procurement scenarios where buyers have budgets and the payment scheme should be carefully designed. After that, Singer and Mittal [14] present constant-competitive truthful mechanisms for maximizing the number of tasks under a budget. Some works [7,28] focus on budget-feasible mechanisms in online scenario where users arrive online and the requester wants to select users for maximizing the value of services under a budget constraint. Singla *et al.* [15] use the approach of regret minimization by combining multi-armed bandits to design budget-feasible mechanisms that achieve near-optimal utility for the requester.

Although Sybil attack has been addressed in some auction scenarios, it has rarely been studied in budget-feasible mechanisms. For example, Terada and Yokoo [16] propose a false-name-proof multi-unit auction protocol. The works [4,17,24] focus on truthful and Sybil-proof mechanisms in combinatorial auctions. Wang *et al.* [19] design mechanisms that detect Sybil attack in dynamic spectrum auctions. Brill *et al.* [2] consider Sybil-proofness for users who may manipulate the recommendation by performing a false-name manipulation in social networks. Yao *et al.* [23] propose a novel Sybil attack detection method based on Received Signal Strength Indicator (RSSI) for Vehicular Ad Hoc Networks (VANETs). For crowdsourcing systems, Lin *et al.* [10] and Zhang *et al.* [26] investigate truthful and Sybil-proof mechanisms in auction-based systems. However, these mechanisms do not take into account the budget constraints of requesters, thus cannot be applied to the procurement scenarios in crowdsourcing systems.

In summary, although the bidding behaviors are well regulated, existing budget-feasible mechanisms in crowdsourcing systems are still vulnerable to Sybil attack. Therefore, it is vital to design budget-feasible mechanisms that are robust in truthfulness and Sybil-proofness.

3   Preliminaries

We consider a crowdsensing system that consists of a platform and n users denoted by $u = \{1, 2, \ldots, n\}$. Users may participate in this system to finish crowdsensing tasks. Denote by T_i the task set user i wants to finish. Let $T = \bigcup_{i \in u} T_i$ denote the whole tasks that can be finished by all users. In addition, each task $t_l \in T$ has a value $v_l > 0$ to the platform. The platform gains value v_l when task t_l is completed.

3.1   Reverse Auction Model

We model the interaction between the platform and users as a reverse auction, where the platform acts as a requester/buyer and users serve as sellers. We take into account the procurement scenario in the crowdsensing systems where the requester wants to procure service from users (sellers) within budget $\mathbb{B}$. We assume that each user i has a cost function $c_i(\mathcal{B})$ to show the cost of finishing all tasks in a bundle $\mathcal{B} \subseteq T$. Following the assumption in [10], the cost function $c_i(\cdot)$ of user i satisfies the following properties:

- $c_i(\emptyset) = 0$ and $c_i(\{t_l\}) = \infty, \forall t_l \in T \backslash T_i$;
- $c_i(\mathcal{B}') \leq c_i(\mathcal{B}''), \forall \mathcal{B}', \mathcal{B}'' \subseteq T$ with $\mathcal{B}' \subseteq \mathcal{B}''$;

- $c_i(\mathcal{B}) \leq c_i(\mathcal{B}') + c_i(\mathcal{B}''), \forall \mathcal{B}', \mathcal{B}'' \subseteq T$ and $\mathcal{B} = \mathcal{B}' \cup \mathcal{B}''$.

These four properties characterize the cost of performing tasks in practice.

Meanwhile, we consider the scenario of *incomplete information* auction where only user herself knows her private information, *e.g.*, task sets and cost function. The budget and value function of the requester, as an auctioneer, are common knowledge. Let $(T_i, c_i(\cdot))$ denote the *task-cost pair* of each user i. Initially, all users would bid their costs. In the auction model, we consider the strategic scenario where each user is selfish and rational and she may misreport her cost for more utilities denoted by $\tilde{c}_i \neq c_i$ or her task set denoted by $\tilde{T}_i \neq T_i$. Similarly, let $(\tilde{T}_i, \tilde{c}_i(\cdot))$ denote the reported *task-cost pair* of each user i and $\vec{\beta} = \{(\tilde{T}_1, \tilde{c}_1), (\tilde{T}_i, \tilde{c}_i), \cdots, (\tilde{T}_i, \tilde{c}_i)\}$ denote the bid profile of all users. Specifically, denote by $\vec{\beta}_{-i}$ the bid profile of all users except user i.

After receiving the bids from users, the requester/buyer selects a subset of users $u_w \subseteq u$ called winners and assign each winner $i \in u_w$ a task set $A_i = T_i$ to finish, and, $A_i = \emptyset$ if $i \notin u_w$. Let $\vec{A} = (A_1, A_2, \ldots, A_n)$ denotes the assignment profile. To stimulate users to participate in the auction, the platform gives payment p_i to each winner i. Note that $p_i = 0$ if $i \notin u_w$. Let $\vec{p} = (p_1, p_2, \ldots, p_n)$ denote the payment profile. We further consider the *budget feasibility* on the buyer's side which requires that the total payments paid to the sellers cannot exceed the budget, *i.e.*, $\sum_{i \in u} p_i \leq \mathbb{B}$. We assume that each user is willing to perform only the whole set T_i following the assumption in [10,22]. We thus define user i's *utility* as the payment minus her cost, *i.e.*,

$$u_i((\tilde{T}_i, \tilde{c}_i(\cdot)), \vec{\beta}_{-i}) = \begin{cases} p_i - c_i(T_i), \; if \; A_i = T_i \\ \quad 0, \qquad otherwise \end{cases} \tag{1}$$

The platform adopts value function $\mathcal{V}(\cdot)$ to calculate the total value over a subset of users. We define the utility of platform as total value procured from winners, *i.e.*, $\mathcal{V}(u_w)$, which is the sum of value of all tasks in the union set of assigned task sets of winners, *i.e.*, $u_b = \mathcal{V}(u_w) = V(\cup_{i \in u_w} T_i) = \sum_{t_i \in \cup_{i \in u_w} T_i} v_i$, where function $V(\cdot)$ denotes the sum of value of all tasks in the subset of T. It is easy to show that the value function $\mathcal{V}(\cdot)$ is a monotone submodular function by the following definition.

Definition 1 *(Monotone Submodular Function): Let G be a finite set. For any $X \subseteq Y \subseteq G$ and $x \in G$, a function $f : 2^G \leftarrow R$ is called submodular if and only if $f(X \cup \{x\}) - f(X) \geq f(Y \cup \{x\}) - f(Y)$ and it is monotone (increasing) if and only if $f(X) \leq f(Y)$.*

3.2 Sybil Attack

We further consider the Sybil attack where a user could submit multiple fictitious identities. As a simple case, user i could submit two task-cost pairs $(\tilde{T}_{i'}, \tilde{c}_{i'})$ and $(\tilde{T}_{i''}, \tilde{c}_{i''}))$ under two identities i' and i'', respectively. This case is sufficient to represent the general Sybil attack.

Assume that user i submits $(\tilde{T}_{i'}, \tilde{c}_{i'})$ and $(\tilde{T}_{i''}, \tilde{c}_{i''})$ under two identities i' and i'', where $\tilde{T}_{i'} \cup \tilde{T}_{i''} = T_i$. Let $A_{i'}$ and $A_{i''}$ denote the assigned task set for user i' and i'', respectively. Similarly, denote by $p_{i'}$ and $p_{i''}$ the corresponding payments for them. As user i is willing to perform only the whole set T_i, her utility $\tilde{u}_i$ under Sybil attack will be zero if the union assigned task set among generated identities is not equal to T_i. Thus, $\tilde{u}_i = p_{i'} + p_{i''} - c_i(T_i)$ if $A_{i'} \cup A_{i''} = T_i$ and otherwise $\tilde{u}_i = 0$. When $\tilde{u}_i > u_i$, user i has an incentive to conduct Sybil attack.

3.3   Properties

The goal of this paper is to design a budget-feasible and Sybil-proof mechanism maximizing the utility of platform under the crowdsensing model above and guaranteeing the following desired properties:

(1) **Individual Rationality:** Each user i has a non-negative utility when bidding her true task-cost pair, *i.e.*, $u_i((T_i, c_i(\cdot)), \vec{\beta}_{-i}) \geq 0$.
(2) **Truthfulness:** Reporting true cost function is user i's dominant strategy, *i.e.*, $u_i((T_i, c_i(\cdot)), \vec{\beta}_{-i}) \geq u_i((\tilde{T}_i, \tilde{c}_i(\cdot)), \vec{\beta}_{-i})$.
(3) **Budget Feasibility:** The total payment cannot exceed the budget of requester, *i.e.*, $\sum_{i \in u} p_i \leq \mathbb{B}$.
(4) **Sybil-proofness:** Any user's utility is maximized when bidding her true task-cost pair using a single identity, *i.e.*, $\tilde{u}_i \leq u_i$.
(5) **Computational Efficiency:** The mechanism terminates in polynomial time.

4   Sybil Attack on Budget-Feasible Mechanisms

In this section, we discuss the vulnerability to Sybil attack in truthful incentive mechanisms. As discussed in Sect. 2, many existing budget-feasible mechanisms do not take into account the threat of Sybil attack. Thus, we present a detailed example showing how Sybil attack increases a dishonest user's utility.

In budget-feasible mechanisms [3,13], the proportional share allocation rule is widely used to generate budget-feasible allocations and elicit the truthfulness. Denoted by m_i or $m_i(S) = \mathcal{V}(S \cup \{i\}) - \mathcal{V}(S)$ the marginal value of a user i with respect to set S, users are sorted according to their non-decreasing order of the cost relative to marginal contributions, *i.e.*, $i + 1 = \text{argmin}_{j \in u} \frac{c_j}{m_j(S_i)}$ where $S_i = \{1, 2, \ldots, i\}$, and selected as winners if $\frac{c_i}{m_i(S_{i-1})} \leq \frac{\mathbb{B}/2}{\mathcal{V}(S_i)}$. In addition, user i in the winner set is rewarded $m_i(S_{i-1}) \cdot \frac{\mathbb{B}}{\mathcal{V}(u_w)}$ to guarantee the truthfulness.

> **Proportional share allocation rule:** (1) Sort all the users according to their non-decreasing costs relative to marginal contribution. (2) Allocate user i to winner set u_w if $\frac{c_i}{m_i(S_{i-1})} \leq \frac{\mathbb{B}/2}{\mathcal{V}(S_i)}$.

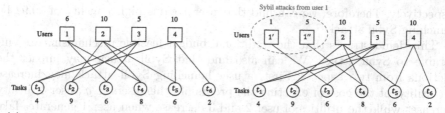

(a) Example of budget-feasible mecha-   (b) Example of Sybil attack in budget-
nisms.   feasible mechanisms.

Fig. 1. Example of the sybil-attack.

Next, we illustrate an example to show Sybil attack in Fig. 1. In this example, we set the budget of the requester (buyer) at $\mathbb{B} = 50$. We use squares to denote users (sellers) while circles represent tasks. The edge between a user and a task means that this task is in this user's task set. Each user owns a task set T_i and the number above user i is her cost for task set T_i. The number below task t_j is her value v_j to the buyer. There are four users $u = \{1, 2, 3, 4\}$, and corresponding task sets: $T_1 = \{t_3, t_4\}, T_2 = \{t_2, t_4, t_5\}, T_3 = \{t_1, t_2, t_3\}, T_4 = \{t_1, t_5, t_6\}$. In addition, their costs are $c_1 = 6, c_2 = 10, c_3 = 5, c_4 = 10$, and the value of these tasks are $v_1 = 4, v_2 = 9, v_3 = 6, v_4 = 8, v_5 = 6, v_6 = 2$, respectively. Let $\mathcal{V}(S) = \sum_{t_j \in \cup_{i \in S} T_i} v_j$ denote the value function given the user subset S.

According to proportional share allocation rule, user 3 with the minimum cost per marginal value $\frac{c_3}{V(T_3)} = \frac{c_3}{v_1+v_2+v_3} = \frac{5}{19} \leq \frac{\mathbb{B}/2}{\mathcal{V}(\{3\})} = \frac{25}{19}$ is first selected as a winner. Then, user 2 with the minimum cost per marginal value among remaining users $\{1, 2, 4\}$ is selected as the second winner, *i.e.*, $\frac{c_2}{V(T_2 \setminus T_3)} = \frac{c_2}{v_4+v_5} = \frac{10}{14} \leq \frac{\mathbb{B}/2}{\mathcal{V}(\{3,2\})} = \frac{25}{33}$. Last, user 4 has the minimum cost per marginal value $\frac{c_4}{V(T_4 \setminus (T_2 \cup T_3))} = \frac{c_4}{v_6} = \frac{10}{2}$, but exceeds the threshold $\frac{\mathbb{B}/2}{\mathcal{V}(\{3,2,4\})} = \frac{25}{35}$. Thus, we have the winner set $\{3, 2\}$. According to the payment scheme, we have the payment $p_1 = 0, p_2 = (v_4 + v_5) \cdot \frac{\mathbb{B}}{\mathcal{V}(\{3,2\})} \approx 21.21, p_3 = (v_1 + v_2 + v_3) \cdot \frac{\mathbb{B}}{\mathcal{V}(\{3,2\})} \approx 28.79, p_4 = 0$. The utilities of these four users are $u_1 = 0, u_2 = 11.21, u_3 = 23.79, u_4 = 0$, respectively.

Now, we assume that user 1 generates two identities: user $1'$ with task set $T_{1'} = \{t_3\}$ and cost $c_{1'} = 1$, and user $1''$ with task set $T_{1''} = \{t_4\}$ and cost $c_{1''} = 5$, as shown in Fig. 1(b).

In such a scenario, user $1'$ is selected as the first winner with the minimum cost per marginal value $\frac{c_{1'}}{v_3} = \frac{1}{6} \leq \frac{\mathbb{B}/2}{\mathcal{V}(\{1'\})} = \frac{25}{6}$. Then, user 3 is selected as the second winner since $\frac{c_3}{v_1+v_2} = \frac{5}{13} \leq \frac{\mathbb{B}/2}{\mathcal{V}(\{1',3\})} = \frac{25}{19}$. After that, user $1''$ is selected as the third winner by $\frac{c_{1''}}{v_4} = \frac{5}{8} \leq \frac{\mathbb{B}/2}{\mathcal{V}(\{1',3,1''\})} = \frac{25}{27}$. Last, user 4 has the minimum cost per marginal value $\frac{c_4}{v_6} = \frac{10}{8}$ which exceeds $\frac{\mathbb{B}/2}{\mathcal{V}(\{1',3,1'',4\})} = \frac{25}{35}$. Thus, we have the winner set $\{1', 3, 1''\}$. According to the payment scheme, we have the payment $p_{1'} = 11.11, p_{1''} = 14.81, p_2 = 0, p_3 = 24.07, p_4 = 0$. The utilities of these four users in this case are $u_1 = 19.92, u_2 = 0, u_3 = 19.07, u_4 = 0$, respectively. Therefore, we can find that user 1 gains higher utility of 19.92 by launching Sybil attack.

This demonstrates that the traditional budget-feasible mechanisms are vulnerable to Sybil attack. We can also find that Sybil attack may impact the auctions from two aspects: First, a user launching Sybil attack may increase her utility, at the cost of effecting the profits of other users, $e.g.$, user 1' utility increases while the utilities of user 2 and 3 decrease when user 1 generates fake identities. This behaviour may hinder the willingness of other users to participate in the system. Second, Sybil attack can also hurt the platform's utility, $e.g.$, the platform can only achieve utility 27 in Fig. 1(b) rather than 33 in Fig. 1(a). Therefore, this motivates us to design an incentive budget-feasible mechanism that is robust against Sybil attack.

5   Mechanism TBS

In this section, we propose a budget-feasible Sybil-proof mechanism **TBS** (**T**ruthful **B**udget-feasible and **S**ybil-proof mechanism).

The main idea of Mechanism TBS is as follows. In order to detect Sybil attack, we first group all the users by the task size and sort all the groups in the decreasing order of their users' task size. Mechanism TBS consists of two phases: winner selection and payment determination. In winner selection scheme, we scan these groups to select winners starting from the group with largest task size. Within each group, we iteratively select the user with the lowest bid per marginal value until the specified threshold set to guarantee budget feasibility and truthfulness is violated. In payment determination scheme, we find a threshold payment, above which bids cannot be selected as winners.

Next, we introduce more details of Mechanism TBS as shown in Algorithm 1. Considering the budget constraint, we use $B = \frac{\mathbb{B}}{2}$ as virtual budget to select winners. We first group all the users by the task size $|T_i|$, $i.e.$, users in the same group have the same task size, and sort these groups in the decreasing order of task size, $i.e.$, $\mathcal{G}_1, \mathcal{G}_2, \ldots, \mathcal{G}_l$, and start from the largest task size group. Assume that we are now considering group $\mathcal{G}_h$. We find the user with the lowest bid per marginal value $i = \mathrm{argmin}_{j \in \mathcal{G}_h} \frac{b_j}{v_j(R_j)}$ where set R_j denotes the union set of all assigned task sets before j and $v_j(R_j)$ denotes the marginal value of user j given the task set R_j, $i.e.$, $v_j(R_j) = V(R_j \cup T_j) - V(R_j)$. Suppose that i is the i-th lowest user in this group. Let q denote the *maximum bid per marginal value* among winners in the previous groups, $i.e.$, $q = \max_{j \in \mathcal{G}_l \cap u_w, \forall l < h} \frac{b_j}{v_j(R_j)}$. User i will be selected as a winner if it satisfies

$$\frac{b_i}{v_i(R_i)} \leq \frac{B}{V(R_i \cup T_i)}, q \leq \frac{B}{V(R_i \cup T_i)} \tag{2}$$

and

$$b_i \leq v_i(R_i), V(R_i) + v_i(R_i) \leq \mathbb{B}. \tag{3}$$

Based on the strategies above, the mechanism can control winners' cost per marginal value and elicit the budget-feasibility as well as the Sybil-proofness.

Algorithm 1: Mechanism **TBS**

Input: Sensing task set T, budget $\mathbb{B}$, user set u, bidding profile $\vec{\beta}$.
Output: Assignment profile $\vec{A}$, payment profile $\vec{p}$.

1   $R \leftarrow \emptyset, q \leftarrow 0, p_i \leftarrow 0, A_i \leftarrow \emptyset, u_w \leftarrow \emptyset$;
2   Group users by the task set size, and sort these groups according to the decreasing order of task size, *i.e.*, $\mathcal{G}_1, \mathcal{G}_2, ..., \mathcal{G}_l$;
3   $k \leftarrow 1, q \leftarrow 0, B \leftarrow \mathbb{B}/2, q_0 \leftarrow 0, R_0 \leftarrow \emptyset$;
4   **while** $k \leq l$ **do**
5   // **Winner Selection**;
6      $\mathcal{G}' \leftarrow \mathcal{G}_k, i \leftarrow \text{argmax}_{j \in \mathcal{G}'} \frac{b_j}{v_j(R)}$;
7   **while** $\frac{b_i}{v_i(R)} \leq \frac{B}{V(R \cup T_i)}$ **and** $q \leq \frac{B}{V(R \cup T_i)}$ **and** $b_i \leq v_i(R)$ **and** $\mathcal{V}(R) \leq \mathbb{B}$
       and $\mathcal{G}' \neq \emptyset$ **do**
8        $\mathcal{G}' \leftarrow \mathcal{G}' \setminus \{i\}, q \leftarrow \max\{q, \frac{b_i}{v_i(R)}\}, A_i \leftarrow T_i, R \leftarrow R \cup T_i, u_w \leftarrow u_w \cup \{i\}$;
9        $i \leftarrow \text{argmax}_{j \in \mathcal{G}'} \frac{b_j}{v_j(R)}$;
10   **end**
11      $q_k \leftarrow q, R_k \leftarrow R$
12   // **Payment Determination**;
13   **for** $i \in \mathcal{G}_k, A_i \neq \emptyset$ **do**
14        $\mathcal{G}' \leftarrow \mathcal{G}' \setminus \{i\}, R' \leftarrow R_{k-1}, q' \leftarrow q_{k-1}$;
15        $i_j \leftarrow \text{argmax}_{j \in \mathcal{G}'} \frac{b_j}{v_j(R)}$;
16   **while** $\frac{b_{i_j}}{v_{i_j}(R')} \leq \frac{B}{V(R' \cup T_{i_j})}$ **and** $q' \leq \frac{B}{V(R' \cup T_{i_j})}$ **and** $b_{i_j} \leq v_{i_j}(R')$ **and**
         $V(R') \leq \mathbb{B}$ **and** $\mathcal{G}' \neq \emptyset$ **do**
17          $p_i \leftarrow \max\{p_i, \min\{v_i(R') \cdot \min\{\frac{b_{i_j}}{v_{i_j}(R')}, \frac{B}{V(R' \cup T_i)}\}, v_i(R')\}\}$;
18          $\mathcal{G}' \leftarrow \mathcal{G}' \setminus \{i\}, q' \leftarrow \max\{q', \frac{b_i}{v_i(R')}\}, R' \leftarrow R' \cup T_{i_j}$;
19          $i_j \leftarrow \text{argmax}_{j \in \mathcal{G}'} \frac{b_j}{v_j(R')}$;
20   **end**
21   **end**
22      $k \leftarrow k + 1$;
23   **end**
24   **return** $u_w, \vec{A}, \vec{p}$

The process repeats until the bid per marginal value of user $\frac{b_i}{v_i(R_i)}$ or the value of q exceeds the threshold $\frac{B}{V(R_i \cup T_i)}$, or the total value is higher than the budget in this group $V(R_i) + v_i(R_i) > \mathbb{B}$, or user's submitted cost is higher than the marginal value $b_i > v_i(R_i)$.

Then we calculate the payment p_i for each winner i in this group $\mathcal{G}_h$. Following the general rule in [11,13], to elicit the truthfulness, the payment should be set as the *threshold payment* by bidding which the winner can replace one of the virtual winners as the winner. Thus, to find the threshold payment, we select virtual winners from the same group without the winner herself using the same winner selection scheme as follow. Given users $\mathcal{G}_h \setminus \{i\}$, we similarly execute the winner selection scheme to select new virtual winning users denoted by $u^h_{w,-i}$ and we assume that $|u^h_{w,-i}| = K_i$. Let i_j denote the selected user in the j-th iteration. User i can be selected as a winner instead of user i_j when her reported cost $\tilde{c}_i(T_i)$ satisfies:

$$\frac{\tilde{c}_i(T_i)}{v_i(R_{i_j})} \leq \frac{b_{i_j}}{v_{i_j}(R_{i_j})}, \frac{\tilde{c}_i(T_i)}{v_i(R_{i_j})} \leq \frac{B}{V(R_{i_j} \cup T_i)}, \tilde{c}_i(T_i) \leq v_i(R_{i_j}), \tag{4}$$

simultaneously. Thus, to replace the virtual winner $i_j \in u^h_{w,-i}$, the bid of user i is at most the minimum of three values: $\theta_{i(j)} = \min\{\delta_{i(j)}, \rho_{i(j)}, v_i(R_{i_j})\}$ where $\delta_{i(j)} = v_i(R_{i_j}) \cdot \frac{b_{i_j}}{v_{i_j}(R_{i_j})}$ and $\rho_{i(j)} = v_i(R_{i_j}) \cdot \frac{B}{V(R_{i_j} \cup T_i)}$. Moreover, we have K_i bids since the size of set $u^h_{w,-i}$ is K_i and the last virtual winner is i_{K_i}. To replace one of these virtual winners, user i should report at most the maximum among these values and the marginal value of user i after K_i iteration:

$$p_i = \max \left\{ \max_{i_j \in u^h_{w,-i}} \theta_{i(j)}, v_i(R_{i_{K_i}} \cup T_{i_{K_i}}) \right\} \tag{5}$$

which will be set as the final payment for winner i in this group.

After considering the current group, we process the next group $\mathcal{G}_{h+1}$. This will repeat until no users can be selected as winners.

5.1   Theoretical Analysis on Desired Properties

Next, we analyze the properties of mechanism TBS.

Lemma 1. *Mechanism TBS is computationally efficient.*

Proof. The running time of Mechanism TBS is dominated by the loop in the winner selection phase (lines 8–12) and payment determination phase (lines 18–31). In the winner selection process, the running time is at most $O(n^2)$ because finding the minimum price-per-value user will take $O(n)$ time and the number of winners is at most n. In the payment scheme phase, the running time is $O(n^3)$ since the select scheme will be executed n times. Therefore, the total computational complexity of Mechanism TBS is $O(n^4)$ since at most n groups need to be processed.

Lemma 2. *Mechanism TBS is individually rational.*

Proof. To simplify the notation, we neglect the label of group and let user i denote the i-th winner in group $\mathcal{G}_h$. Recall that user i_j is the j-th virtual winner in payment determination phase which selects virtual winners by excluding user i herself. We assume that the order of virtual winners in the payment determination phase is $i_1, i_2, \ldots, i_{[i]}, \ldots, i_{K_i}$ where $[i]$ denotes the place where user i should be selected in the winner selection phase if it was involved. It is obvious that previous $i - 1$ sellers, i.e., from i_1 to $i_{[i]-1}$, are still selected as winners in the winner selection phase. According to (2), we have

$$\begin{cases} \frac{c_i}{v_i(R_i)} \leq \frac{B}{V(R_i \cup \{T_i\})} \\ c_i \leq v_i(R_i) \end{cases} \tag{6}$$

since user i is the winner in the truthful case. For the virtual winner $i_{[i]}$, we have

$$\frac{c_i}{v_i(R_i)} \leq \frac{c_{i_{[i]}}}{v_{i_{[i]}}(R_{i_{[i]}})} \tag{7}$$

since user i is selected as winner rather than user $i_{[i]}$ in the winner selection phase. Recall that $\delta_{i_{[i]}} = v_i(R_{i_{[i]}}) \cdot \frac{c_{i_{[i]}}}{v_{i_{[i]}}(R_{i_{[i]}})}$ and $\rho_{i_{[i]}} = v_i(R_{i_{[i]}}) \cdot \frac{B}{V(R_{i_{[i]}} \cup \{T_i\})}$. By combining (6) and (7), we have

$$\begin{cases} c_i \leq \frac{v_i(R_i) \cdot c_{i_{[i]}}}{v_{i_{[i]}}(R_i)} = \frac{v_i(R_{i_{[i]}}) \cdot c_{i_{[i]}}}{v_{i_{[i]}}(R_{i_{[i]}})} = \delta_{i_{[i]}} \\ c_i \leq \frac{v_i(R_i) \cdot B}{V(R_i \cup T_i)} = \frac{v_i(R_{i_{[i]}}) \cdot B}{V(R_{i_{[i]}} \cup T_i)} = \rho_{i_{[i]}} \\ \qquad c_i \leq v_i(R_i) = v_i(R_{i_{[i]}}). \end{cases}$$

Thus, it is obvious that $c_i \leq \theta_{i([i])}$. According to Eq. (5), we have $p_i \geq \max \theta_{i(j)} \geq \theta_{i([i])} \geq c_i$. Therefore, TBS guarantees the individual rationality.

Before analyzing Mechanism TBS's truthfulness, we first introduce a general rule for verifying truthfulness:

Theorem 1 *(Monotone theorem, [11,13]). In single parameter domains, an auction mechanism is truthful iff:*

- *The selection rule is monotone: If user i wins the auction by bidding b_i, it also wins by bidding $b'_i \leq b_i$;*
- *Each winner is paid the critical value, which is the smallest value such that user i would lose the auction if it bids higher than this value.*

Lemma 3. *Mechanism TBS is truthful.*

Proof. We first prove that user i cannot improve her utility by submitting a false task set. We assume that user i submits a false task set $\tilde{T}_i \neq T_i$. If $\tilde{T}_i \subset T_i$, the utility of i is zero according to Eq. (1). If $T_i \subset \tilde{T}_i$, user i can not finish all the tasks as a winner, thus fails to get payment. Therefore, users have to submit her true task set for utility maximization.

Next, we prove that user i cannot improve her utility by bidding a false cost. Following the general rule in Theorem 1, we show that the designed mechanism is monotone and the payment to each winning seller is the critical value.

Monotonicity: Assume that a winner user i in group $\mathcal{G}_h$ bids a lower cost $b'_i < c_i$. Since user i is a winner, we have

$$\begin{cases} \frac{c_i}{v_i(R_i)} \leq \frac{B}{V(R_i \cup T_i)} \\ \quad q \leq \frac{B}{V(R_i \cup T_i)} \\ \quad c_i \leq v_i(R_i) \end{cases} \tag{8}$$

Suppose that user i converts to the j-th $(j < i)$ lowest price per marginal value after bidding the lower bid. Recall that R_i denotes the total value of winners before user i. It is obvious that the total value of winners before i and j satisfies $R_j \leq R_i$. Thus, we have $\frac{b_i'}{v_i(R_j)} \leq \frac{c_i}{v_i(R_i)}$ due to $b_i' \leq c_i \leq v_i(R_i) \leq v_i(R_j)$. In addition, we have $\frac{b_i'}{v_i(R_j)} \leq \frac{B}{V(R_j \cup T_i)}$ since $\frac{c_i}{v_i(R_i)} \leq \frac{B}{V(R_i \cup T_i)} \leq \frac{B}{V(R_j \cup T_i)}$. Additionally, note that $q \leq \frac{B}{V(R_i \cup T_i)} \leq \frac{B}{V(R_j \cup T_i)}$. According to (2) and (3), user i is still selected as a winner. Thus, mechanism guarantees monotonicity.

Threshold Payments: Now we consider each winner's payment. Assume that user i_{K_i} is the last winner in the payment determination scheme which processes the users without i herself. Recall that

$$p_i = \max \left\{ \max_{1 \leq j \leq K_i} \{\theta_{i(j)}\}, v_i(R_{i_{K_i}} \cup T_{i_{K_i}}) \right\} \tag{9}$$

If user i bids a higher cost $b_i > p_i$, we have $b_i > \max_{1 \leq j \leq K_i}\{\theta_{i(j)}\}$ which means that user i will not be selected as a winner before K_i iterations in this group. We also have $p_i > v_i(R_{i_K} \cup T_{i_{K_i}})$, thus i will not be selected after K_i iterations since any user will not be selected as winners if her bid is higher than it marginal value due to (3). Thus, p_i is the threshold payment and any winner will not be selected as winner if her bid is higher than the payment p_i. Therefore, users have no incentive to submit false bid.

In summary, no sellers can increase her utility by submitting a false task-cost pair. Therefore, Mechanism TBS guarantees truthfulness.

Before starting to consider budget feasibility, we first introduce a useful lemma inspired by [3].

Lemma 4. *Consider any set $S \subset T \subseteq \mathcal{G}_h$ in one group and $i = \text{argmin}_{j \in T \setminus S} \frac{c_j}{v_j(S)}$. Then*

$$\frac{c(T) - c(S)}{V(T) - V(S)} \geq \frac{c_i}{v_i(S)}. \tag{10}$$

Proof. Assume that $\frac{c(T)-c(S)}{V(T)-V(S)} < \frac{c_i}{v_i(S)}$ which implies $\frac{c(T)-c(S)}{V(T)-V(S)} < \frac{c_t}{v_t(S)}$ for any $t \in T \setminus S$. After adding all inequalities, we have

$$\frac{c(T) - c(S)}{V(T) - V(S)} < \frac{\sum_{t \in T \setminus S} c_t}{\sum_{t \in T \setminus S} v_t(S)}$$

$$= \frac{c(T) - c(S)}{\sum_{t \in T \setminus S} v_t(S)}$$

which means $V(T) - V(S) > \sum_{t \in T \setminus S} v_t(S)$ contradicting to the submodularity.

Lemma 5. *Mechanism TBS is budget-feasible.*

Proof. We prove the budget feasibility by showing that Mechanism TBS satisfies two properties: $\sum_{i \in u_w} \max_{1 \leq j \leq K} \theta_{i(j)} \leq \mathbb{B}$ and $\sum_{i \in u_w} v_i(R_{i_{K_i}} \cup T_{i_{K_i}}) \leq \mathbb{B}$. Assume that user K in group k is the last winner in mechanism TBS. We focus on payment determination phase and let i_j denote j-th user in the group $e(e \leq k)$ without user i. We consider two cases:

(1) Consider $\theta_{i(j)}$: Recall that $\theta_{i(j)} = \min\{\delta_{i(j)}, \rho_{i(j)}, v_i(R_{i_j})\}$ and $\delta_{i(j)} = v_i(R_{i_j}) \cdot \frac{c_{i_j}}{v_{i_j}(R_{i_j})}$. For each user $i_j < i_{[i]}$, we have $\frac{c_{i_j}}{v_{i_j}(R_{i_j})} \leq \frac{c_i}{v_i(R_{i_j})}$ since user i_j is selected as winner instead of user i. Thus $\delta_{i(j)} \leq c_i$ which means $\theta_{i(j)} \leq c_i$.

In group $\mathcal{G}_h$, for each user $i_j \geq i_{[i]}$, we assume that the set of winners we have chosen before i_j is $\mathcal{S}$. Suppose user i replaces i_j as a winner by bidding $\theta_{i(j)}$. Thus we have $\mathcal{S} \cup \{i\} \subset u_w \cup \mathcal{S}$. In addition, we have

$$\frac{\theta_{i_j}}{v_i(\mathcal{S})} \leq \frac{c(u_w \cup \mathcal{S}) - c(\mathcal{S} \cup \{i\})}{V(u_w \cup \mathcal{S}) - V(\mathcal{S} \cup \{i\})} \tag{11}$$

Recall that $\frac{c_i}{v_i(R_i)} \leq \frac{B}{V(u_w)}$ which means that $\sum_{i \in u_w} c_i \leq B$. Thus, we have

$$\begin{aligned} \frac{V(u_w) - V(\mathcal{S} \cup \{i\})}{B} &\leq \frac{V(u_w) - V(\mathcal{S} \cup \{i\})}{\sum_{i \in u_w} c_i} \\ &\leq \frac{V(u_w \cup \mathcal{S}) - V(\mathcal{S} \cup \{i\})}{c(u_w \cup \mathcal{S}) - c(\mathcal{S} \cup \{i\})} \end{aligned} \tag{12}$$

Assume that $\theta_{i(j)} > v_i(R_i) \cdot \frac{\mathbb{B}}{V(u_w)}$, we have

$$\frac{v_i(\mathcal{S})}{\theta_{i(j)}} < \frac{v_i(\mathcal{S})}{v_i(R_i)} \cdot \frac{V(u_w)}{\mathbb{B}} \leq \frac{V(u_w)}{\mathbb{B}} \tag{13}$$

where the second inequality is due to $R_i \subseteq \mathcal{S}$. By combining (11) and (13), we have

$$\frac{\mathbb{B}}{V(u_w)} < \frac{c(u_w \cup \mathcal{S}) - c(\mathcal{S} \cup \{j\})}{V(u_w \cup \mathcal{S}) - V(\mathcal{S} \cup \{j\})} \tag{14}$$

According to (12) and (14), we have $V(u_w) < 2V(\mathcal{S} \cup \{i\})$ due to $\mathbb{B} = 2B$. However, since $\frac{b_{i_j}}{v_{i_j}(R_{i_j})} \leq \frac{B}{V(\mathcal{S} \cup \{i_j\})}$, we have

$$\begin{aligned} \theta_{i(j)} \leq \delta_{i(j)} &= \frac{v_i(R_{i_j}) \cdot b_{i_j}}{v_{i_j}(R_{i_j})} \leq \frac{v_i(R_{i_j}) \cdot B}{V(\mathcal{S} \cup \{i_j\})} \\ &\leq v_i(R_i) \cdot \frac{\mathbb{B}}{V(u_w)} \end{aligned} \tag{15}$$

which contradicts to the assumption $\theta_{i(j)} > v_i(R_i) \cdot \frac{\mathbb{B}}{V(u_w)}$. Thus, we have $\theta_{i(j)} \leq v_i(R_i) \cdot \frac{\mathbb{B}}{V(u_w)}$.

(2) Consider $v_i(R_{i_K})$: It is obvious that $v_i(R_{i_K}) \leq v_i(R_i)$. Thus, we have $v_i(R_{i_K}) \leq v_i(R_i) \leq v_i(R_i) \cdot \frac{\mathbb{B}}{V(u_w)}$ since $V(u_w) \leq \mathbb{B}$ due to (3).

Therefore, Mechanism TBS guarantees budget feasibility.

Before proving Sybil-proofness of Mechanism TBS, we introduce the following general rules introduced by [10]:

Theorem 2. *A mechanism is Sybil-proof if it satisfies the following two conditions:*

- *If any user i pretends two identities i' and i'', and both i' and i'' are selected as winners, then i should be selected as a winner while using only one identity;*
- *If any user i pretends two identities i' and i'', the payment to i should not be less than the summation of the payments to i' and i''.*

Lemma 6. *Mechanism TBS is Sybil-proof.*

Proof. We first prove TBS satisfies the first condition. Assume that user i generate two identities i' and i'' which implies that $T_{i'} \subset T_i$ and $T_{i''} \subset T_i$, and both of them are selected as winners. Let $\mathcal{R}', \mathcal{R}''$ denote the union assigned task set before i', i'' are selected as winners respectively. Similarly, denote by $\mathcal{R}$ the union assigned task set before user i is selected as the winner. W.l.o.g, suppose the group of i' is ahead of the group of i''. Since user i' and i'' are all winners, according to conditions of being winners in (2) and (3), we have

$$b_{i'} \le v_{i'}(\mathcal{R}'), b_{i''} \le v_{i''}(\mathcal{R}'') \tag{16}$$

and

$$\frac{b_{i'}}{v_{i'}(\mathcal{R}')} \le \frac{B}{V(\mathcal{R}' \cup T_{i'})}$$

$$\frac{b_{i''}}{v_{i''}(\mathcal{R}'')} \le \frac{B}{V(\mathcal{R}'' \cup T_{i''})} \tag{17}$$

$$\frac{b_{i'}}{v_{i'}(\mathcal{R}')} \le \frac{B}{V(\mathcal{R}'' \cup T_{i''})}$$

where the reason for the third inequality is that the maximum bid per marginal value q among winners before i'' is higher than $\frac{b_{i'}}{v_{i'}(\mathcal{R}')}$ and it must be not higher than the current average price-per-value $\frac{B}{V(\mathcal{R}'' \cup T_{i''})}$ according to the selection rule (2). Also note that

$$v_{i'}(\mathcal{R}') \le v_{i'}(\mathcal{R}), v_{i''}(\mathcal{R}'') \le v_{i''}(\mathcal{R}) \tag{18}$$

since the groups of user i', i'' will follow the groups of user i. Combining (16) and (18), we have

$$\begin{aligned} b_i = c_i \le c_{i'} + c_{i''} &= b_{i'} + b_{i''} \\ &\le v_{i'}(\mathcal{R}') + v_{i''}(\mathcal{R}'') \\ &\le v_{i'}(\mathcal{R}) + v_{i''}(\mathcal{R}) \\ &\le v_i(\mathcal{R}) \end{aligned} \tag{19}$$

where the reason for the first and last inequality is because $T_i = T_{i'} \cup T_{i''}$. By combining (17) and (19), it holds that

$$\begin{aligned} \frac{c_i}{v_i(\mathcal{R})} &\le \frac{b_{i'} + b_{i''}}{v_{i'}(\mathcal{R}') + v_{i''}(\mathcal{R}'')} \le \frac{B}{V(\mathcal{R}'' \cup T_{i''})} \\ &\le \frac{B}{V(R \cup T_i)}. \end{aligned} \tag{20}$$

Thus user i will still be selected as a winner without generating multiple identities.

Next, we consider the second condition. It is obvious that the payment of user i is at least $v_i(R_{i_{K_i}})$ according to (5). Recall that the order of winning seller in the payment determination phase is $i_1, i_2, \ldots, i_{[i]}, \ldots, i_{K_i}$ and $[i]$ denotes the place where user i should be selected in the winner selection phase if it was involved. Thus, we have $b_{i_j} \leq v_{i_j}(R_{i_j})$ where i_j is selected in K_i iterations. For the user after $i_{[i]}$, we have $\min\{\delta_{i(r)}, \rho_{i(r)}, v_i(R_{i_r})\} \leq v_i(R_{i_r}) \leq v_i(R_i)$ where $i_r \geq i_{[i]}$. For the user i_j before $i_{[i]}$, we have

$$\min\{\delta_{i(j)}, \rho_{i(j)}, v_i(R_{i_r})\} \leq \delta_{i(j)} = v_i(R_{i_j}) \cdot \frac{b_{i_j}}{v_{i_j}(R_{i_j})}$$
$$\leq v_i(R_{i_j}) \cdot \frac{b_i}{v_i(R_{i_j})} = b_i \qquad (21)$$
$$\leq v_i(R_i)$$

where the reason for the second inequality is that user i_j is selected as winner instead of user i. Furthermore, after K_i iterations, we have $v_i(\mathcal{R}_{K_i}) \leq v_i(R_i)$. Hence, we have $p_i \leq v_i(R_i)$. Similarly, we have $p_{i'} \leq v_{i'}(R_{i'})$ and $p_{i''} \leq v_{i''}(R_{i''})$. Thus we have

$$p_{i'} + p_{i''} \leq v_{i'}(R_{i'}) + v_{i''}(R_{i''})$$
$$\leq v_{i'}(\mathcal{R}_{i_{k'}}) + v_{i'}(\mathcal{R}_{i_{k'}}) \qquad (22)$$
$$\leq v_i(\mathcal{R}_{i_{K'}}) \leq p_i$$

Hence, the second condition is satisfied.

Therefore, Mechanism TBS guarantees Sybil-proofness.

6   Performance Evaluation

In this section, we conduct extensive simulations to validate the performance of Mechanism TBS. We first verify the desired properties (truthfulness and robustness against Sybil attack) of Mechanism TBS. Then, we validate the performance of Mechanism TBS in terms of various optimization metrics, $e.g.$, $payment$ (the payment for target value), $platform\ utility$ (the total value procured from users), and $average\ user\ utility$ (the sum of users' utilities over the number of sellers). Under these metrics, we take the Sybil-proof Mechanism SPIM-S proposed in [10] as the benchmark algorithm. Although Mechanism SPIM-S cannot work in the procurement scenario with budget constraint, we can achieve the comparison by enumerating the inputs of Mechanism SPIM-S.

Simulation Setup. In our evaluation, we assume that the task size of each user is uniformly distributed over $[1, 5]$, and value of each task is uniformly distributed over $[1, 20]$. The users' costs for each task is uniformly distributed over $[1,10]$. In default, we set the number of users, the number of tasks and budget at 150, 200 and 200, respectively. To evaluate the impact of number of total users on the performance of platform utility and average user utility, we

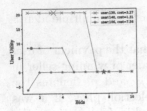 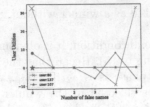 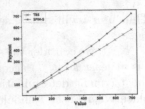

Fig. 2. The impact of untruthful bids on TBS.

Fig. 3. The impact of Sybil attack on TBS.

Fig. 4. The impact of Sybil attack on TBS.

vary the number of users from 40 to 140 with the increment of 20. Similarly, to evaluate the impact of number of total tasks on the performance of platform utility and average user utility, we let the number of tasks vary from 80 to 200 in increment of 20. Furthermore, to evaluate the impact of total budget on the performance of platform utility and average user utility, we change the budget from 120 to 400 with the increment of 40. All the results are averaged over 100 instances.

6.1   Evaluation of Desired Properties

The properties like individual rationality and budget feasibility of Mechanism TBS can be easily verified. In this part, we mainly validate the truthfulness and Sybil-proofness of Mechanism TBS by letting users submit false bids or launch Sybil attack unilaterally, and monitoring the corresponding utilities. In the simulation, we fix the number of tasks at 150 and the number of users at 200, respectively.

Figure 2 shows the impact of (untruthful) bids on user utilities for Mechanism TBS. To validate truthfulness, we let each of these users unilaterally change her bid in $[1, 10]$. We randomly select three users: 139, 140, and 106 of TBS. User 106 is a loser with real cost 7.56 while users 139 and 140 are winners with real costs 3.27 and 1.31, respectively. Specifically, in Fig. 2, we use larger markers to indicate the utilities for truthful bids and the smaller ones to indicate those of untruthful bids. We observe that these users cannot achieve more utilities after bidding false costs, *e.g.*, user 139 obtains less utilities if her bid is not equal to her real cost 3.27. Therefore, users obtain the maximum utility when bidding real cost, which validates the truthfulness of Mechanism TBS.

Figure 3 shows the impact of Sybil attack on user utilities of Mechanism TBS. To validate Sybil-proofness, we let each of these users create up to 5 false names. For each false name, the submitted task set is a subset of the submitted tasks of the user. We select three users: 80, 137, and 107 of TBS. User 80 and 137 are winners while user 107 is a loser. Moreover, in Fig. 3, we also use larger markers to indicate the utilities when users do not launch Sybil attack. We observe that these users achieve the highest utilities without generating fake identities, *e.g.*, user 137 obtains less utilities after creating more false names. Therefore, a user cannot increase her utility by unilaterally launching Sybil attack which validates the Sybil-proofness of Mechanism TBS.

6.2 Evaluation of Optimization Metrics

Next, we compare Mechanism TBS with Mechanism SPIM-S using aforementioned metrics. Since Mechanism SPIM-S actually cannot be applied to procurement scenarios with budget constraint, we use enumeration method when conducting the comparison. Specifically, to evaluate the platform utility and average user utility, we enumerate the possible outputs of SPIM-S by running multiple rounds of auctions to find the one that uses up the given budget. To evaluate the payment, we enumerate possible outputs until a target total value is procured.

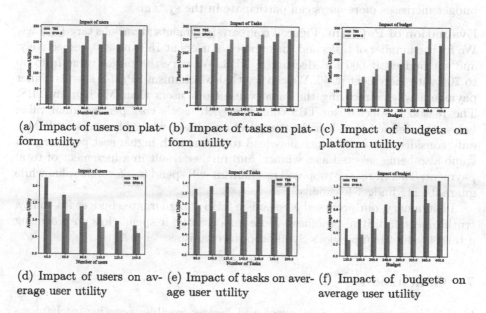

(a) Impact of users on platform utility

(b) Impact of tasks on platform utility

(c) Impact of budgets on platform utility

(d) Impact of users on average user utility

(e) Impact of tasks on average user utility

(f) Impact of budgets on average user utility

Fig. 5. (a)–(c) show the impact of users, tasks and budget on platform utility, while (d)–(f) on average user utility.

Evaluation of Platform Utility. The top part of Fig. 5 shows the impact of users, tasks and budget on platform utility for Mechanism TBS and Mechanism SPIM-S. In Fig. 5(a), Fig. 5(b) and Fig. 5(c), we see that while preserving the budget feasibility, Mechanism TBS can achieve similar platform utility to that of the Mechanism SPIM-S. This demonstrates that a small loss of overall platform utility might be needed to guarantee budget constraint in the procurement scenario. Furthermore, in both Fig. 5(a) and Fig. 5(b), the platform utility increases in all these two mechanisms. The respective reasons are that competition among users has become more fierce which leads to more users with lower costs being winners, and more users can be selected as winners with increments of tasks. However, as a constant budget is set in the experiment, the platform utility is growing slowly in both figures. In addition, the platform utility grows steadily with the increments of budget as shown in Fig. 5(c).

Evaluation of Average User Utility. The bottom part of Fig. 5 shows the performance of user average utility, defined as the total utilities of all uses over the number of users, for Mechanism TBS, and Mechanism SPIM-S. In Fig. 5(d), Fig. 5(e) and Fig. 5(f), we can see that the average user utility of Mechanism TBS is higher than Mechanism SPIM-S. This is because TBS can select the user with smaller cost which improves winner's utility. As shown in Fig. 5(d) and Fig. 5(f), average user utility decreases as the number of users increases, while it rises as budgets increase. This is because 1) as the number of users increases the competition among users become more fierce resulting in lower payments, 2) as budget increases more users can participate in the system.

Evaluation of Payment. Figure 4 compares payments at various target values. We fix the number of tasks and the number of users at 150 and 200, respectively, and set budget at 600 for Mechanism TBS. We vary the target value from 50 to 700 with increment of 50. We can see that Mechanism SPIM-S spends higher payments when procuring the same value from users than Mechanism TBS. The reason is Mechanism TBS considers each user's cost per marginal value below the defined threshold in the selection phase, while Mechanism SPIM-S only considers the costs which may lead to a user with higher cost per marginal value also being selected as a winner, and further result in a increment of total payment. Therefore, our proposed mechanism will spend less for unit value while guaranteeing budget-feasibility.

In summary, our proposed Mechanism TBS can guarantee budget feasibility, truthfulness and Sybil-proofness. More importantly, it spends less on procuring a target value than previous Sybil-proof mechanisms.

7 Conclusion

In this paper, we study Sybil-proof and budget-feasible incentive mechanisms for crowdsensing systems. We design an incentive mechanism TBS and prove that the designed mechanism guarantees the individual rationality, truthfulness, budget-feasibility and Sybil-proofness. We validate the desired properties of the designed mechanism through extensive simulations.

Acknowledgements. The work is supported in part by the National Key Research and Development Program of China under grant No. 2019YFB2102200, National Natural Science Foundation of China under Grant No. 61672154, 61672370, 61972086 and the Postgraduate Research & Practice Innovation Program of Jiangsu Province under grant No. KYCX19_0089.

References

1. Anari, N., Goel, G., Nikzad, A.: Mechanism design for crowdsourcing: an optimal 1-1/e competitive budget-feasible mechanism for large markets. In: 2014 IEEE 55th Annual Symposium on Foundations of Computer Science, pp. 266–275. IEEE (2014)
2. Brill, M., Conitzer, V., Freeman, R., Shah, N.: False-name-proof recommendations in social networks. In: Proceedings of the 2016 International Conference on Autonomous Agents & Multiagent Systems, pp. 332–340. International Foundation for Autonomous Agents and Multiagent Systems (2016)
3. Chen, N., Gravin, N., Lu, P.: On the approximability of budget feasible mechanisms. In: Proceedings of the Twenty-Second Annual ACM-SIAM Symposium on Discrete Algorithms, pp. 685–699. Society for Industrial and Applied Mathematics (2011)
4. Cui, J., et al.: TCAM: a truthful combinatorial auction mechanism for crowdsourcing systems. In: 2018 IEEE Wireless Communications and Networking Conference (WCNC), pp. 1–6. IEEE (2018)
5. Feng, Z., Zhu, Y., Zhang, Q., Ni, L.M., Vasilakos, A.V.: TRAC: truthful auction for location-aware collaborative sensing in mobile crowdsourcing. In: IEEE INFOCOM 2014-IEEE Conference on Computer Communications, pp. 1231–1239. IEEE (2014)
6. Gao, L., Hou, F., Huang, J.: Providing long-term participation incentive in participatory sensing. In: 2015 IEEE Conference on Computer Communications (INFOCOM), pp. 2803–2811. IEEE (2015)
7. Goel, G., Nikzad, A., Singla, A.: Mechanism design for crowdsourcing markets with heterogeneous tasks. In: Second AAAI Conference on Human Computation and Crowdsourcing (2014)
8. Huang, H., Xin, Y., Sun, Y.E., Yang, W.: A truthful double auction mechanism for crowdsensing systems with max-min fairness. In: 2017 IEEE Wireless Communications and Networking Conference (WCNC), pp. 1–6. IEEE (2017)
9. Koutsopoulos, I.: Optimal incentive-driven design of participatory sensing systems. In: 2013 Proceedings IEEE INFOCOM, pp. 1402–1410. IEEE (2013)
10. Lin, J., Li, M., Yang, D., Xue, G., Tang, J.: Sybil-proof incentive mechanisms for crowdsensing. In: IEEE INFOCOM 2017-IEEE Conference on Computer Communications, pp. 1–9. IEEE (2017)
11. Myerson, R.B.: Optimal auction design. Math. Oper. Res. **6**(1), 58–73 (1981)
12. Qiao, Y., Wu, J., Cheng, H., Huang, Z., He, Q., Wang, C.: Truthful mechanism design for multiregion mobile crowdsensing. In: Wireless Communications and Mobile Computing 2020 (2020)
13. Singer, Y.: Budget feasible mechanisms. In: 2010 IEEE 51st Annual Symposium on Foundations of Computer Science, pp. 765–774. IEEE (2010)
14. Singer, Y., Mittal, M.: Pricing mechanisms for crowdsourcing markets. In: Proceedings of the 22nd International Conference on World Wide Web, pp. 1157–1166. ACM (2013)
15. Singla, A., Krause, A.: Truthful incentives in crowdsourcing tasks using regret minimization mechanisms. In: Proceedings of the 22nd International Conference on World Wide Web, pp. 1167–1178. ACM (2013)
16. Terada, K., Yokoo, M.: False-name-proof multi-unit auction protocol utilizing greedy allocation based on approximate evaluation values. In: Proceedings of the Second International Joint Conference on Autonomous Agents and Multiagent Systems, pp. 337–344. ACM (2003)

17. Todo, T., Iwasaki, A., Yokoo, M., Sakurai, Y.: Characterizing false-name-proof allocation rules in combinatorial auctions. In: Proceedings of the 8th International Conference on Autonomous Agents and Multiagent Systems, vol. 1, pp. 265–272. International Foundation for Autonomous Agents and Multiagent Systems (2009)

18. Von Ahn, L., Maurer, B., McMillen, C., Abraham, D., Blum, M.: reCAPTCHA: human-based character recognition via web security measures. Science **321**(5895), 1465–1468 (2008)

19. Wang, Q., et al.: ALETHEIA: robust large-scale spectrum auctions against false-name bids. In: Proceedings of the 16th ACM International Symposium on Mobile Ad Hoc Networking and Computing, pp. 27–36. ACM (2015)

20. Xu, J., Xiang, J., Yang, D.: Incentive mechanisms for time window dependent tasks in mobile crowdsensing. IEEE Trans. Wirel. Commun. **14**(11), 6353–6364 (2015)

21. Yang, D., Xue, G., Fang, X., Tang, J.: Crowdsourcing to smartphones: incentive mechanism design for mobile phone sensing. In: Proceedings of the 18th Annual International Conference on Mobile Computing and Networking, pp. 173–184. ACM (2012)

22. Yang, D., Xue, G., Fang, X., Tang, J.: Incentive mechanisms for crowdsensing: crowdsourcing with smartphones. IEEE/ACM Trans. Network. **24**(3), 1732–1744 (2015)

23. Yao, Y., et al.: Multi-channel based Sybil attack detection in vehicular ad hoc networks using RSSI. IEEE Trans. Mob. Comput. **18**(2), 362–375 (2018)

24. Yokoo, M., Sakurai, Y., Matsubara, S.: The effect of false-name bids in combinatorial auctions: new fraud in internet auctions. Games Econom. Behav. **46**(1), 174–188 (2004)

25. Zhang, X., Xue, G., Yu, R., Yang, D., Tang, J.: Truthful incentive mechanisms for crowdsourcing. In: 2015 IEEE Conference on Computer Communications (INFOCOM), pp. 2830–2838. IEEE (2015)

26. Zhang, X., Xue, G., Yu, R., Yang, D., Tang, J.: Countermeasures against false-name attacks on truthful incentive mechanisms for crowdsourcing. IEEE J. Sel. Areas Commun. **35**(2), 478–485 (2017)

27. Zhao, D., Li, X.Y., Ma, H.: How to crowdsource tasks truthfully without sacrificing utility: online incentive mechanisms with budget constraint. In: IEEE INFOCOM 2014-IEEE Conference on Computer Communications, pp. 1213–1221. IEEE (2014)

28. Zhao, D., Li, X.Y., Ma, H.: Budget-feasible online incentive mechanisms for crowdsourcing tasks truthfully. IEEE/ACM Trans. Network. (TON) **24**(2), 647–661 (2016)

29. Zhu, X., An, J., Yang, M., Xiang, L., Yang, Q., Gui, X.: A fair incentive mechanism for crowdsourcing in crowd sensing. IEEE Internet Things J. **3**(6), 1364–1372 (2016)

Author Index

Printed in the United States
by Baker & Taylor Publisher Services